NUCLEAR FISSION AND FISSION-PRODUCT SPECTROSCOPY

To learn more about the AIP Conference Proceedings, including the Conference Proceedings Series, please visit the webpage **http://proceedings.aip.org/proceedings**

NUCLEAR FISSION AND FISSION-PRODUCT SPECTROSCOPY

3rd International Workshop on Nuclear Fission and Fission-Product Spectroscopy

Cadarache, France   11 – 14 May 2005

EDITORS

Héloïse Goutte
CEA/DAM Ile de France
Bruyères-le-Châtel, France

Herbert Faust
Institut Laue Langevin
Grenoble, France

Gabriele Fioni
CEA/DSM Saclay
Gif-sur-Yvette, France

Dominique Goutte
GANIL
Caen, France

SPONSORING ORGANIZATIONS
CEA/DAM, Bruyères-le-Châtel
CEA/DSM, Saclay
CEA, Cadarache
ILL, Grenoble
GANIL, Caen
Région Provence Alpes Côte d'Azur

Melville, New York, 2005
AIP CONFERENCE PROCEEDINGS ■ VOLUME 798

Editors:

Héloïse Goutte
CEA/DAM Ile de France
DPTA/Service de Physique Nucléaire
BP 12
F-91680 Bruyères-le-Châtel
France
E-mail: heloise.goutte@cea.fr

Gabriele Fioni
DSM/DIR
CEA/Saclay
F-91191 Gif-sur-Yvette
France
E-mail: gabriele.fioni@cea.fr

Herbert Faust
Institut Laue Langevin
BP 156
F-38042 Grenoble Cedex 9
France
E-mail: faust@ill.fr

Dominique Goutte
GANIL
BP 55027
F-14076 Caen Cedex 5
France
E-mail: goutte@ganil.fr

Authorization to photocopy items for internal or personal use, beyond the free copying permitted under the 1978 U.S. Copyright Law (see statement below), is granted by the American Institute of Physics for users registered with the Copyright Clearance Center (CCC) Transactional Reporting Service, provided that the base fee of $22.50 per copy is paid directly to CCC, 222 Rosewood Drive, Danvers, MA 01923, USA. For those organizations that have been granted a photocopy license by CCC, a separate system of payment has been arranged. The fee code for users of the Transactional Reporting Services is: 0-7354-0288-4/05/$22.50.

© 2005 American Institute of Physics

Permission is granted to quote from the AIP Conference Proceedings with the customary acknowledgment of the source. Republication of an article or portions thereof (e.g., extensive excerpts, figures, tables, etc.) in original form or in translation, as well as other types of reuse (e.g., in course packs) require formal permission from AIP and may be subject to fees. As a courtesy, the author of the original proceedings article should be informed of any request for republication/reuse. Permission may be obtained online using Rightslink. Locate the article online at http://proceedings.aip.org, then simply click on the Rightslink icon/"Permission for Reuse" link found in the article abstract. You may also address requests to: AIP Office of Rights and Permissions, Suite 1NO1, 2 Huntington Quadrangle, Melville, NY 11747-4502, USA; Fax: 516-576-2450; Tel.: 516-576-2268; E-mail: rights@aip.org.

L.C. Catalog Card No. 2005934831
ISBN 0-7354-0288-4
ISSN 0094-243X
Printed in the United States of America

CONTENTS

Preface xi
Committees xiii
Sponsors xiv
Program xv
Opening Address xvii
    B. Bigot

CROSS SECTIONS AND RESONANCES

Fission Cross Section Calculations for Pa Isotopes 3
    F.-J. Hambsch, G. Vladuca, A. Tudora, S. Oberstedt, and D. Filipescu
Actinide Fission and Capture Cross Section Measurements at ILL:
The Mini-INCA Project 11
    A. Letourneau, I. Al Mahamid, C. Blandin, O. Bringer, S. Chabod,
    F. Chartier, H. Faust, G. Fioni, Y. Foucher, F. Marie, P. Mutti, and
    C. Veyssiere
Determination of Minor Actinides Fission Cross Sections by Means
of Transfer Reactions 19
    B. Jurado, M. Aiche, G. Barreau, S. Boyer, S. Czajkowski, D. Dassié,
    C. Grosjean, A. Guiral, B. Haas, B. Osmanov, M. Petit, E. Berthoumieux,
    F. Gunsing, L. Perrot, C. Theisen, E. Bauge, F. Michel-Sendis,
    A. Billebaud, J. N. Wilson, I. Ahmad, J. P. Greene, and R. V. F. Janssens
Recent Results on the Neutron-Induced Fission Cross-Section
of ^{231}Pa 27
    A. Oberstedt, S. Oberstedt, E. Birgersson, F.-J. Hambsch, V. Fritsch,
    G. Lövestam, G. Vladuca, A. Tudora, and N. Kornilov
Evaluations of Photonuclear Cross Sections for Actinides 31
    M.-L. Giacri-Mauborgne, M. B. Chadwick, J.-C. David, D. Doré,
    D. Ridikas, A. Van Lauwe, and W. B. Wilson

FISSION AT HIGHER ENERGIES—I

Distribution of Nuclides Produced in the Collision of 1 AGeV
^{238}U-Ions on p 37
    M. Bernas, P. Napolitani, F. Rejmund, C. Stephan, J. Taieb, L. Tassan-Got,
    P. Armbruster, T. Enqvist, M. V. Ricciardi, K.-H. Schmidt, J. Benlliure,
    E. Casajeros, J. Pereira, A. Boudard, R. Legrain, S. Leray, and C. Volant
Nuclide Cross-Sections of Fission Fragments in the Reaction ^{208}Pb+p
at 500 A MeV 45
    C. Volant, B. Fernandez-Dominguez, P. Armbruster, L. Audouin,
    J. Benlliure, M. Bernas, A. Boudard, E. Casarejos, S. Czajkowski,
    J. E. Ducret, T. Enqvist, B. Jurado, R. Legrain, S. Leray, B. Mustapha,
    J. Pereira, M. Pravikoff, F. Rejmund, M. V. Ricciardi, K.-H. Schmidt,
    C. Stéphan, J. Taieb, L. Tassant-Got, and W. Wlazlo

Experimental Cross Sections and Velocities of Light Nuclides
Produced in the Proton-Induced Fission of ^{238}U at 1 GeV 49
    M. V. Ricciardi, K.-H. Schmidt, F. Rejmund, T. Enqvist, P. Armbruster,
    J. Benlliure, M. Bernas, B. Mustapha, L. Tassan-Got, C. Stéphan,
    A. Boudard, S. Leray, C. Volant, S. Czajkowski, and M. Pravikoff

Transmutation and Neutron Flux Studies with Fission Chambers
in the MEGAPIE Target 57
    S. Chabod, C. Blandin, F. Chartier, G. Fioni, Y. Foucher, A. Letourneau,
    F. Marie, and J. C. Toussaint

NEA Nuclear Data Services: EXFOR, JANIS, and the JEFF Project 61
    Y. Rugama and H. Henriksson

FISSION: MASS AND CHARGE YIELDS

Low Energy Fission: Dynamics and Scission Configurations 69
    H. Goutte, J.-F. Berger, D. Gogny, and W. Younes

Microscopic Calculations of Potential Energy Surfaces: Fission
Barriers, Fission and Fusion Properties 77
    L. Bonneau and P. Quentin

Microscopic Description of Fission in Odd Nuclei 85
    S. Pérez and L. M. Robledo

Self-Consistent Study of Fission Barriers of Even-Even Superheavy
Nuclei 93
    A. Staszczak, J. Dobaczewski, and W. Nazarewicz

LIGHT PARTICLES AND CLUSTER EMISSION

Microscopic Description of Cluster Emission with the Gogny Force 103
    L. M. Robledo and J. L. Egido

New Results on the Ternary Fission of ^{243}Cm 111
    J. Heyse, C. Wagemans, S. Vermote, O. Serot, P. Geltenbort, T. Soldner,
    and J. Van Gils

Studies on Particle-Accompanied Fission of ^{252}Cf(sf) and ^{235}U(n_{th}, f) 115
    Y. N. Kopatch, V. Tishchenko, M. Speransky, M. Mutterer, F. Gönnenwein,
    P. Jesinger, A. M. Gagarski, J. von Kalben, I. Kojouharov, E. Lubkiewics,
    Z. Mezentseva, V. Nezvishevsky, G. A. Petrov, H. Schaffner, H. Scharma,
    W. H. Trzaska, and H.-J. Wollersheim

Sudden Emission of Nucleons at Scission 123
    N. Carjan, P. Talou, D. Strottman, O. Serot, and H. Goutte

SPECTROSCOPY OF NEUTRON RICH NUCLEI

Nuclear Structure Studies of Neutron-Rich Cu and Zn Isotopes
Produced by Means of Proton-Induced Fission of ^{238}U 131
    J.-C. Thomas, H. De Witte, M. Gorska, M. Huyse, K. Kruglov,
    Y. Kudryavtsev, D. Pauwels, N. V. S. V. Prasad, K. Van de Vel, P. Van

Duppen, J. Van Roosbroeck, S. Franchoo, J. Cederkall, H. O. U. Fynbo, U. Georg, O. Jonsson, U. Köster, L. Weissman, W. F. Mueller, V. N. Fedoseyev, V. I. Mishin, D. Fedorov, A. De Maesschalck, N. A. Smirnova, The IS365 Collaboration, and the ISOLDE Collaboration

Recent Results and Future Prospects for Nuclear Structure Studies at the ILL 137
    G. S. Simpson, J. Genevey, J. A. Pinston, I. Tsekhanovich, and W. Urban

Neutron-Rich In and Cd Isotopes Close to the Doubly-Magic ^{132}Sn.......... 145
    A. Scherillo, J. Genevey, J. A. Pinston, A. Covello, H. Faust, A. Gargano, R. Orlandi, G. S. Simpson, and I. Tsekhanovich

Shape Coexistence in Odd and Odd-Odd Nuclei in the A~100 Region 149
    J. A. Pinston, J. Genevey, G. Simpson, and W. Urban

Disposition of Legacy Materials at LBNL and Reuse of Valuable Items in Target Preparation 157
    I. Al Mahamid, D. A. Shaughnessy, and R. Sudowe

RESONANCES, BARRIERS, AND FISSION TIMES

Fission Modeling, Data Evaluation, and Integral Data Testing at Los Alamos 167
    M. B. Chadwick, P. G. Young, R. E. MacFarlane, P. Talou, and T. Kawano

Fission Barriers of Exotic Nuclei 178
    A. Kelić and K.-H. Schmidt

Determination of the ^{243,246,248}Cm Thermal Neutron Induced Fission Cross Sections 182
    O. Serot, C. Wagemans, S. Vermote, J. Heyse, T. Soldner, and P. Geltenbort

Neutron Cross Section Data for Pd-105, Ag-109, Xe-131, and Cs-133 190
    Y. D. Lee and Y. O. Lee

What Can we Learn from Fission Times? 194
    M. Morjean

FRAGMENT EXCITATION AND NEUTRON EMISSION

Current Status of the Search for Scission Neutrons in Fission and Estimation of their Main Characteristics 205
    G. A. Petrov

Influence of Fission Fragment Excitation Energy on Prompt Fission Neutron Observables 213
    S. Lemaire, P. Talou, T. Kawano, M. B. Chadwick, and D. G. Madland

Distributions for Excitation Energy and Kinetic Energy in Nuclear Fission.. 221
    H. R. Faust and Z. Bao

Kinetic Energy Distributions in Thermal Neutron Induced Fission of ^{245}Cm 232
    B. Weiss, H. Faust, and N. Bessolaz

MASS AND ENERGY DISTRIBUTIONS

Statistical Approaches to the Even-Odd Effect in Fission. 239
    K-.H. Schmidt, A. V. Ignatyuk, F. Rejmund, A. Kelić, and M. V. Ricciardi

Prompt Fission Neutron Multiplicity and Spectrum Evaluations 247
    A. Tudora, G. Vladuca, B. Morillon, F.-J. Hambsch, and S. Oberstedt

Prompt Neutron Emission from Fragments in Spontaneous Fission of ^{244,248}Cm and ^{252}Cf 255
    A. S. Vorobyev, V. N. Dushin, F.-J. Hambsch, V. A. Jakovlev,
    V. A. Kalinin, A. B. Laptev, B. F. Petrov, and O. A. Shcherbakov

Isotopic Yields of Fission Fragments from Transfer-Induced Fission 263
    F. Rejmund, G. Barreau, B. Jurado, K.-H. Schmidt, A. Kelic,
    M. V. Ricciardi, E. Casarejos, C.-O. Bacri, L. Tassant-Got, C. Schmitt,
    J. Benlliure, T. Enqvist, N. Alahari, M. Rejmund, J. Frankland,
    M. Morjean, H. Savajols, and W. Mittig

NEEDS FOR NUCLEAR DATA AND NEW FACILITIES—I

Status of High Intensity Laser Experiments for Nuclear Fission Investigations. 269
    J. Galy and J. Magill

The New Pulsed Mono-Energetic Neutron Source at the IRMM and the Shape Isomer Search in ^{239}U 273
    S. Oberstedt, G. Lövestam, C. Chaves de Jesus, T. Gamboni, W. Geerts,
    and R. J. Tornin

Delayed Neutrons from High Energy Fission-Spallation Reactions. 277
    D. Ridikas, P. Bokov, J.-C. David, D. Doré, M.-L. Giacri, X. Ledoux,
    A. Plukis, R. Plukiene, and A. Van Lauwe

IAEA Coordinated Research Project on Fission Product Yield Data for Minor Actinides up to 150 MeV 285
    M. Lammer and A. L. Nichols

ANGULAR MOMENTA AND FISSION AT HIGHER ENERGIES—II

Scission Configurations and the Spin of Fission Fragments 297
    L. Bonneau, P. Quentin, and I. N. Mikhailov

Capture and Fusion-Fission Processes in Heavy Ion Induced Reactions 305
    M. G. Itkis, S. Beghini, B. R. Behera, A. A. Bogatchev, V. Bouchat,
    L. Corradi, O. Dorvaux, E. Fioretto, A. Gadea, F. Hanappe, I. M. Itkis,
    M. Jandel, J. Kliman, G. N. Knyazheva, N. A. Kondratiev, E. M. Kozulin,
    L. Krupa, A. Latina, V. G. Lyapin, T. Materna, G. Montagnoli,
    Y. T. Oganessian, I. V. Pokrovsky, E. V. Prokhorova, N. Rowley,
    V. A. Rubchenya, A. Y. Rusanov, R. N. Sagaidak, F. Scarlassara,
    C. Schmitt, A. M. Stefanini, L. Stuttge, S. Szilner, M. Trotta,
    W. H. Trzaska, and V. M. Voskresenski

ISOLDE Beams of Neutron-Rich Zinc Isotopes: Yields, Release, Decay Spectroscopy 315
    U. Köster, T. Behrens, C. Clausen, P. Delahaye, V. N. Fedoseyev,
    L. M. Fraile, R. Gernhäuser, T. J. Giles, A. Ionan, T. Kröll, H. Mach,
    B. Marsh, M. Seliverstov, T. Sieber, E. Siesling, E. Tengborn, F. Wenander,
    and J. Van de Walle

NEW FACILITIES—II

High Intensity Beams of Fission Fragments at SPIRAL 2 327
    M. Lewitowicz

Accelerator Studies for an ADS within the European Project EUROTRANS 334
    A. C. Mueller

POSTERS

Major Actinide Diffusion (U, Pu) in Oxidised Zirconium Application to Nuclear Fuel Cladding Tubes 345
    N. Bérerd, Y. Pipon, N. Moncoffre, A. Chevarier, H. Faust, and H. Catalette

Binary Fission Fragment Yields from the Reaction ^{251}Cf(n_{th}, f) 349
    E. Birgersson, S. Oberstedt, A. Oberstedt, F.-J. Hambsch, D. Rochman, and
    I. Tsekhanovitsch

The Neutron Induced Fission Cross-Section of ^{240}Pu, ^{243}Am, and ^{nat}W in the Energy Range 1–200 MeV 353
    A. B. Laptev, A. Y. Donets, A. V. Fomichev, A. A. Fomichev, R. C. Haight,
    O. A. Shcherbakov, S. M. Soloviev, Y. V. Tuboltsev, and A. S. Vorobyev

A Novel High-Resolution Time-of-Flight Spectrometer with Tracking Capabilities for Photo-Fission Fragments and Beams of Exotic Nuclei 357
    N. Nankov, E. Grosse, A. Hartmann, A. R. Junghans, K. Kosev,
    K. D. Schilling, M. Sobiella, and A. Wagner

On the Double and Triple-Humped Fission Barriers and Half-Lives of Actinide Elements 361
    G. Royer and C. Bonilla

Experimental Study of Energy Dependence of Proton Induced Fission Cross Sections for Heavy Nuclei in the Energy Range 200–1000 MeV 365
    A. A. Kotov, Y. A. Gavrikov, L. A. Vaishnene, V. G. Vovchenko,
    V. V. Poliakov, O. Y. Fedorov, T. Fukahori, Y. A. Chestnov, and
    A. I. Shchetkovskiy

Development of Digital Technique for the Determination of Fission Fragments and Emitted Prompt Neutron Characteristics 369
    N. Varapai, F.-J. Hambsch, S. Oberstedt, O. Serot, G. Barreau, N. Kornilov,
    and S. Zeinalov

Calculations of Fission Fragment Yields at Low and Intermediate Energy Fission 373
    S. Yavshits and O. Grudzevich

List of Participants..377
Author Index..383

PREFACE

The present book contains the proceedings of the third workshop of a series of workshops previously held in Seyssins in 1994 and 1998. About 100 scientists attended the conference in the friendly working atmosphere of the Castle of Cadarache in the heart of the Provence.

In his opening address, Prof. B. Bigot, the French High Commissioner for Atomic Energy, outlined France's energy policy for the next few decades. He emphasized the continuing progress of nuclear fission in both technical and economic terms, allowing it to contribute to the energy needs of the planet even more in the future than it does today.

We believe that such progress implies a very strong link between fundamental and applied research based on experimental and theoretical approaches. Precisely with this idea in mind, we tried to gather in this workshop the different nuclear communities studying the fission process, including the following topics:

- nuclear fission experiments,
- spectroscopy of neutron rich nuclei,
- fission data evaluation,
- theoretical aspects of nuclear fission,
- and innovative nuclear systems and new facilities.

During the workshop the most recent achievements in these different fields were discussed.

The meeting clearly profited from the exchange of information and from discussions among these different communities.

The scientific program was suggested by the International Advisory Committee and we would like to thank our colleagues for this crucial help. Among them let us spare a thought for Dave Warner and express here our deep sadness at his passing.

This workshop would not have been possible without the efficient organization from Ms. Christine Lemaitre and Ms. Estelle Radufe and the kind hospitality of the CEA/Cadarache and of the Castle of Cadarache.

This workshop was sponsored by the research centers of the Commissariat à l'Energie Atomique located in Bruyères-le-Châtel, in Cadarache and in Saclay, by the Institut Laue-Langevin (ILL) in Grenoble, by the Grand Accélérateur National d'Ions Lourds (GANIL) in Caen, and by the Regional Council of Provence Alpes Côte d'Azur.

Heloïse Goutte
Herbert Faust
Gabriele Fioni
Dominique Goutte

ORGANIZING COMMITTEE

Héloïse GOUTTE
CEA/DAM Ile de France
Bruyères le Châtel
France
Herbert FAUST
ILL Grenoble
France
Gabriele FIONI
CEA/DSM Saclay
France
Dominique GOUTTE
GANIL Caen
France

INTERNATIONAL ADVISORY COMMITTEE

M.B. Chadwick,
Los Alamos National Laboratory,
USA
F. Gönnenwein,
University of Tübingen,
Germany
D. Gogny,
Lawrence Livermore National Laboratory,
USA
M. Itkis,
JINR Dubna,
Russia
R. Jacqmin,
CEA/DEN Cadarache,
France
J. Lachkar,
CEA Paris,
France
P. Ledermann,
CEA/DEN Saclay,
France
K.H. Schmidt,
GSI-Darmstadt,
Germany
J.-L. Sida,
CEA/DAM Ile de France,
France
C. Wagemans,
University of Gent,
Belgium
D.D. Warner,
University of Surrey-Daresbury,
United Kindom

Supported by :

- CEA / DAM, Bruyères-le-Châtel
- CEA, Cadarache
- CEA/DSM, Saclay
- ILL, Grenoble,
- GANIL, Caen,
- Conseil Régional Provence Alpes-Côte d'Azur.

PROGRAMME

Wednesday 11		Thursday 12		Friday 13		Saturday 14	
		III. Fission: Mass and Charge yields		VII. Fragment Excitation and Neutron Emission		X. Angular Momenta and Fission at Higher Energies 2	
		8:30 to 10:30	H. Goutte / L. Bonneau / S. Heinrich / S. Perez / A. Staszczak	8:30 to 10:30	G. Petrov / S. Lemaire / H. Faust / B. Weiss	9:00 to 10:30	P. Quentin / P. Romain / M. Itkis / U. Köster
		10:30-11:00	COFFEE BREAK	10:30-11:00	COFFEE BREAK	10:30-10:45	COFFEE BREAK
		IV. Light Particles and Cluster Emissions		VIII. Mass and Energy Distributions		XI. New Facilities 2	
		11:00 to 12:30	L. Robledo / J. Heyse / F. Gönnenwein / N. Carjan	11:00 to 12:30	K-H. Schmidt / A. Tudora / A. Vorobyev / F. Rejmund	10:45-11:10	M. Lewitowicz
						11:10-11:35	A. Mueller
						11:35-12:15	Round table session of chairmen
13:30-14:00	OPENING ADDRESS	12:30-14:00	LUNCH	12:30-14:00	LUNCH	12:15	End of the workshop SANDWICH LUNCH
I. Cross Sections and Resonances		V. Spectroscopy of Neutron Rich Nuclei					
14:00 to 16:00	F. Hambsch / A. Letourneau / B. Jurado / A. Oberstedt / M. Giacri	14:00 to 16:00	J-C. Thomas / G. Simpson / J. Genevey / J-A. Pinston / I. Al Mahamid	14:00 to 17:00	VISIT OF TORE SUPRA AT CADARACHE RESEARCH CENTER		
16:00-16:30	COFFEE BREAK	16:00-16:30	COFFEE BREAK	17:00-17:30	COFFEE BREAK		
II. Fission at Higher Energies 1		VI. Resonances, Barriers and Fission Times		IX. Needs for Nuclear Data and New Facilities 1			
16:30 to 18:30	M. Bernas / C. Volant / M. Ricciardi / S. Chabod / Y. Rugama	16:30 to 18:30	M. Chadwick / A. Kelic / O. Serot / Y. Lee / M. Morjean	17:30 to 19:15	D. Iracane / J. Galy / S. Oberstedt / D. Ridikas / A. Nichols		
18:30-20:00	Cocktail and Poster session	18:30-20:00	Apéritif				
20:00	DINNER	20:00	DINNER	20:00	DINNER After dinner talk F. Gönnenwein		

Opening of the "FISSION 2005" Workshop at Cadarache, on 11 May 2005
By B. Bigot, High Commissioner for Atomic Energy

Ladies and Gentlemen, Colleagues,

I am particularly happy to address you on the occasion of this the Third International Workshop on Nuclear Fission and Fission Product Spectroscopy.

Happy, first of all, given that this is the third of the series launched in Biarritz in 1996 and continued in Seyssins in 1998. For, in this field, continuity and tenacity are essential.

Secondly, I am all the more happy that this workshop is being held in Cadarache, which has become a veritable temple for energy generation by nuclear fission, given the number and quality of research reactors sited here over the last fifty years; these facilities have established Cadarache's renown within the scientific and industrial community. But now, beyond the restricted circle of nuclear specialists, Cadarache is becoming known throughout the world, following France's proposal to the European Union that the commune be selected to site the major experimental set-up to be developed under the international ITER programme on magnetically confined thermonuclear fusion. The 16 months of difficult negotiations between the 6 partners to the project – China and Korea, the United States and Japan, Russia and the European Union – in search of a consensus on the physical location of the main facilities, have brought to the world's attention both the challenges of fusion and the Cadarache site itself. Indeed, the scientific prestige of this site can only be enhanced by the fact that, last Thursday in Geneva, the European Union and Japan reached a technical agreement opening the way to a final decision on the launching of ITER in the weeks to come. Your decision, therefore, to hold your meeting here in Cadarache, in the midst of these events concerning this new chapter of research with such promise for the future of nuclear power generation, clearly reflects your desire to see fission develop its full potential in the course of, and indeed beyond, this century, as we await the advent of controlled fusion.

I believe with you that the future of the planet depends on the continuing evolution, in both technical and economic terms, of nuclear fission, allowing it to contribute even more than it does today (at 17% of world electricity use) to the energy needs of the planet.

Given the solid advantages of this mode of production - (and here I am thinking of the presence across 5 continents of natural resources of fissile and fertile material with the potential to match our needs for large-scale energy production in a sustainable manner, offering the highest possible guarantees in terms of acceptable energy dependence: no greenhouse gas emissions generating serious risks of climate change with long-lasting consequences for the stability of our planet, no emission of significant volumes of gaseous effluents difficult to contain and impacting negatively on human health or the environment) - given such advantages therefore, there is strong justification for actively pursuing our R&D efforts across all fission production systems, in order to ensure that the risks associated with them are mastered to the fullest extent possible and the likelihood of their occurrence reduced to a minimum.

These risks include, as you are well aware, the potential for radiological contamination of workers and local communities associated with the normal operation of the constituents of the industry (reactors, the plant and installations within the nuclear fuel and waste management cycles), and the risks of incident or accident. The sustainable development of nuclear energy is entirely dependent on engendering confidence within the general public; the demonstration that we can fully master these risks is an essential requirement for this. This in turn supposes that all those involved share a common culture of safety, and that the technology they depend on is both robust technically, and tried and tested scientifically.

You are in the forefront when it comes to designing the technology and establishing the safety culture, and this is both a demanding and fascinating responsibility. It is a long-term mission, leaving no quarter for imprecision or rigidity. It will take all your expertise, energy, creativity and will to succeed, to achieve the aims we have set in good time.

France, for its part, has defined its priorities clearly. The challenge is first to maintain, or improve wherever possible, the performance of its reactor and fuel cycle facilities, whilst ensuring their total safety through the course of their life-time, and then also to prepare for their future; in both cases, we will need to develop procedures for the safe management of all the nuclear waste generated by their operation.

It is clear, in the light of the requirements foreseeable over the next few decades in scientific, technical and industrial terms, that at least half of our current plant will have to be renewed between 2015/2020 and 2035/2040. This is envisageable only through the construction of new EPR generation reactors. For the remaining facilities, the question is whether France's industrial sector will be able to count on the services of a new-generation fast-neutron reactor in the period beyond 2040 - i.e. whether France will be sufficiently well placed, at that time, to start exploiting efficiently the entire stock of uranium and plutonium she will have accumulated by then, or whether, alternatively, banking on the continued availability, in sufficient quantities, of natural uranium at a cost capable of maintaining the competitivity of thermal technology, she chooses to postpone the fast-neutron scenario, in other words, putting it aside until there is a need to renew the EPRs at the end of the century. The first of these options would clearly avoid the risks of a uranium market under pressure (the first signs of which are already visible) and would enable us to accumulate industrial experience ensuring, for several centuries to come, a supply of energy impacting only weakly on the climate and environment. It would also be the logical development of France's current decision on the processing of spent UOX fuel and the storage of spent MOX fuel for later processing, with the ultimate objective of partitioning the minor actinides they contain and transmuting most if not all of these, in conjunction with what is currently being done for plutonium.

Concretely, in terms of reactor R&D, this means that between now and 2015 ambitious R&D programmes will have to be established, based on the following:
- an equal effort on both the SFR and the GFR systems, bringing together all those in the field of research capable of contributing at national and international level to overcome the clearly identified scientific and technical obstacles they pose, to ensure that we are well prepared to make the final choice in 2015/2020 between the two reactor types, thus ensuring that by 2040 our country possesses an innovative industrial reactor meeting the objectives set by the Generation IV Forum;
- a sustained R&D effort on the production of hydrogen using energy from the nuclear cycle;
- the pursuit of R&D efforts into thermal neutron reactors, aimed at minimal lifetimes of between 40 and 50 years for the majority of them, and at achieving operating conditions throughout this period that improve their overall performance (improved specific burn-up, fuel conversion and effective availability, etc.);
- R&D efforts on two specific aims (accelerator, spallation target), as a prerequisite for the decision on the construction of a prototype ADS "transmuter" reactor from 2015 onwards.

On fuel and waste R&D, we need:

- a clear demonstration that the formal requirements are being met with regard to the disposal in clay geological formations of high-level (HA) waste and intermediate-level long-lived (MA-VL) waste as they are now being produced, with a view to obtaining a decision on the creation of a site in 2020/2025 optimizing disposal conditions. This should principally be based on compliance with the technical standards that could be set by our country, i.e. that the radiological dose rate to which future generations could be exposed be limited to no more than the level we receive today in the vicinity of our power stations, and which in our consideration presents no risk to the public and the environment. The fundamental principle is to provide for barriers between man and the waste sufficient in number to ensure that any release into the biosphere (whether of the radionuclides held in the original waste or of their decay products) occurs on a time scale allowing their eventual radiological impact to be below, or at the very most equal to, what is today considered capable of exerting an acceptable impact on the environment and the general health of the population;
- a demonstration of the feasibility of introducing spent fuel processing systems compatible with the project to partition minor actinides for transmutation in fast-neutron reactors;
- a demonstration of the feasibility of manufacturing fuel suitable for use in fast reactors, and of reprocessing this fuel, in line with the aim of producing ultimate waste containing only fission products and less than 0.1% uranium, plutonium, americium and neptunium.

In the light of these aims, a meeting of this type, uniting over 100 participants from 13 different countries and jointly organised by the different divisions of the CEA and two major international laboratories, is particularly welcome. It provides the occasion for scrutiny, in a convivial setting, of the respective available data and the most recent models and theoretical developments; it will also incite debate on the new instrumentation that will be needed if new data is to be acquired.

This is all the more important as the fission process brings into play the more intricate elements within the structure of the nucleus and the most comprehensive reaction mechanisms determining the nature and the consequences of the neutron's interaction with the nucleus – in other words both the static and the dynamic components within the nuclear edifice. It is indeed essential, if we are to satisfy our curiosity through a deeper understanding of the universe we live in, or to make full use of the energy potential within fissile nuclei.

We have already assembled a considerable set of data. We must go even further, if the promises of fission are to be realised fully.

I sincerely hope that this meeting provides the conditions you need to make as efficient progress as possible to this end. We are confident that you will succeed, both in your research and in this workshop.

Thank you for your attention.

CROSS SECTIONS AND RESONANCES

Fission Cross Section Calculations for Pa Isotopes

F.-J. Hambsch[†], G. Vladuca[¶], A. Tudora[¶], S. Oberstedt[†], D. Filipescu[¶],

[†] *EC - JRC IRMM, B-2440 Geel, Belgium*
[¶] *Faculty of Physics, Bucharest University, RO-76900 Bucharest, Romania*

Abstract. Based on the recently measured cross-section values for the neutron-induced fission of ^{231}Pa and our experience gained with other isotopes, new self consistent neutron cross section calculations for n+^{231}Pa have been performed up to 30 MeV. The results are quite different to the existing evaluations, especially above the first chance fission threshold.

Keywords: Protactinium-231, Protactinium-233, neutron-induced fission, fission cross-section, statistical model calculations, Th-U fuel cycle.
PACS: 24.10.Pa, 25.85.Ec, 28.20.-v

INTRODUCTION

Besides the importance of ^{231}Pa for basic fission studies it is also of interest in the field of future reactor design [1,2,3] in which it plays a role in the generation of ^{232}U in the ^{232}Th/^{233}U breeder cycle, contemplated to be able to provide *"clean"* fission energy, according to the following successive processes:

$$^{232}_{90}\text{Th} \xrightarrow{(n,2n)} {}^{231}_{90}\text{Th} \xrightarrow{\beta^-(25.5h)} {}^{231}_{91}Pa \xrightarrow{(n,\gamma)} {}^{232}_{91}Pa \xrightarrow{\beta^-(1.31d)} {}^{232}_{92}U \xrightarrow{(n,\gamma)} {}^{233}_{92}U.$$

Many experimental data of the neutron-induced fission of ^{231}Pa exist but the scatter is rather large. Also, the two existing evaluated files (ENDF/B-VI [4] and JENDL3.3 [5]) arrive at quite different results for the ^{231}Pa(n,f) cross section.

Based on the newly available IRMM experimental data of the ^{231}Pa(n,f) fission cross-section in the incident neutron energy range 0.76 to 3.46 MeV [6], a new, self-consistent, neutron cross-section calculation for n+^{231}Pa, in the incident neutron energy range 0.01-30.0 MeV has been performed.

This calculation involves also the inclusion of lighter Pa isotopes (^{228,229,230}Pa). In view of that, also the previous calculation of n+^{233}Pa up to 20 MeV (see Ref. [7]) could be extended up to 50.0 MeV.

The fission cross section of the involved neutron reactions were calculated in competition with all other possible processes. In the studied incident neutron energy range, the neutron interaction takes place through both direct and compound nucleus mechanisms. The direct mechanism has been treated with the coupled channel method and the compound nucleus mechanism with the statistical model.

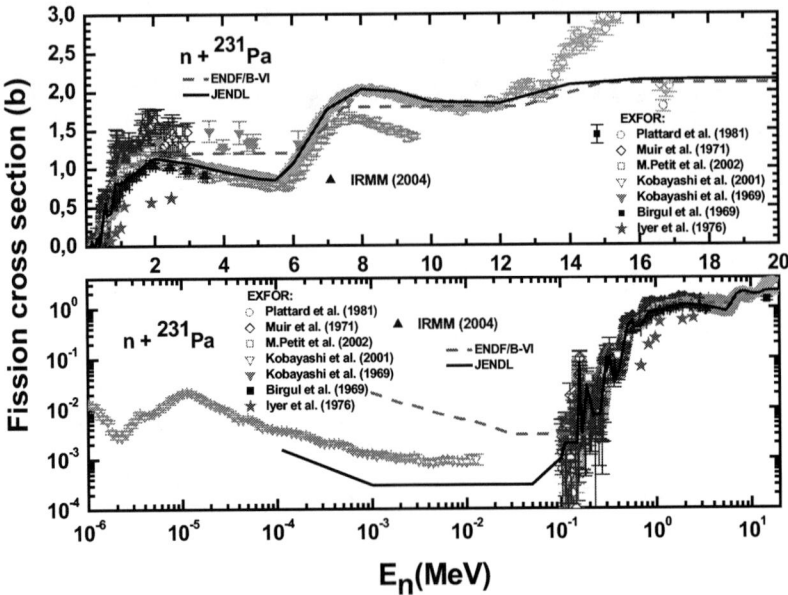

FIGURE 1. Experimental database (from EXFOR [8]) for the fission cross section of ^{231}Pa, including the most recent IRMM measurements [6].

CALCULATIONAL PROCEDURE

One important ingredient in the statistical model calculations and, in fact the first step, is to have a correct coupled channel calculation. This calculation provides the total cross-section (σ_t), the direct elastic ($\sigma_{n,n}^{DI}$) and individual inelastic cross-sections (σ_{n,n_i}^{DI}) for each coupled level i, the neutron transmission coefficients $T_{nlj}^{J\pi}$ for different partial waves and moments (lj) for each possible spin and parity ($J\pi$) of the compound nucleus and the compound nucleus cross-section (σ^{CN}), which is calculated with the neutron transmission coefficient.

From the coupled channel calculation it turns out that:

$$\sigma^{CN} = \sigma_t - \sigma_{n,n}^{DI} - \sigma_{n,inel}^{DI} \qquad (1)$$

where:

$$\sigma_{n,inel}^{DI} = \sum_{i=1}^{N} \sigma_{n,n_i}^{DI} \qquad (2)$$

is the total direct inelastic cross-section and N is the number of coupled levels.

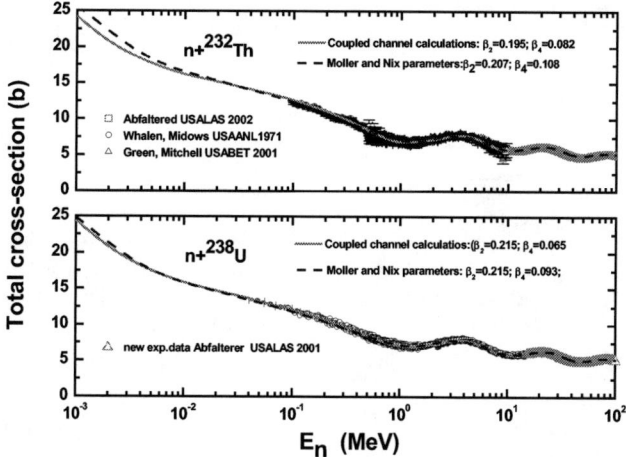

FIGURE 2. Total cross section of ^{232}Th and ^{238}U obtained by coupled channel calculations up to 100 MeV.

Also, the compound cross-section σ^{CN} given by the direct interaction calculation has to be equal with all competitive cross-sections undergoing compound nucleus formation. For instance, in the incident neutron energy where only the first fission chance is possible, this cross-section will be:

$$\sigma^{CN} = \sigma_{n,f} + \sigma_{n,\gamma} + \sigma_{n,n}^{CN} + \sigma_{n,inel}^{CN} \tag{3}$$

From the above considerations the importance of coupled channel calculations is obvious. Also the parameterization for the phenomenological deformed optical potential that must be used for the nuclei involved is crucial.

We extended our parameterization for each actinide given in Ref. [9] up to 100 MeV incident neutron energy with the same coupling levels and the same deformation parameters. The agreement of the coupled channel calculation, performed with the ECIS95 code [10], with the existing total cross-sections of neighboring isotopes ^{232}Th and ^{238}U is very good, as it can be seen in Fig. 2. Therefore, confidence in the calculations for Pa is justified, although for Pa isotopes no experimental data exist. In these figures the deformation parameters given in Ref. [9] and used in the coupled channel calculation as well as the Möller and Nix parameters [11] are given.

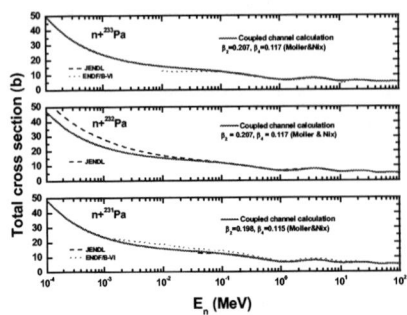
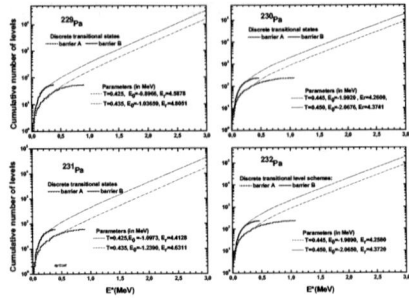

FIGURE 3. Total cross section of ^{231,232,233}Pa obtained by coupled channel calculations using deformation parameters of Möller and Nix in comparison with the ENDF/B-VI and JENDL 3.3 evaluation (broken lines).

FIGURE 4. Cumulative number of transitional states at the saddle points for some Pa isotopes considered in the calculations.

Since the isotopes ^{232}Th and ^{238}U are close to the protactinium isotopes we conclude that the Möller and Nix deformation parameters can be used for the protactinium isotopes, too. The total cross-sections for n+^{231,232,233}Pa, from 0.1 keV up to 100 MeV are compared with the ENDF and JENDL evaluations in Fig. 3.

The elastic and individual inelastic cross-sections for the coupled levels are the sum of the direct and compound nucleus contributions. For the other inelastic levels (discrete and continuum) only the compound nucleus contribution was considered. The compound nucleus mechanism was treated in the framework of the statistical model. For the involved Pa-isotopes we used the discrete level spectrum (if discrete levels exist) and the continuum spectrum. For instance in ^{231}Pa a discrete level spectrum up to 329.10 keV was assumed consisting of 22 levels taken from the Reference Input Parameter Library (RIPL) [12]. For the nucleus ^{230}Pa the level scheme is completely unknown, only the ground state level with spin and parity $I^{\pi} = 2^{-}$ is provided [12]. Therefore, a rotational band with $I^{\pi} = 2^{-},3^{-},4^{-}$ and $\hbar^2/2\Im = 6.84\ keV$ and a discrete level spectrum up to 95.75 keV was constructed consisting of 3 levels. For ^{229}Pa, 10 discrete levels are given in RIPL but some of them without spin and/or parity. In consequence, we identified for this isotope some band-heads, and assigned the spin and parity for the unknown levels. Finally, a discrete level spectrum up to 241.90 keV consisting of 9 levels has been attributed for this nucleus. Similar problems arose for the other Pa-isotopes involved in this calculation e.g. ^{227,228}Pa. From the above discussion it is obvious that the continuum spectrum is very important for the present calculation. This spectrum is described by the composite level density function of Gilbert and Cameron [13]. Because of lack of experimental data for the neutron s-wave average level spacing at the neutron binding energy, B_n, for the Pa-isotopes (except for ^{232}Pa for which $D_{exp}(B_n) = (0.45+/-0.05)$ eV) the level density parameter, a, was obtained theoretically using the Gilbert and Cameron [13] recipe. The parameters are given in Table 1.

Practically the same procedure was used for the transitional level spectrum (discrete and continuum) at the deformations of the saddle points. Fig. 4 shows the dependence of the cumulative number of transitional states at the saddle points on the excitation energy of four isotopes of Pa considered in the calculations. A very smooth crossing of the experimental levels with the calculated one is observed.

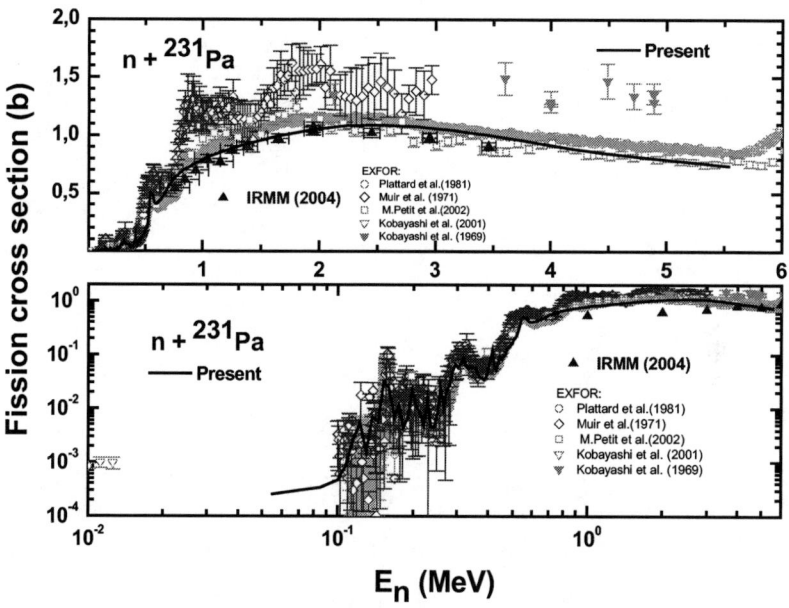

FIGURE 5. The experimental database (from EXFOR [8]) for the fission cross section of ^{231}Pa, including the most recent IRMM measurements [6] is compared to the calculation up to 5.5 MeV incident neutron energy.

Parameters	$^{228}_{91}Pa$	$^{229}_{91}Pa$	$^{230}_{91}Pa$	$^{231}_{91}Pa$	$^{232}_{91}Pa$
a (MeV^{-1})	28.3636	28.6770	28.7811	28.7595	28.7761
B$_n$ (MeV)	5.9810	7.0915	5.7948	6.8170	5.5540
Δ (MeV)	0.0000	0.7900	0.0000	0.6000	0.0000
θ (MeV)	0.3975	0.3947	0.3937	0.3936	0.3933
E$_0$ (MeV)	-1.2852	-0.4998	-1.2894	-0.6859	-1.2834
Er (MeV)	3.1772	3.9633	3.1696	3.7658	3.1621
Dcalc(eV)	0.1884	0.1060	0.2260	0.1212	0.5110

Table 1. Parameters entering the Gilbert-Cameron level density calculations.

The fission cross section was calculated with the STATIS code [7]. In this code every transitional state is represented by a double-humped fission barrier. Each fission barrier is constructed from inverted parabolas for the inner (A) and outer barrier (B) and a parabolic barrier for the isomeric well (I). The transmission coefficients for absorption and direct fission are calculated in the JWKB approximation using the optical model for fission. The direct, indirect, and isomeric fission processes are taken into account, as well as sub-barrier effects. For the transition states both the discrete and continuum part of the spectrum are considered.

Compared to the results published in Ref. [7], the parameters of the fundamental double humped fission barrier as well as the temperatures $T_{A,B}$ were slightly modified to have best agreement with the recent IRMM data [6] on ^{231}Pa(n,f). A comparison of the present calculation with the available experimental data is given in Fig. 5. Here one should emphasize the very good agreement with the data of Plattard et al. [14] and the most recent surrogate experiment of Petit et al. [15]. In the lower part of Fig. 5 the resonance structure is emphasized. Also here the present calculations (full line) are in very good agreement with the data of Plattard et al. [14].

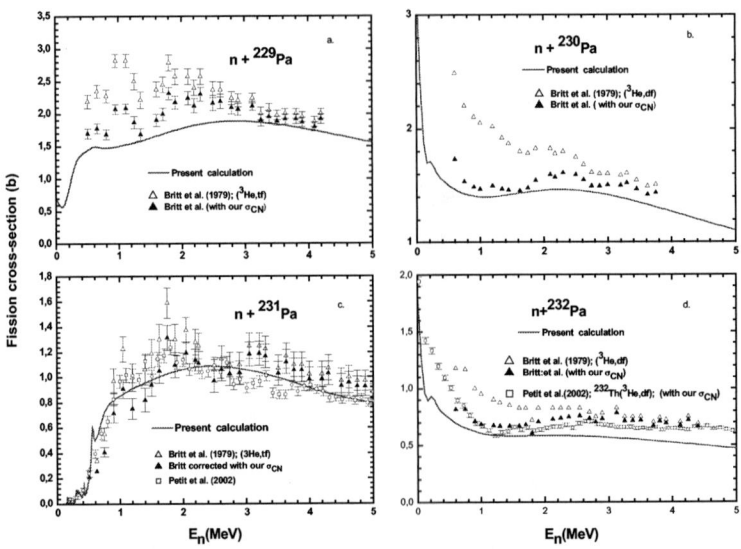

FIGURE 6. Calculated fission cross sections for the involved Pa isotopes in comparison with surrogate fission cross section data obtained by Britt and Wilhelmy [16].

The first chance fission cross-section for n+^{229,230,231,233}Pa given in Fig. 6 are very important for the ^{231}Pa(n,f) calculation up to 30 MeV where higher fission chances are involved. In the reaction ^{231}Pa(n,f) the ^{230}Pa(n,f) (Fig. 6b) is very important in this calculation being the second chance and, hence, is important in the neutron energy range from about 6 MeV up to 14 MeV. In this neutron energy range, as can be seen in the upper part of Fig. 1 important differences exist between the Plattard et al. [14] and the Petit et al. [15] experimental data. The calculation in Fig. 6 is compared to data from surrogate experiments by Britt and Wilhelmy [16]. Fig. 7 shows the full calculation of the fission cross section of ^{231}Pa up to 30 MeV incident neutron energy.

It is obvious from this figure that large discrepancies exist in the available experimental data above the second chance fission threshold. Here the spread amounts to more than 100%. Our new calculation favours more the tendency given by the surrogate experiment of Petit et al. [15] and the singular point of Birgul et al. [17]. Also the evaluations are quite apart from our calculation from 7 MeV to 17 MeV.

CONCLUSION

In this paper we have reported about self-consistent neutron cross section calculations for ^{231}Pa in the neutron incident energy range up to 30 MeV. For the first time the resonance structure in the threshold of the fission cross section could be reproduced in the calculation in reasonable agreement with the experimental values. Although not shown beside the fission cross section also all other reaction cross sections have been calculated in addition.

The new calculation shows best agreement with the most recent experimental data [6, 15] and deviates quite drastically from the evaluated files and older measurements especially above the emissive fission threshold. Additional experimental investigations are necessary to pinpoint down the higher chance fission cross sections.

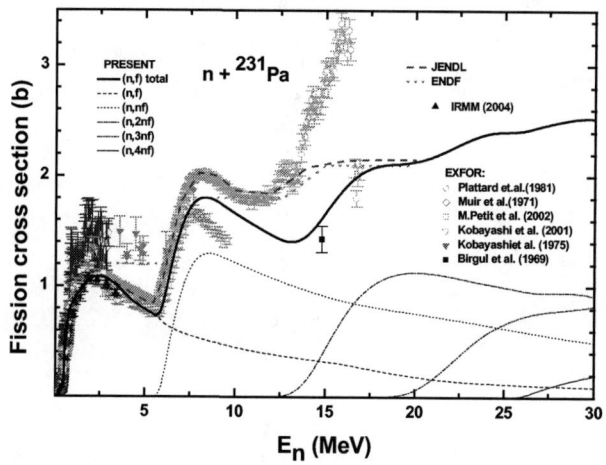

FIGURE 7. Calculated fission cross section up to 30 MeV in comparison with the evaluated files and the available experimental data from EXFOR [8] and the recent IRMM experimental data [6].

REFERENCES

1. Rubbia, C., Rubio, J. A., Buono, S., Carminati, F., Fiétier, N., Galvez, J., Gelès, C., Kadi, Y., Klapisch, R., Mandrillon, P., Revol J.P., and Roche, C., *CERN report No. CERN/AT/***95-44**(*ET*), (1995).

2. IAEA, CRP Evaluated nuclear data for the Th-U fuel cycle, http://www.iaea.org/programmes/ripc/nd/crps/eval_nucldata_th_u.htm
3. B. D. Kuzminov, V. N. Manokhin, Nuclear Constants, Issue No. 3-4, 41 (1997), (in Russian). English translation printed in *IAEA report INDC(CCP)-416*, (1998).
4. ENDF/B-VI evaluated nuclear data library IAEA-NDS (2003), ed.3, 1990, MAT=9137, MF=3, MT=18
5. JENDL3.3 evaluated nuclear data library IAEA-NDS (2003), rev.2, 2002, MAT=9137, MF=3, MT=18
6. A. Oberstedt, S. Oberstedt, E. Birgersson, F.-J. Hambsch, V. Fritsch, G. Vladuca, A. Tudora, this conference.
7. Vladuca, G., Hambsch, F.-J., Tudora, A., Oberstedt, S., Oberstedt, A., Tovesson, F., and Filipescu, D., *Nucl. Phys. A* **740**, 3-19, (2004).
8. EXFOR nuclear data library, IAEA-NDS (2004), target nucleus Pa-231, reaction (n,f) quantity: cross-section.
9. G. Vladuca, A. Tudora, M. Sin, Rom. J. Phys. 41 (1996) 515; and Reference Input Parameter Library (RIPL),segm IV, id600.
10. J. Raynal, 1994, Notes on ECIS94, CEA-N-2772 and private communication.
11. Reference Input Parameter Library, RIPL, Segment I Deformations, recommended file (Möller P., Nix J.R), 1998.
12. Reference Input Parameter Library, RIPL, Segment II, Discrete Level Schemes, recommended file, 1998.
13. A. Gilbert, A.G.W.Cameron, Can. J. Phys. 43 (1965) 1446.
14. Plattard, S., Auchampaugh, G. F., Hill, N. W., de Saussure, G., Harvey, J. A., and Perez, R. B., *Phys. Rev. Lett.* **46**, 633 (1981).
15. Petit, M., Aiche, M., Barreau, G., Boyer, S., Carjan, N., Czajkowski, S., Dassié, D., Grosjean, C., Guiral, A., Haas, B., Karamanis, D., Misicu, S., Rizea, C., Saintamon, F., Andriamonje, S., Bouchez, E., Gunsing, F., Hurstel, A., Lecoz, Y., Lucas, R., Theisen, Ch., Billebaud, A., Perrot, L., and Bauge, E., *Nucl. Phys. A* **735**, 345–371 (2004).
16. H.C. Britt, J.B.Wilhemy, Nucl.Sci.Eng 72 (1979)222.
17. O. Birgul and S. J. Lyle, J. Radiochimica Acta 11 (1969) 108.

ns at ILL: the Mini-INCA project

A. Letourneau[1*], I. Al Mahamid[5], Ch. Blandin[2], O. Bringer[1], S. Chabod[1], F. Chartier[3], H. Faust[4], G. Fioni[1], Y. Foucher[1], F. Marie[1], P. Mutti[4], Ch. Veyssiere[1]

[1]*CEA/Saclay/DSM/DAPNIA – Gif-sur-Yvette, France. *Email: aletourneau@cea.fr*
[2]*CEA/Cadarache/DEN/DER/SPEX – Saint-Paul-lez-Durances, France*
[3]*CEA/Saclay/DEN/DPC/SECR – Gif-sur-Yvette, France*
[4]*Laue-Langevin Institut, Grenoble, France*
[5]*Lawrence Berkeley National Laboratory, E.H.&S division – USA (CA)*

Abstract. Fission cross section of short-lived minor actinides is of prime importance for the incineration of minor actinides in high and thermal neutron fluxes. But due to the shortness of their half-lives, measurements are difficult to handle on these isotopes and the existing data present some large discrepancies. An original method has been developed, in the framework of the Mini-INCA project at ILL, to measure the fission and capture cross sections of minor actinides with low error bars associated even for short-lived isotopes. This method lies on a quasi on-line alpha- and gamma-spectroscopy of irradiated samples and on the use of fission micro-chambers. Coupled to a very powerful Monte-Carlo simulation, both microscopic information on nuclear reactions (total and partial cross sections for neutron capture and/or fission reactions) and macroscopic information on transmutation and incineration potentials could be gathered. In this paper, the method is explained in its originality and some recent results are given and compared with existing measurements and evaluated data libraries.

Keywords: Neutron physics, fission, transmutation
PACS: 25.40.Lw, 25.85.Ec, 28.41.Kw, 29.40.Cs

INTRODUCTION

Fission is a key process when dealing with the incineration of Minor Actinides (MA) responsible for the long term radiotoxicity of nuclear wastes. The major contributors have very low fission cross sections in the thermal energy region and their incineration is mainly envisaged into fast neutron fluxes with the advantage of reducing the production of higher mass isotopes. Another envisaged option is to use high thermal neutron fluxes [1] with the advantage of higher fission cross sections and thus a gain in the incineration process of at least a factor 1000. In the latter option, high neutron fluxes are needed to transmute first the fertile actinide into fissile one and to make the neutron-induced fission process competitive with the natural decay process. Unfortunately, for some of the major fissile actinides the half-lives are short ranging from 26 mn (^{244m}Am) to 2.117 days (^{238}Np). Neutron fluxes above 10^{15} n/cm^2/s are thus needed to induce the fission before the fissile isotope could decay. In this context and within the idea to study the best conditions of neutron fluxes to optimize the transmutation-incineration process, it is clear that precise measurements on the formation probabilities of the fissile isotopes and on their disappearance probabilities (fission and capture cross sections, β- and α-decay probabilities) are obviously needed. But due to their short half-lives, measurements on these fissile isotopes are extremely difficult to handle, explaining the rarity of experimental data and the observed discrepancies between existing ones and between evaluated ones. As a consequence, nuclear data do not have the quality required to asses the optimal performances of a system dedicated to the transmutation of MA and it could even result in wrong conclusions. For instance, when considering the transmutation chain of ^{237}Np, the formation probability of the fissile ^{238}Np is know with an error of 12% due to the two existing values of the ^{237}Np(n,γ)^{238}Np cross section of 161.7 b for JENDL3.2 and 181 b for ENDF-B6 and JEFF 3.0. The ^{238}Np(n,F) cross section has been measured by Abramovitch et al. [2] and Danon et al. [3]. They found respectively a value of (2110±740) b and (2638±58) b. The ^{238}Np(n,γ)^{239}Np has been recently measured by Hirada et al. who found a value of (479±24) b whereas ENDF-B6 gives 203 b and JENDL3.2 gives 450 b.

This situation has motivated the development of a very ambitious experimental program, namely Mini-INCA, dedicated to the study of the nuclear parameters (total and partial capture cross sections, fission cross sections, half-lives) of MA isotopes and their incineration potential in different neutron fluxes, by the mean of integral measurements. In the framework of this program, an original method has been developed to measure the fission cross sections of short-lived fissile isotopes involved in the transmutation-incineration process of MA.

EXPERIMENTAL METHOD AND APPARATUS

The originality of the method and the accuracy obtained on the data lie within the coupling of different precise nuclear techniques of measurement, precise Monte Carlo modelisations and the availability of high neutron fluxes. The experimental method could be described basically into three steps. The first step consist to characterize precisely the neutron flux in which samples are irradiated. Then the second step is to characterize the transmutation chain (capture and combustion cross sections) of the fertile actinide to finally measure the fission cross section of the fissile one.

The Mini-INCA experimental set-up is installed at the High Flux Reactor (HFR) of the Laue Langevin Institut (Grenoble - France). The HFR is a 58 MW reactor with an enriched fuel element surrounded by heavy water and providing thermal neutrons with intensities as high as $1.5 \cdot 10^{15}$ n/cm^2/s. The Mini-INCA set-up uses two irradiation channels, namely H9 and V4 (fig.1) which provide thermal neutrons with 2% and 15% of epithermal neutrons, respectively, and with an intensity of about $6 \cdot 10^{14}$ n/cm^2/s for H9 and a maximum intensity of $1.5 \cdot 10^{15}$ n/cm^2/s for V4.

The H9 channel is coupled to the Mini-Inca chamber which allows accurate alpha- and gamma-spectroscopy measurements [4] just after the irradiation and even between two successive irradiations of the same sample. Samples are automatically transferred from the irradiation point in H9 to the Mini-INCA chamber and positioned in front of the detection system with an absolute precision of 0.1 mm. The detection system is composed by a high purity coaxial Ge detector with a resolution of 1.8 keV at 1.17 MeV and by a Passivated Implanted Planar Silicon (PIPS) with a resolution of 11.8 keV for 5.486 MeV alpha particles. Both of them can move on their axis to choose the most suitable counting rates. The chamber and the detectors are shielded by a 5 cm thick lead wall and by a combination of borated polyethylene and B_4C to reduce the gamma and neutron background.

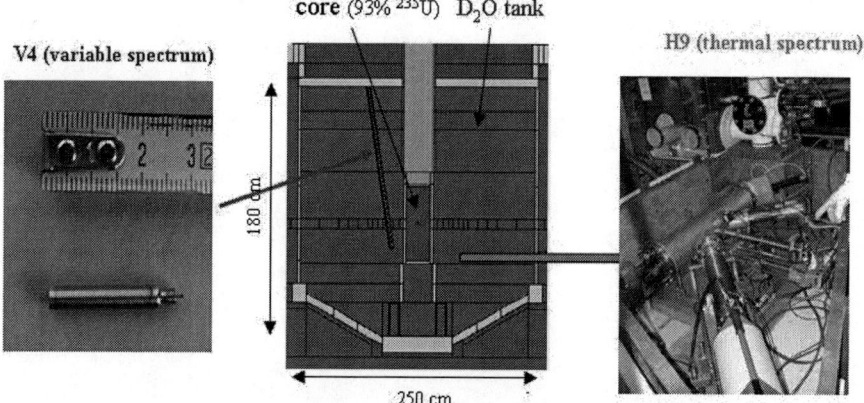

FIGURE 1. The Mini-INCA installation at ILL. The core reactor is shown with the two irradiation channels: the vertical one V4 starting at 10 cm from the fuel element, and the horizontal one H9. Also are shown the alpha and gamma-spectroscopy station on H9 (right photo) and the fission micro-chamber (left photo) that instruments the V4 channel.

The V4 channel could be equipped either with an irradiative system allowing the irradiation of samples for off-line spectrometry, for isomer production [13], neutron induced damage studies on material..., or by fission micro-chambers (FC). These 4 mm in diameter and 2 cm in length detectors have been designed and developed [4] to operate in high neutron fluxes and to measure on-line the fission probabilities of the deposited isotope. Thanks to a small gap between the anode and the cathode, space charge effects due to the very high number of fissions are reduced as compared to a bigger geometry.

The complementarities of these two experimental apparatus and the high intensity neutron fluxes are essential for the success of the experiments: whereas H9 is more dedicated to the capture and combustion cross sections and half-life measurements at thermal energy, V4 gives information on fission cross sections and on all the transmutation chain parameters in different neutron energy spectra.

NEUTRON FLUX CHARACTERISATION

One of the main critical point in integral measurements is the characterization of the neutron flux in which samples are irradiated in order to extract usable data. We have used the MCNP code to modelise the simple cylindrical geometry of the ILL core [7] and to simulate the spectral shape of the neutron energy distributions (fig. 2) available at the different irradiation positions (V4 positions and H9 channel). This approach has been validated by integral measurements using the $^{59}Co(n,\gamma)^{60}Co$ and $^{93}Nb(n,\gamma)^{94}Nb$ standard reactions and the $^{235}U(n,F)$ standard reaction with FC [6].

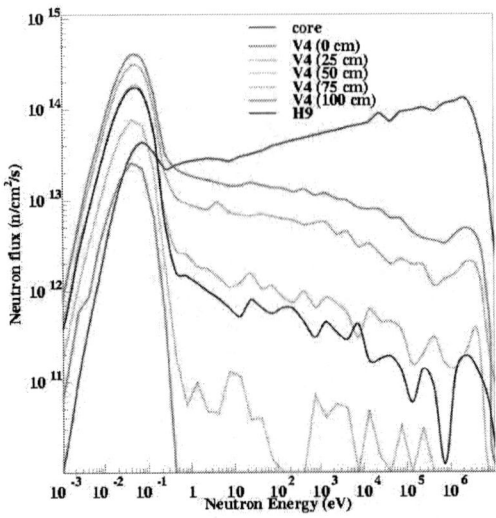

FIGURE 2. Neutron flux energy spectra as simulated with MCNP at different positions of irradiation at the HFR.

TABLE 1. Measured and calculated (MCNP) neutron fluxes (n/cm²/s) at different positions of the V4 channel and H9 channel. The fluxes were measured with Co and Nb activated foils and with the fission micro-chambers (FC).

Irradiation positions	Simulated	Co/Nb foils	FC
0 cm	15.7±0.5	16.0±0.7	17.1
25 cm	13.0±0.4	12.7±0.9	13.6
50 cm	7.0±0.2	6.87±0.48	7.3
H9	6.0±0.2		6.78
75 cm		2.78±0.19	2.89
100 cm		0.90±0.06	0.96

Results obtained experimentally and calculated are compared in table 1 and show a quite good agreement, validating by the way the use and the functioning of the fission micro-chambers in intense neutron fluxes.

Measured cross sections strongly depend on the shape of the modelised neutron flux and resonances could play a non negligible role for some isotopes due to the non-thermal contributions of the neutrons. In order to estimate these corrections and to extract the point-like cross section at 25.3 meV from the measured data, a mean cross section is calculated by integrating the differential data library cross section over the modelised neutron flux. This value is compared to the point-like value in the library and the measured value corrected from the ratio of these two quantities as explained in [8]. In that sense the role play by the resonances is taken into account and errors estimated in consequence.

CAPTURE CROSS SECTIONS

Within the objectives to characterize the transmutation chain of MA and in particular the formation probabilities of fissile MA isotopes, the capture cross sections are of course of prime importance. Measurements already performed on H9 mainly concern the capture cross sections of isotopes which contribute significantly to the mass inventory of the nuclear waste as ^{237}Np and ^{241}Am or to their transmutation chain. All these measurements have been done with approximately 10 μg of pure actinide, deposited on a thin Ni-backing. The diameter of the deposit was less than 5 mm, centered on the target. The typical irradiation times were between 2 hours for the short irradiations dedicated to the capture cross sections determination, and 2 days for the long irradiations dedicated to the combustion cross section determination. Each actinide samples were accompanied with an Al-0.1%Co monitor to measure in the same irradiation conditions the neutron flux.

Results are presented in Table 2 and compared with evaluated data. All the results was obtained with the Mini-INCA chamber coupled to the H9 beam tube [10] except for ^{241}Am [9].

TABLE 2. Reaction cross sections obtained in this work and compared to evaluated data libraries

Reaction	This work	ENDF-BVI	JENDL3.2	JEFF2.2
^{241}Am (n, γ)^{242}Am Branching Ratio	(696 ± 48) b 0.914 ± 0.007 [9]	618.65	639.49	618.15
^{242gs}Am (n, γ)^{243}Am	(330 ± 50) b [9]	252.09	219.02	5500
^{243}Am(n, γ)^{244}Am	(81.8 ± 3.9) b [10][11]	75.101	78.55	76.02
^{243}Am(n, γ)^{244gs}Am	(5.2 ± 1.7) b [10][11]			
^{242}Pu (n, γ)^{243}Pu	(22.5 ± 1.1) b [10][11]	19.27	18.80	18.52
^{237}Np (n, γ)^{238}Np	(180 ± 5) b [12]	181.02	161.73	181.02
^{238}Np combustion	(3200 ± 240) b [12]	2229.73	2520	

A global comparison shows some discrepancies between our results and evaluated data libraries. In particular, the new ^{242gs}Am $(n,\gamma)^{243}$Am cross section is 16 times less than JEF2.2 whereas it is compatible with ENDF-B6 and JENDL3.2 data. Less spectacular is the ^{243}Am$(n,\gamma)^{244}$Am cross section that we found 9% higher than ENDF-B6 but compatible with JENDL3.2. The ^{243}Am$(n,\gamma)^{244gs}$Am cross section is compatible with previous data whereas ^{242}Pu $(n,\gamma)^{243}$Pu cross section is 16% higher than library values. The new ^{237}Np $(n,\gamma)^{238}$Np cross section is compatible with ENDF-B6, JEFF2.2 but 9% higher than JENDL3.2. The ^{238}Np combustion cross section is 27% and 44% higher than the ENDF and JENDL values respectively.

FISSION CROSS SECTIONS

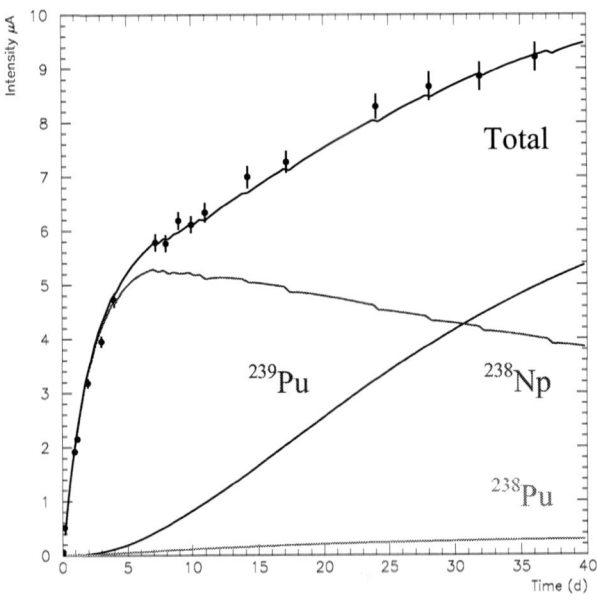

FIGURE 3. Current measured (symbols) with a TDFC containing initially ^{237}Np as a function of the irradiation time. The irradiation was done during 40 days. Also is shown the total adjusted evolution function, included all the contributors to the fission current (continuous line named Total) and the partial contribution of each main isotopes.

Fission cross sections are measured with FC irradiated in V4. Single Deposit Fission Chambers (SDFC) with ^{235}U as active deposit have been tested at ILL [6] in 2002 and used to characterize the neutron flux and to follow its temporal variations. In 2004 we have developed and tested Triple Deposit FC (TDFC) based on three electrically independent SDFC sharing the same gas: one chamber without deposit dedicated to background measurements, another with ^{235}U deposit to monitor the neutron flux and the last one with actinide deposit dedicated to the fission rate and incineration potential measurements. The first measurement was done in 2004 on ^{237}Np to determine the fission cross section of ^{238}Np and its evolution in a high thermal neutron flux. A TDFC containing 42±3.1 μg of ^{237}Np in one chamber, 4±0.02 μg of ^{235}U in the second one has been irradiated during 40 days in a neutron flux of 10^{15} n/cm^2/s with about 90% of neutrons having energies less than 1 eV.

The measured current as a function of time is shown on Fig.3. On the figure are also shown the global adjustment done with the solutions of the Bateman equations and the contributions of each isotope to the fission rate as a function of time. It is clear that during the first days only ^{238}Np contribute to the fission rate. The analysis is still in progress and the results presented in this paper are still preliminary. We found a value of 2197±33 b (relative error) for the ^{238}Np(n,F) cross section that is 10% higher than evaluated one. This value has been obtained by adjusting the measured intensity as a function of time with the solutions of the Bateman equations and by fixing the capture cross section of ^{237}Np and the combustion cross section of ^{238}Np to the values measured on H9. The capture cross section of ^{238}Np that could be determined by the same method is less sensitive than the fission cross section and its value determination needs a more precise analysis.

CONCLUSION

The Mini-INCA set-up installed at the ILL High Flux Reactor is well suited for transmutation studies in high thermal neutron fluxes. The two dedicated apparatus are complementary and allow microscopic as well as macroscopic measurements even on rare and short-lived isotopes. The recent capture cross sections that we have measured show good agreements with the most recent measurements but show sometimes great divergences from evaluated data libraries. The new FC (TDFC) developed for high neutrons fluxes seems to be a very promising tool for transmutation and incineration studies. They allow to follow the evolution of an actinide in a given flux and also to monitor the variations of the neutron flux. These TDFC have been used to measure on-line the fission cross section of ^{238}Np. The program will be extended soon to actinides heavier than Am (Cm, Bk, Cf, ...). Moreover, taking benefit of the FC, the same method will be applied to characterize the neutron flux inside the MEGAPIE [14] spallation target and to measure its incineration potential onto ^{241}Am and ^{237}Np.

REFERENCES

1. F. Lelièvre, *PhD Thesis*, Université d'Orsay - Paris XI, 1998 (in French).
2. S.Abramovich et al., 13.Meeting on Physics of Nucl.Fission, Obninsk (1995) 303.
3. Y.Danon,et al, Nuclear Science and Engineering 124 (1996) 482.
4. O. Deruelle, *PhD Thesis*, Université d'Orsay - Paris XI, 2002 (in French).
5. M. Fadil et al., *Proceedings of the Int. Conf. on Neutron Field Spectrometry in Science, Technology and Radiation Protection*, Pisa, June 2000, *Nucl. Inst.. Meth.* **A476** 313 (2002).
6. M. Fadil, *PhD Thesis*, INPG - Grenoble, 2003 (in French).

7. D. Ridikas et al., *Proceedings of the 6th OECD/NEA Information Exchange Meeting on Actinide and Fission Partitioning and Transmutation*, 11-13 December 2000, Madrid, Spain.
8. A. Letourneau, F. Marie, D. Ridikas, *Internal report*, DAPNIA-03-243 (2003).
9. G. Fioni, M. Cribier, F. Marie et al., *Nucl. Phys.* **A 693** 546-564 (2001).
10. O. Deruelle, F. Marie et al., *Proceedings of the 11th International Symposium on Capture Gamma-Ray Spectroscopy and Related Topics*, 2-6 september 2002, Prague, Czech Republic.
11. F. Marie et al., to be submitted.
12. A. Letourneau et al., to be submitted.
13. O.Roig et al., *Nucl. Instr. Meth.* A**521**, 5-11 (2004).
14. S. Chabod et al., this proceeding.

Determination Of Minor Actinides Fission Cross Sections By Means Of Transfer Reactions

B. Jurado[1], M. Aiche[1], G. Barreau[1], S. Boyer[1], S. Czajkowski[1],
D. Dassié[1], C. Grosjean[1], A. Guiral[1], B. Haas[1], B. Osmanov[1],
M. Petit[1], E. Berthoumieux[2], F. Gunsing[2], L. Perrot[2], Ch. Theisen[2],
E. Bauge[3], F. Michel-Sendis[4], A. Billebaud[5], J. N. Wilson[4,5], I. Ahmad[6],
J.P. Greene[6] and R.V.F. Janssens[6]

1 CENBG (UMR 5795 CNRS/IN2P3-Univ. Bordeaux 1), Le Haut Vigneau, 33175 Gradignan, France
2 CEN Saclay, DSM/DAPNIA/SPhN, 91191 Gif-sur-Yvette cedex, France
3 CEA, SPhN, BP12 91680 Bruyères-le-Châtel, France
4 IPN, 15 rue G. Clémenceau, 91406 ORSAY CEDEX, France
5 LPSC, 53 Avenue des Martyrs, 38026 Grenoble cedex, France
6 ANL, 9700 S. Cass Avenue, Argonne, IL 60439.

Abstract. We present an original method that allows to determine neutron-induced cross sections of very short-lived minor actinides. This indirect method, based on the use of transfer reactions, has already been applied with success for the determination of the neutron-induced fission and capture cross section of ^{233}Pa, a key nucleus in the ^{232}Th-^{233}U fuel cycle. A recent experiment using this technique has been performed to determine the neutron-induced fission cross sections of $^{242, 243, 244}$Cm and ^{241}Am which are present in the nuclear waste of the current U-Pu fuel cycle. These cross sections are highly relevant for the design of reactors capable to incinerate minor actinides. The first results will be illustrated.

Keywords: Fission induced by transfer reactions. Minor actinides. Neutron-induced fission cross sections.
PACS: 25.85.–w, 25.40.Hs, 25.45.Hi

INTRODUCTION

Minor actinides (Np, Am and Cm isotopes) are produced by successive neutron captures, alpha and beta decays starting from ^{238}U in the current U-Pu cycle. These nuclei represent one of the most harmful types of nuclear waste as they are strong neutron and alpha emitters with specific activities, in some cases of the order of 10^9 Bq/μg. At present, two different strategic approaches are proposed for minor actinides waste disposal: direct disposal without any reprocessing and spent fuel reprocessing with the aim to optimise the extraction of minor actinides and to then incinerate them. Incineration amounts for the transmutation of minor actinides into less radiotoxic or short-lived species obtained by neutron-induced fission reactions.

The reliable design of reactors for incineration requires an accurate knowledge of minor actinides cross sections in a fast neutron flux. However, in the particular case of the Cm isotopes, the available data are rather scarce. For instance, the only available data for ^{242}Cm are the cross-section measurements for fission induced by neutrons with energies of 0.1-1.4 MeV performed by Vorotnikov et al. in 1984 [1]; no data are available for other decay channels such as ^{242}Cm(n,γ), ^{242}Cm(n,n'), ^{242}Cm(n,2n), etc. The reason for this lack of data is the short half-life of ^{242}Cm (162.94 days) which makes it very difficult to produce and to manipulate targets of this isotope. The experimental technique we present here allows to overcome these difficulties. This indirect method, which was developed in the 70s by Cramer et al. [2] consists in measuring the decay probability of a compound nucleus (e.g. fission or radiative capture) produced via a few-nucleon transfer reaction. The transfer reaction chosen is such that the resulting nucleus has the same mass A and charge Z as the compound nucleus that would be formed if a neutron would be directly absorbed by the minor actinide. The neutron-induced fission cross section is then deduced from the product of the measured fission probability and the compound nucleus cross section obtained from optical model calculations[3]. One may wonder whether the compound states populated by neutron absorption and transfer reactions are the same. In fact, model calculations [4] show that, for neutron energies below 2 MeV, the average value of the angular momentum distributions populated via neutron absorption is approximately one or two units lower than the average angular momentum induced in transfer reactions [5], this difference becoming smaller with increasing neutron energy. This means that, in neutron-induced fission reactions, low-lying states might be populated that are not accessible via transfer reactions and vice-versa. Therefore, one would expect to find some difference between fission induced by these two reactions mechanisms at the fission threshold. However, the comparison between experimental neutron-induced fission cross sections and transfer-induced fission cross sections shows a very good agreement at the fission threshold. This concordance may be due to a lack of resolution. At higher excitation energies, the fission probability is anyway rather insensitive to the angular momentum as illustrated by the scaling laws of the ratio Γ_n/Γ_f determined by Vandenbosch [6] and Gavron [7].

Our group has already successfully applied this technique to determine the neutron-induced fission [8] and capture [9] cross sections of ^{233}Pa via the transfer reaction ^{232}Th(^{3}He, p)^{234}Pa. ^{233}Pa plays a fundamental role in the Th/U cycle, but up to this measurement the only data available for this nucleus were rather inaccurate. In a recent experiment, we have applied the transfer reaction method to determine the neutron-induced fission cross sections of $^{242, 243, 244}$Cm and ^{241}Am up to 10 MeV neutron energy. Besides the relevance of these fission data for the design of reactors capable to incinerate minor actinides, the comparison of the measured cross sections with model calculations will enable the determination of fundamental fission parameters and the investigation of low-lying transition states which are not well known for these short-lived nuclei. In addition, once the model parameters are fixed, cross sections that are hardly measurable, such as (n, γ) and (n,n') can be predicted.

EXPERIMENT

In this experiment, the access to neutron-rich Cm isotopes via few-nucleon transfer reactions with a light projectile such as ^{3}He implied the use of a 30μg ^{243}Am target. This target was prepared at the Argonne National Laboratory and was deposited on a 75 μg/cm^2 carbon backing. The ^{3}He beam at 30 MeV was provided by the Tandem accelerator at the IPN Orsay. The ^{3}He-induced transfer reactions on the ^{243}Am target lead to the production of various fissioning nuclei. The nature (mass and charge) of the light particle ejected indicated the formation of a given heavy residue. Table 1 lists the different transfer channels available in the present experiment and the corresponding neutron-induced reactions that this indirect method will allow to "reconstruct".

TABLE 1. Transfer channels investigated in the reaction ^{3}He + ^{243}Am and the corresponding neutron-induced fission reactions.

Transfer channel	Equivalent neutron-induced reaction
^{243}Am(^{3}He,p)^{245}Cm	^{244}Cm(n,f)
^{243}Am(^{3}He,d)^{244}Cm	^{243}Cm(n,f)
^{243}Am(^{3}He,t)^{243}Cm	^{242}Cm(n,f)
^{243}Am(^{3}He,^{4}He)^{242}Am	^{241}Am(n,f)

Table 1 illustrates the advantage of the transfer reaction technique with respect to the standard direct method: the simultaneous access to several transfer channels allows one to determine neutron-induced fission cross sections of various nuclei from just one projectile-target combination.

Experimental Set-up

The detection set-up used to determine the fission probability of the different compound nuclei formed after the transfer reaction is displayed in figure 1. Two Si telescopes placed at 90 and 130 degrees with respect to the beam axis served to identify the light charged particles emerging from the transfer reaction. If the corresponding heavy residue would fission, one of the fission fragments was detected in coincidence with the light particle by means of a fission-fragment multi-detector. This multi-detector was designed to achieve a large efficiency for fission fragment angular distribution measurements. It consisted of 15 photovoltaic cells distributed among 5 units, each unit consisting of 3 cells placed vertically one above the other. All the units were situated normal to the reaction plane defined by the two telescopes and the beam axis. Four units were placed in the forward direction with a covering angle from 14 to 125 degrees. In order to add a point at backward angles to the measured angular distribution, the fifth unit was positioned at 180 degrees from the foremost unit. The solid angle subtended by this fission multi-detector is about 48% of 4π. More details on the experimental set-up can be found in reference [8].

Fission Probability

The Si telescopes allowed identification of the light charged particles and determination of their kinematics parameters (energy and angle). With this information and the related Q-values, we could determine the excitation energy E^* of the corresponding compound nuclei. Figure 2a) shows the identification plot representing the energy loss versus the residual energy measured in one of the telescopes. The typical hyperbolas corresponding to the different light-charged particles can be easily distinguished. By selecting one type of light particle, for example tritons t, we can construct the spectrum represented by the solid line in figure 2b), the so-called "singles" spectrum. This spectrum represents the number of tritons, i. e. the number of ^{243}Cm nuclei, N_{SING}, as a function of their excitation energy. The broad resonances at the highest excitation energies stem from transfer reactions on the carbon backing and on ^{16}O impurities in the target. The singles spectrum has been extrapolated under these peaks, introducing an additional source of uncertainty.

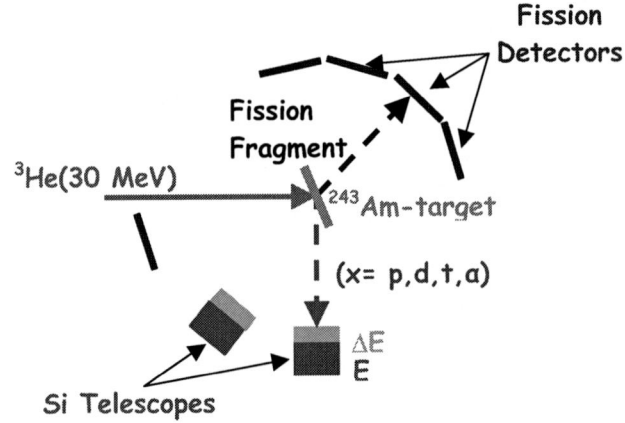

FIGURE 1. Experimental set-up for fission probability measurements of compound nuclei formed via transfer reactions.

If we now select the tritons detected in coincidence with a fission event, we obtain the spectrum represented by the dashed line in figure 2b) which represents the number of ^{243}Cm nuclei that have undergone fission, N_{COIN}. For each excitation energy bin we can then determine the ratio between the fission events spectrum (dashed line) and the compound nucleus spectrum (full line). This ratio, corrected for the fission detector efficiency $Eff(E^*)$, gives the fission probability of ^{243}Cm as a function of the excitation energy:

$$P_f(E^*) = \frac{N_{COIN}(E^*)}{N_{SING}(E^*) \cdot Eff(E^*)} \qquad (1)$$

As was stated before, the geometrical efficiency of the fission detector is approximately 48%. However, for fission induced by neutrons or light charged particles, the fission-fragment angular distributions can be forward peaked. The anisotropy depends on the recoil angle of the system undergoing fission, and thus, on its excitation energy. Therefore, we should actually talk about an effective detection efficiency that includes not only the geometrical effects of the fission detector, but also the fragment angular distribution effects. Our fission detector allows for measuring the angular distribution anisotropy; with this information and a Monte-Carlo simulation it is possible to calculate the effective efficiency.

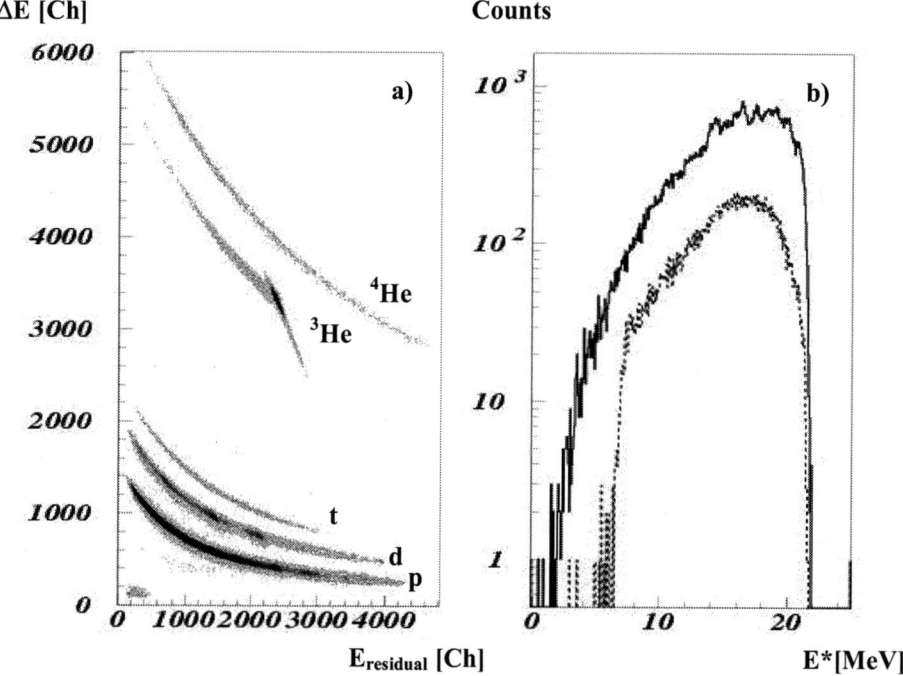

FIGURE 2. a) Identification plot representing the energy loss versus residual energy in one of the Si telescopes. b) Number of tritons detected in the Si detector located at 130 degrees as a function of the excitation energy of the corresponding ^{243}Cm nucleus. The full line represents the total number of tritons detected. The dashed line represents the number of tritons detected in coincidence with the fission detector.

The effect of the angular anisotropy on the detector efficiency has not yet been determined; thus our data have only been corrected for the geometrical efficiency. Nevertheless, the angular distribution effect is expected to be smaller than 10%. Up to now we have analysed the deuteron, triton and alpha channels which correspond to the fission probabilities of ^{244}Cm, ^{243}Cm and ^{242}Am, respectively. The preliminary results are shown in figure 3. Error bars in these spectra represent statistical errors as well as uncertainties due to the subtraction of the contaminant peaks.

As shown in figure 3, all our measurements extend up to the onset of second-chance fission. None of the existing neutron-induced fission data for ^{242}Cm go as high in excitation energy. The ^{244}Cm fission probability will allow to reconstruct the neutron-induced cross section ^{243}Cm(n,f). ^{243}Cm is a fissile nucleus and the neutron binding energy Bn of the compound nucleus ^{244}Cm is higher than its fission barrier. Therefore, neutron-induced fission of this nucleus doesn't allow exploring the fission threshold. As shown in figure 3b), the transfer reaction used makes the fission threshold of ^{244}Cm accessible. The fission probability of ^{242}Am is of great importance as the existing data for the corresponding neutron-induced fission cross section ^{241}Am(n,f) will serve to validate our method.

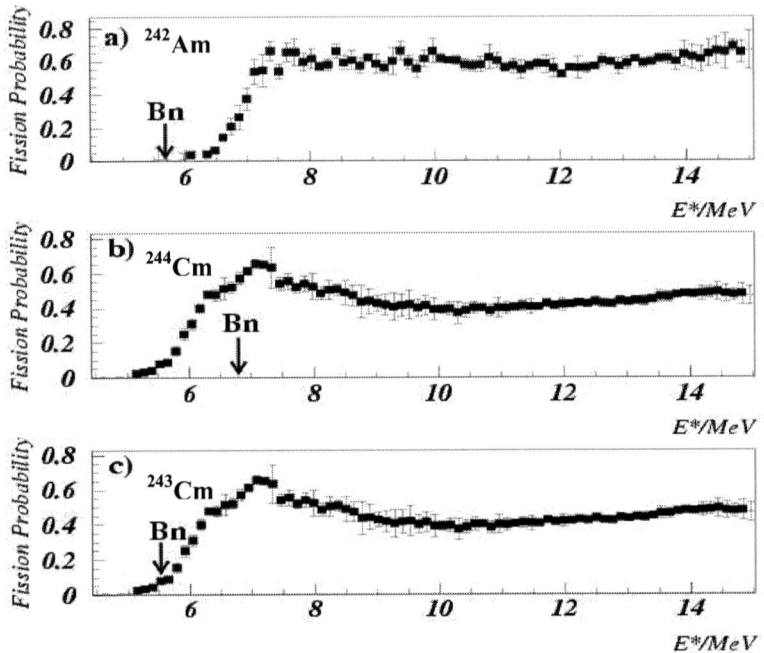

FIGURE 3. Preliminary fission probabilities of various compound nuclei as a function of the excitation energy. The arrows indicate the neutron binding energy of the fissioning nucleus.

MODEL CALCULATIONS

Once the neutron-induced cross sections will be reconstructed, they will be compared with model calculations. This comparison will allow one to extract fundamental fission parameters like the fission barrier heights and curvatures, several parameters related to the level densities (e. g. pairing gaps over the two fission barriers) and the energies of the low-lying transition states. The latter are particularly interesting as for the Cm isotopes under investigation here they are not yet well known. Moreover, once the model parameters are fixed, the calculations can be used to predict the neutron-induced cross sections of several competing decay channels such as (n, γ) and (n,n') which are very difficult to measure.

Our calculations are based on the statistical model. After absorption of a neutron of incident energy E_n the compound nucleus is formed in a certain nuclear state characterised by a given total angular momentum J and parity π. This nucleus can then decay by fission, neutron and gamma emission. Recently, the model has been extended to take into account second-chance fission and two-neutron emission. The different neutron-induced decay cross sections can be written as:

$$\sigma_i(E_n) = \sum_{J,\pi} \sigma_{CN}(E_n, J, \pi) \cdot P_i(E_n, J, \pi) \qquad (2)$$

where $\sigma_{CN}(E_n, J, \sigma)$ represents the compound nucleus formation cross section and $P_i(E_n, J, \pi)$ is the probability for a certain decay channel i. The probabilities for fission, neutron and gamma decay channels are calculated by dividing the corresponding partial decay width Γ_i by the total partial decay width Γ_T. Each partial decay width is obtained by summing up over the available nuclear states of the fissioning/residual nucleus. The integration is done over both the low-energy discrete nuclear states and the nuclear states in the continuum. In the particular case of the fission decay width, the traversing of the double-humped fission barrier is taken into account. Moreover, for even compound nuclei, fission may occur through low-lying discrete states located on the top of the barriers. The model has been recently modified to include the contribution to fission from these states. Besides, the probability for second-chance fission P_{nf} and two-neutron emission P_{2n} has been incorporated in the following way:

$$P_{nf} = \left(\frac{\Gamma_n}{\Gamma_T}\right)_A \cdot \left(\frac{\Gamma_f}{\Gamma_T}\right)_{A-1} \qquad (3)$$

$$P_{2n} = \left(\frac{\Gamma_n}{\Gamma_T}\right)_A \cdot \left(\frac{\Gamma_n}{\Gamma_T}\right)_{A-1} \qquad (4)$$

where Γ_f and Γ_n represent the fission-decay width and the neutron-decay width, respectively.

The determination of the partial widths requires the computation of level densities at different deformations of the fissioning systems (ground state and barriers) as well as at the ground state of the residual nuclei. For this purpose, we have used the phenomenological prescription of Ignatyuk et al. [10]. The key parameter in the level density description is the level density parameter a. For this quantity we have used the description of reference [11] which includes the dependence of the level density parameter with deformation, shell effects and excitation energy. The effect of collective enhancement in the level density is included as established by Bohr and Mottelson [12]. More details on the model calculations can be found in references [8, 13]. As shown in ref. [8], the experimental neutron-induced fission cross section of ^{233}Pa was compared to the present model. This comparison enabled the determination of previously unknown fission barrier parameters of ^{234}Pa and the prediction of the ^{233}Pa(n,γ) cross section. Later, this neutron-induced capture cross section was experimentally determined in reference [9] using the same transfer reaction technique. The good agreement between the experimental capture cross section and the model prediction confirms the quality of the model calculations.

CONCLUSION

The design of nuclear reactors capable to incinerate minor actinides requires a good knowledge of neutron-induced cross sections of Cm and Am isotopes. However, the enormous specific activity of these nuclei considerably complicates the direct measurement of these cross sections. Recently, we have performed an experiment to determine the neutron induced-fission cross sections of $^{242, 243, 244}$Cm and ^{241}Am using an indirect technique. This technique is based on the use of transfer reactions and its validity has been confirmed by previous experiments. The preliminary results for the fission probability of $^{243, 244}$Cm and ^{242}Am compound nuclei have been presented. The final data will be compared to model calculations based on the statistical model. This comparison will allow one to determine several fundamental fission parameters and predict neutron-induced cross sections such as (n,γ), (n,n'), which are otherwise very complicate to measure.

ACKNOWLEDGMENTS

We thank the Orsay tandem accelerator staff for their great support during the experiment. This work was supported by the GEDEPEON program, the Conseil Régional d'Aquitaine, and by the U.S. Department of Energy under contract W-31-109-ENG-38. The authors are also indebted for the use of ^{243}Am to the Office of Basic Energy Sciences, U.S. Department of Energy, through the transplutonium element production facilities at Oak Ridge National Laboratory.

REFERENCES

1 Vorotnikov, P. E. et al., *J,YF,* 40,(5),1141,8411
2 Cramer, J. D. and Britt, H. C., *Nucl. Sci. Eng.* **41**, 177 (1970)
3 Bauge, E., Private Communication
4 Delaroche, J. P. et al., Private Communication
5 Back, B. B. et al. *Phys. Rev. C* **9**, 1924 (1974)
6 Vandenbosch, R. et al., Nuclear Fission, Academic Press New York and London (1973)
7 Gavron, A. et al. *Phys. Rev. C* **13**, 2374 (1976)
8 Petit, M. et al., *Nucl. Phys. A* **735**, 345 (2004)
9 Boyer, S., PhD Thesis, Université Bordeaux (2004)
10 Ignatyuk, A. V. et al., *Sov. J. Nucl. Phys.* **21**, 255 (1975)
11 Junghans, A. R. et al., Nucl. Phys. A **629**, 635 (1998)
12 Bohr, A. and Mottelson, B. R., Nuclear Structure, World Scientific, Singapore (1998)
13 Grosjean, C., PhD Thesis, Université Bordeaux (2005)

Recent results on the neutron-induced fission cross-section of ^{231}Pa

A. Oberstedt*, S. Oberstedt[†], E. Birgersson*[†], F.-J. Hambsch[†], V. Fritsch[†], G. Lövestam[†], G. Vladuca[¶], A. Tudora[¶], N. Kornilov[†]

Dept. of Natural Sciences, Örebro University, SE-70182 Örebro, Sweden
[†] *EC - JRC IRMM, B-2440 Geel, Belgium*
[¶] *Faculty of Physics, Bucharest University, RO-76900 Bucharest, Romania*

Abstract. The cross-section for the neutron-induced fission of ^{231}Pa has recently been measured from the threshold to E_n = 3.5 MeV. The experimental results are described in terms of extended statistical model calculations.

Keywords: Protactinium-231, neutron-induced fission, fission cross-section, statistical model calculations, Th-U fuel cycle.
PACS: 24.10.Pa, 25.85.Ec, 28.20.-v

INTRODUCTION

In this paper we present recent cross-section data from direct, energy resolved measurements of the neutron-induced fission of ^{231}Pa. The experimental results are compared to other experimental data as well as previously known values from evaluated data bases and discussed in terms of new calculations in the frame work of the statistical model code STATIS. The results may serve as input data for the design of so-called accelerator driven systems (ADS) [1, 2], using ^{232}Th and ^{233}U as fuel. With an average cross-section value of about 1 b for fast neutrons, ^{231}Pa plays an important role for the simulation of the balance of nuclei in and the reactivity of the reactor. Besides, with a half-life of $3.276 \cdot 10^4$ years, which is similar to the one of ^{239}Pu ($2.411 \cdot 10^4$ years), ^{231}Pa has a comparable radiotoxicity in the Th-U cycle with respect to the handling of the radioactive waste as ^{239}Pu has in the U-Pu fuel cycle.

EXPERIMENTS AND DATA TREATMENT

The experiments were performed at the IRMM in Geel, Belgium, by irradiating the ^{231}Pa sample (mass 1190 µg, diameter 28 mm) with quasi-monoenergetic neutrons, produced in the reaction T(p, n)^{3}He in the energy range E_n = 0.8 - 3.5 MeV. The relative widths of the resulting neutron energy distributions are less than a few percent (FWHM). The protons (max 30 µA) were accelerated with a 7 MV Van de Graaff accelerator and impinged on the solid tritium target (TiT). Due to the high α-activity of the sample, the diameter of the ^{231}Pa sample had to be collimated to 6 and 8 mm, respectively. The corresponding effective masses were determined by α-spectroscopy to 53.4±1.3 μg and 87.9±2.7 μg. The measurements were carried out using a twin Frisch-grid ionization chamber, which is described in Ref. [3]. The detector was operated in back-to-back geometry, with the ^{231}Pa sample mounted on one side of the chamber and with a known ^{237}Np sample (182.2 μg (±1,7%), d = 28 mm) on the other. Because both sides of the chamber have the same geometry and efficiency, the fission cross-section of ^{231}Pa may be determined by counting fission events on both sides and by using the rather well-known fission cross-section of ^{237}Np [4].

An unambiguous identification of the fission fragments is achieved by plotting the sum of the pulse heights at both anode and grid, giving information about the range and direction of the particles, as a function of the anode signal. The left-hand side of Fig. 1 shows events from ^{231}Pa, while the right-hand side represents events from ^{237}Np. The events in the lower left corners are caused by piled-up α-particles, whereas the upper right corners contain signals from a pulse generator. The majority of data in the centre of both plots corresponds to fission fragments. For the ^{231}Pa measurement, events parallel to the sum axis at low pulse height appear because of the simultaneous detection of piled-up α-particles and a fission fragment. The interaction of the α-particles with the counting gas is rather weak, leading to a weak anode signal. However, a fission fragment giving a signal at the grid, might not reach the anode before a α-particle does, which creates the stop signal for the data acquisition. Hence, for these events the pulse height at the anode is low, while the grid signal, and thus the sum of both, is not. Obviously, these events, not appearing for the ^{237}Np measurement due to the much lower α-activity, have to be included, too. This is indicated by the chosen regions-of-interest. Finally, the amount of fission events was corrected for dead time (< 4%). The accuracy, with which the individual ^{231}Pa(n, f) cross-section values were determined, was in the order of 5-6%.

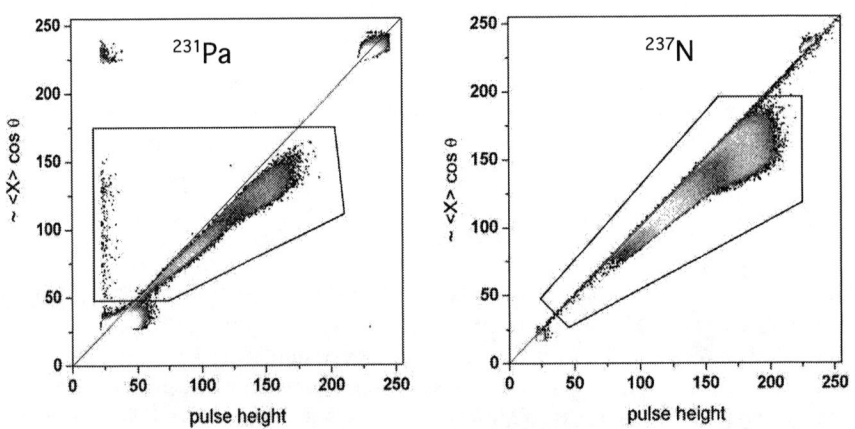

FIGURE 1. Two-dimensional presentation of the sum of anode and grid signal versus pulse height at the anode for ^{231}Pa (left-hand side) and ^{237}Np (right-hand side) in arbitrary units. The polygons correspond to the regions-of-interest for the counting of fission fragments (see text for details).

RESULTS AND DISCUSSION

In this chapter we present the results for the cross-section for neutron-induced fission as a function of neutron energy for ^{231}Pa (Fig. 2), together with a new theoretical description based on calculations in the framework of the extended statistical model code STATIS [5, 6]. They are compared to previous experimental data as well as previous evaluations, ENDF/B-VI [7] and JENDL-3.3 [8]. The data obtained in this work are shown as large open circles and compared to recent data deduced from the reaction ^{232}Th(^{3}He, tf) ^{232}Pa [9]. In addition, some relevant older data from Refs. [10-12], are shown together with the result of evaluations in ENDF/B-VI and JENDL-3.3. It is apparent that our data confirm the values obtained by the mentioned particle exchange reaction. In contrast to the findings for the neutron-induced fission cross-section in ^{233}Pa [6], both direct and indirect cross-section measurement lead here to a similar result. This seems to support the assumption, that only in cases, where ingoing and outgoing particle are similar, particle-transfer reactions give results that are in agreement with those obtained from direct compound nuclear reactions [6]. Our results are in the same order of magnitude as most of the other reported cross-section data, while the data from Ref. [12] have to be ruled out. The position of the fission threshold at about 600 keV could be confirmed as well. New STATIS calculations have been performed [13], following the concept presented in Refs. [5, 6] and based on the new experimental results. The good agreement with these data, even in the energy region of vibrational resonances around the fission threshold, is satisfactory indeed.

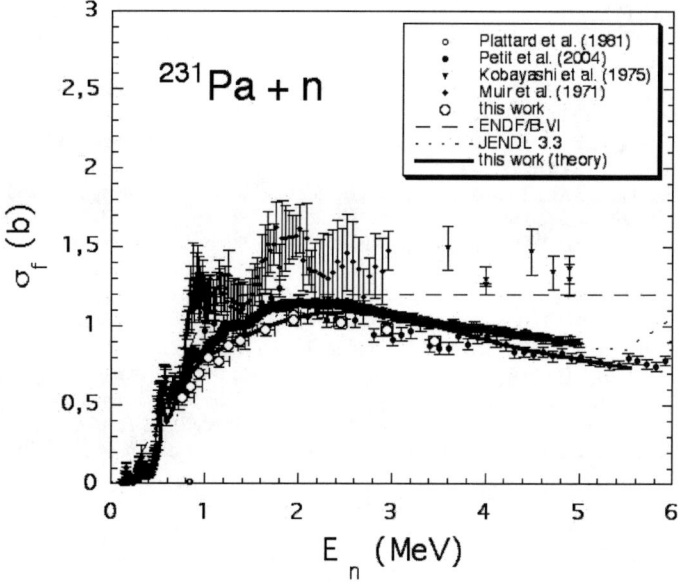

FIGURE 2. Compilation of both experimental and theoretical cross section data for neutron-induced fission of ^{231}Pa. The results obtained in this work are shown as large open circles and full drawn line.

SUMMARY AND CONCLUSION

In this paper we have reported about direct cross-section measurements for the neutron-induced fission of ^{231}Pa. The experiments were performed with quasi-monoenergetic neutrons covering the regime for first chance fission. Using a double-sided Frisch-grid ionization chamber for the unambiguous detection of fission fragments has been proven to be an excellent choice to tackle problems related to high backgrounds of α-particles. The experimental technique of a direct measurement as presented here was shown to be well suited for cross-section studies, since there is no model dependence involved in the data analysis. The experimental results are described well by the calculations with the statistical model code STATIS and provide valuable input data for the modelling of ADS.

We may conclude that the precision, with which the cross-sections were determined, meets the requirements posed by the IAEA in order to serve as reliable input data for the modelling of advanced nuclear reactor systems involving the thorium-uranium fuel cycle [14]. Based on this knowledge, the study of the neutron-induced fission of ^{231}Pa is being continued to neutron energies up to 20 MeV.

REFERENCES

1. Rubbia, C., Rubio, J. A., Buono, S., Carminati, F., Fiétier, N., Galvez, J., Gelès, C., Kadi, Y., Klapisch, R., Mandrillon, P., Revol J.P., and Roche, C., *CERN report No. CERN/AT/95-44(ET)* (1995).
2. Carminati, F., Klapisch, R., Revol, J.P., Roche, C., Rubio, and J. A., Rubbia, C., *CERN report No. CERN/AT/93-47(ET)* (1993).
3. Budtz-Jørgensen, C., Knitter, H.-H., Straede, Ch., Hambsch, F.-J., and Vogt, R., *Nucl. Inst. and Meth. A* **258**, 209 (1987).
4. Tovesson, F., Hambsch, F.-J., Oberstedt, A., Fogelberg, B., Ramström, E., and Oberstedt, S., *Phys. Rev. Lett.* **88**, 062502-1 (2002).
5. Vladuca, G., Hambsch, F.-J., Tudora, A., Oberstedt, S., Tovesson, F., Oberstedt, A., and Filipescu, D., *Phys. Rev. C* **69**, 021604(R) (2004).
6. Vladuca, G., Hambsch, F.-J., Tudora, A., Oberstedt, S., Oberstedt, A., Tovesson, F., and Filipescu, D., *Nucl. Phys. A* **740**, 3-19 (2004).
7. Cross Section Evaluation Working Group, "ENDF/B-VI Summary Documentation", *National Nuclear Data Center, Brookhaven National Laboratory Report No. BNL-NCS-17541 (ENDF-201)*, edited by Rose, P. F. (1991).
8. Shibata, K., Kawano, T., Nakagawa, T., Iwamoto, O., Katakura, J., Fukahori, T., Chiba, S., Hasegava, A., Murata, T., Matsunobu, H., Ohsawa, T., Nakajima, Y., Yoshida, T., Zukeran, A., Kawai, M., Baba, M., Ishikawa, M., Asami, T., Watanabe, T., Watanabe, Y., Igashira, M., Yamamuro, N., Kitazawa, H., Yamano, N., and Takano, H., J., *Nucl. Sci. Technol.* **39**, 1125 (2002).
9. Petit, M., Aiche, M., Barreau, G., Boyer, S., Carjan, N., Czajkowski, S., Dassié, D., Grosjean, C., Guiral, A., Haas, B., Karamanis, D., Misicu, S., Rizea, C., Saintamon, F., Andriamonje, S., Bouchez, E., Gunsing, F., Hurstel, A., Lecoz, Y., Lucas, R., Theisen, Ch., Billebaud, A., Perrot, L., and Bauge, E., *Nucl. Phys. A* **735**, 345–371 (2004).
10. Plattard, S., Auchampaugh, G. F., Hill, N. W., de Saussure, G., Harvey, J. A., and Perez, R. B., *Phys. Rev. Lett.* **46**, 633 (1981).
11. Kobayashi, K., *Tech. report Res. Reactor Inst. Kyoto University, KURRI-TR-***8**, 10 (1975).
12. Muir, D. W., and Vessor, L. R., *Internal report Los Alamos National Laboratory LA-***4648** (1971).
13. Hambsch, F.-J., Vladuca, G., Oberstedt, S., and Filipescu, D., *contribution to this conference*.
14. Pronyaev, V. G.. *IAEA report No. INDC(NDS)-***408** (1999).

Evaluations of Photonuclear Cross Sections for Actinides

M.-L. Giacri-Mauborgne[a], M. B. Chadwick[†], J.-C. David[a], D. Doré[a], D. Ridikas[a], A. Van Lauwe[a] and W. B. Wilson[†]

[a]*CEA Saclay, DSM/DAPNIA/SPhN, 91191 Gif/Yvette, France*
[†]*Los Alamos National Laboratory (T-16), Los Alamos, NM 87545, USA*

Abstract. This article presents calculations of photonuclear reaction cross sections for actinides by using different reaction codes, namely an improved version of HMS-ALICE and GNASH. The GDR parameters from the RIPL2 data library are used to describe the total photo-absorption cross section. In general, our results show satisfactory agreement with experimental data and IAEA evaluations. We plan to include our GNASH evaluations for ^{235}U, ^{238}U, ^{237}Np, ^{239}Pu, ^{240}Pu and ^{241}Am into a new release of the ENDF-B/VII data files.

Keywords: photonuclear activation file, photofission.
PACS: 85.Jg, 25.20-x

INTRODUCTION

A renewed interest in photonuclear process is motivated by a number of different applications where progress in high-intensity electron accelerators was awaited [1]. Major problems in modeling photonuclear reactions are the lack of photonuclear data on corresponding cross sections despite the huge effort of the IAEA [2], where data are available for 164 isotopes only. Up to now no material evolution code including photonuclear reactions is available. Therefore, in close collaboration with LANL, we have been working on the development of a photonuclear activation file (PAF) to be included into the material evolution code CINDER'90 [3]. The reaction codes HMS-ALICE [4] and GNASH [5] were used to calculate photonuclear reactions for more than 600 isotopes. More information on this effort can be found in [6].

In the case of actinides, the PAF includes nine actinides evaluated by IAEA [2]. We complete the library by calculating the remaining cross sections with the HMS-ALICE reaction code. In this work we discuss in detail our results of HMS-ALICE calculations for ^{235}U and ^{238}U, where separate reaction channels as (γ,n), (γ,2n), and (γ,fiss) are compared with the IAEA evaluations.

For some high priority actinides we do more accurate evaluations using the GNASH reaction code, which was routinely employed to calculate neutron induced reactions on actinides. The GNASH evaluations for ^{237}Np, ^{240}Pu, ^{241}Am are performed for the first time, while the re-evaluations for ^{235}U, ^{238}U and ^{239}Pu are also made. All these cross sections agree well with experimental data. These GNASH results will be included in the future release of ENDF-B/VII.

RESULTS OF HMS-ALICE

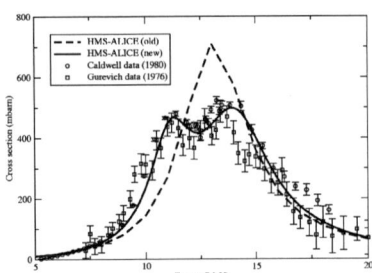

FIGURE 1. Comparison of different parameterizations of the total photo-absorption cross sections in HMS-ALICE for ^{235}U with corresponding data sets.

Fig. 1 presents an improvement we made in modeling the total photo-absorption cross section with HMS-ALICE. In brief, the sum of two Lorentzians is indispensable to reproduce correctly experimental data for deformed nuclei [7, 8]. The GDR parameters from the RIPL data library are used to describe the total photo-absorption cross section in the above parameterization.

Equally, the improvement in predicting the partial photonuclear reaction cross sections is evident as presented in Fig. 2 (upper part – for ^{238}U and lower part – for ^{235}U).

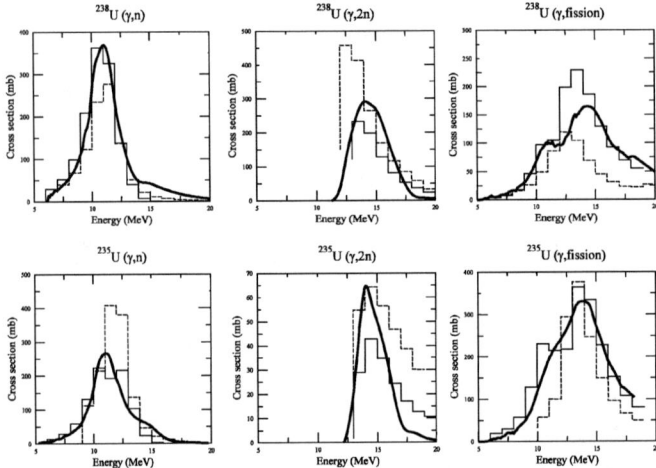

FIGURE 2. Comparison of HMS-ALICE predictions (dash line: old version, solid line: new version) and IAEA evaluations (bold line) for partial photonuclear cross sections in the case of ^{238}U and ^{235}U.

Note that the IAEA evaluations are based on experimental data and serve as a reference in this case. Similar quality of the predictive power of the improved HMS-ALICE was observed for other nine actinides provided by the IAEA evaluated data library. Therefore, we used HMS-ALICE to predict photonuclear cross sections of actinides for which no measurements-evaluations existed at all. For other nuclei such as ^{235}U, ^{238}U, ^{239}Pu, ^{237}Np, ^{240}Pu and ^{241}Am, we performed new evaluations with GNASH.

GNASH CALCULATIONS

The major advantage of making the evaluations with GNASH is that the cross sections become available in the ENDF format and can be used in transport codes. For more details on the photo-nuclear cross section evaluations with GNASH we refer the reader to Ref. [9].

Results for ^{235}U, ^{238}U and ^{239}Pu

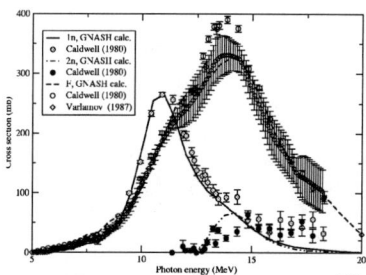

FIGURE 3. GNASH evaluations for ^{235}U.

GNASH evaluations for partial cross sections on ^{235}U are given Fig. 3. Comparable results were also obtained for ^{238}U and ^{239}Pu. We conclude that GNASH can be successfully used to evaluate photonuclear cross sections both for actinides [9] and non-actinides [10]. We also noticed that the GNASH results were quite sensitive to the fission barrier. In brief, the experimental data could not be well reproduced by using the same parameters as for neutron induced reactions.

Predictions for ^{240}Pu, ^{237}Np and ^{241}Am

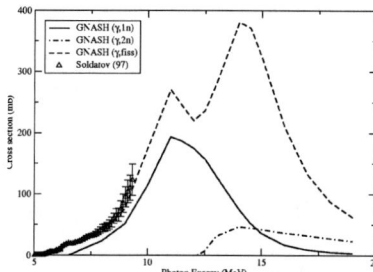

FIGURE 4. GNASH evaluations for ^{240}Pu.

Photonuclear cross section data for ^{240}Pu are very scarce. We found the only experiment on photo-fission up to 11 MeV [11]. In addition, no integral measurements were done for this nucleus.

Again as for other nuclei, we used input parameters from neutron induced reactions on ^{239}Pu adjusting the fission barrier to reproduce the available data as shown in Fig. 4. For the second chance fission we employed the same input values as for the first chance fission of ^{239}Pu.

Therefore, even with the lack of partial data we think that GNASH should provide reasonable predictions for the neutron production channels as (γ,n) and (γ,2n) as we experienced in the case of ^{239}Pu.

TABLE 1. Comparison of relative photo-fission rates: ^{237}Np versus ^{238}U.

Energy	11.5 MeV	17 MeV	20 MeV
Data	2.71±0.08	2.39±0.1	2.40±0.11
GNASH	2.55	2.24	2.11

TABLE 2. Comparison of relative photo-fission rates: ^{241}Am versus ^{238}U.

Energy	11.5 MeV	14.5 MeV	20 MeV
Data	2.45±0.07	2.42±0.06	2.18±0.12
GNASH	2.54	2.51	2.34

As it is presented in Fig. 5 for photo-fission of ^{237}Np, the two sets of experimental data disagree. The GNASH evaluation reproduces perfectly low energy region and for higher energies stands between two sets of data. Note that for the remaining channels as (γ,n) and (γ,2n) the evaluation reproduces well the data (not shown here).

Another validation of the GNASH predictions for the photo-fission cross section is to compare them with integral fission rates of ^{237}Np measured in [12, 13] with different end point Bremsstrahlung energies. Again a reasonable agreement is obtained for all energies considered as shown in Table 1.

Similarly as for ^{237}Np, in Fig. 6 and Table 2 we give the GNASH results for ^{241}Am. Unfortunately, there is no measurement for photo-neutron production in this case. The photo-fission cross section was measured by [14, 15]. The integral fission rates [12, 16] presented in Table 2 are rather well reproduced for all energies, i.e. the photo-fission cross section presented in Fig. 6 should be quite accurate.

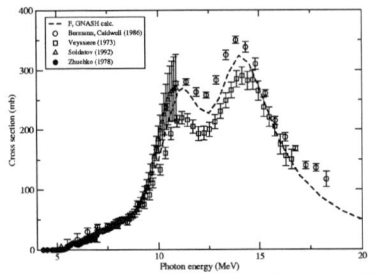 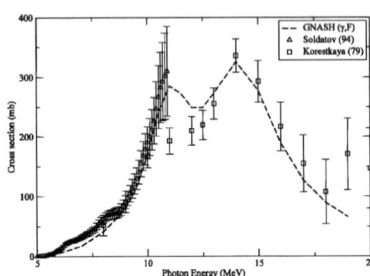

FIGURE 5. GNASH evaluation for photo-fission of ^{237}Np.

FIGURE 6. GNASH evaluation for photo-fission of ^{241}Am.

CONCLUSIONS

The photonuclear cross section evaluations for ^{237}Np, ^{240}Pu and ^{241}Am were performed for the first time. The GNASH reaction code was successfully used for this purpose. These new results in addition to the re-evaluations for ^{235}U, ^{238}U and ^{239}Pu will be included in the new release of ENDF-B/VII.

The photonuclear activation file (PAF), based on compilation of the IAEA evaluations, HMS-ALICE predictions, and GNASH evaluations, by now includes the cross section for more than 600 nuclei. The release of the first version of PAF is planned by the end of 2005. We also add that new measurements of photonuclear cross sections for ^{240}Pu and ^{241}Am are urgently needed to confirm our evaluations.

REFERENCES

1. D. Ridikas et al., *Proc. of the Int. Conf. AccApp/ADTTA '03, San Diego, California, USA*, 2003.
2. M. B. Chadwick et al., *IAEA TECHDOC*, **1178** (2000).
3. W. B. Wilson, Proc. of the Int. Conf. *GLOBAL '95, Versailles, France*, 1995.
4. M. Blann, *Physical Review C*, **34 4**, 1341 (1996).
5. P. Young, E. Arthur, and M. Chadwick, *Proc. of the IAEA Workshop on Nuclear Reaction Data and Nuclear Reactors-Physics, Design, and Safety, Trieste, Italy*, 1996.
6. M.-L. Giacri-Mauborgne et al, *Proc. of the Int. Conf. Nuclear Data 2004, Santa Fe, USA*, 2004.
7. J. T. Caldwell et al., *Physical Review C*, **21 4**, 1215 (1980).
8. G. Gurevich et al. *Nuclear Physics A*, **273**, 326 (1976).
9. M.-L. Giacri, M. B. Chadwick, D. Ridikas, submitted to *Nuclear Science and Engineering*.
10. M. Chadwick et al. *Nuclear Science and Engineering*, **144**, 157 (2003).
11. A. S. Soldatov et al., *Physics of Atomic Energy*, **63 1**, 31 (2000).
12. M. B. Alexandrov et al., *Yadernaya Fizika*, **43**, 290 (1986).
13. J. R. Huizenga, J. E. Gindler, R. B. Duffield, *Physical Review*, **95 4**, 1009 (1954).
14. A. S. Soldatov, G. N. Smirenkin, *INDC (CCP)*, 379 (1994).
15. I. S. Koretskaya et al., *Yadernaya Fizika*, **30**, 910 (1979).
16. Yu. A. Vinogradov et al ., *Yadernaya Fizika*, **24**, 686 (1976).

FISSION AT HIGHER ENERGIES—I

DISTRIBUTION OF NUCLIDES PRODUCED IN THE COLLISION OF 1 AGeV ^{238}U-IONS ON p

M. BERNAS, P. NAPOLITANI, F. REJMUND, C. STEPHAN, J. TAIEB,
L. TASSAN-GOT*, P. ARMBRUSTER, T. ENQVIST, M.-V. RICCIARDI,
K.-H. SCHMIDT[†], J. BENLLIURE, E. CASAJEROS, J. PEREIRA** and
A. BOUDARD, R. LEGRAIN, S. LERAY, C. VOLANT[‡]

*I.P.N. d'Orsay, F-91406 Orsay Cedex France E-mail: bernas@ipno.in2p3.fr
[†]G.S.I. Darmstadt, D-64291 Darmstadt, Germany
**Univ. of Santiago de Compostela, E-15706 Santiago, Spain
[‡]DAPNIA/SPhN CEA/Saclay F-91191 Gif-sur-Yvette Cedex, France

Abstract. Production cross sections and kinematical properties of the complete set of fission fragment residues from the reaction ^{238}U (1 A.GeV) + p have been obtained. Isotopic distributions are measured for all elements from O (Z = 8) to W (Z = 74). Fission velocities and production cross sections are shown as a function of Z, the charge and N, the number of neutrons of the fragments. The very asymmetric pairs of fragments can be attributed to excited fissioning parent nuclei of charge Z, 88< Z < 92.

Keywords: NUCLEAR REACTION p(238U,x),E = 1 GeV/nucleon; Measured fission cross sections of 283 isotopes from Gd to Re; Measured fission fragments velocities; Inverse-kinematics method; In-flight separation by high resolution magnetic spectrometer; Identification in Z and A by ToF and energy-loss measurements; Relevance for accelerator-driven subcritical reactors and for production of radioactive beams
PACS: PACS 24.75.+i:25.40.Sc:25.85.Ge:28.50.Dr:29.25.Rm

INTRODUCTION

Proton-induced spallation in the 1 GeV range is important for many future technological applications. ISOL-separators world-wide use the proton on ^{238}U reaction since 35 years, and future radioactive-beam facilities producing neutron-rich isotopes count on it, but a solid base for the primary isotope production is missing. With 1 GeV protons on ^{238}U two main channels are opened; either fission of a more or less excited spallation residue or evaporation of nucleons [1]. The total cross section of 1.97 b divides in (1.53 ± 0.15)b for fission leading to all elements between nitrogen and rhenium [2] and (0.44 ± 0.06)b for evaporation residues [3].

Evaporation of nucleons is the main source of fragments close to U and down to Z = 74. After a rapid fall, within the 10 mass units close to U covering 0.24 b of cross-section, the main part of the evaporation-cross section, the mass distribution stays at 5 mb to 4 mb down to mass losses of 50 mass units. Finally a break-down by a factor of 10 to the level of 0.4 mb is observed for mass loss of 65, where the evaporation meets the fission cross sections.

Fission leads mainly to fragments which cover the domain of Z = 30 to 60. The distribution of velocities and of isotopic yields of fission fragments have been reported [2]. In this region of the chart, fragment velocities in the emitting-source frame and isotopic yields were shown to be consistent with a binary break-up of a parent nucleus with charge number $88 < Z_0 < 92$. Even for the lightest fragments, $Z < 25$, the shape of the measured velocity distribution revealed that they arise from a binary break-up [4]. The element distribution of the fission fragments drops down from Z = 45 to Z = 17 (N = 20), forms a plateau until Z = 14 and increases again for smaller values of Z and N [4]. The fission partners of these very light elements are expected in the range of $Z > 74$.

In this presentation we concentrate on the region intermediate between fission fragments (FF) and evaporation residues (EVR).

EXPERIMENT

The experimental study of ^{238}U fragments induced by 1 GeV protons has been performed by using inverse kinematics. The 1 A GeV U beam produced by the GSI accelerator facility collides the nuclei from a H$_2$ liquid target contained in a Ti cell. Forward emitted fragments, fully stripped, are momentum analysed with the high resolution FRagment Separator, FRS [5], and identified in Z with an ionisation chamber at the end of their trajectory in the FRS. The mass number A is deduced from the time of flight in the second half of the FRS-system.

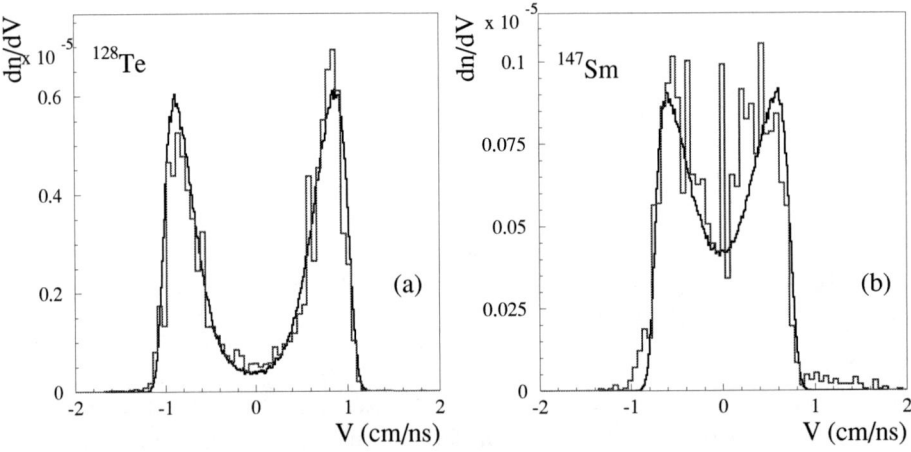

FIGURE 1. Examples of fission fragments velocity distributions in the rest frame measured for ^{128}Te and ^{147}Sm. The fission velocity is deduced from the distance between the external sides of the distributions. It decreases with the mass of the fragment. The simulations reproduce the spectra.

The velocity for each fragment is precisely determined by its magnetic rigidity, once they are identified. The FRS identification-method is precisely explained in the contribution of M.-V. Ricciardi to this conference.

Within a momentum window of 3 %, fragments separated in the FRS, are detected and identified in Z and A. The successive scannings of the FRS magnetic rigidity allow to reconstruct the velocity spectra of each fragment however truncated by the FRS angular acceptance of ± 15 mr. The rest frame velocity spectra of fission fragments is obtained by applying the Lorenz transformation to the lab. spectra.

A bunch of 3 to 4 isotopes of the 36 elements populated by fission are simultaneously measured. This coherence minimizes the relative uncertainties. The transmission through the FRS increases towards 100% for fragments with masses A > 150. The production cross sections are obtained by integrating the velocity distributions and accounting for the transmission in the FRS.

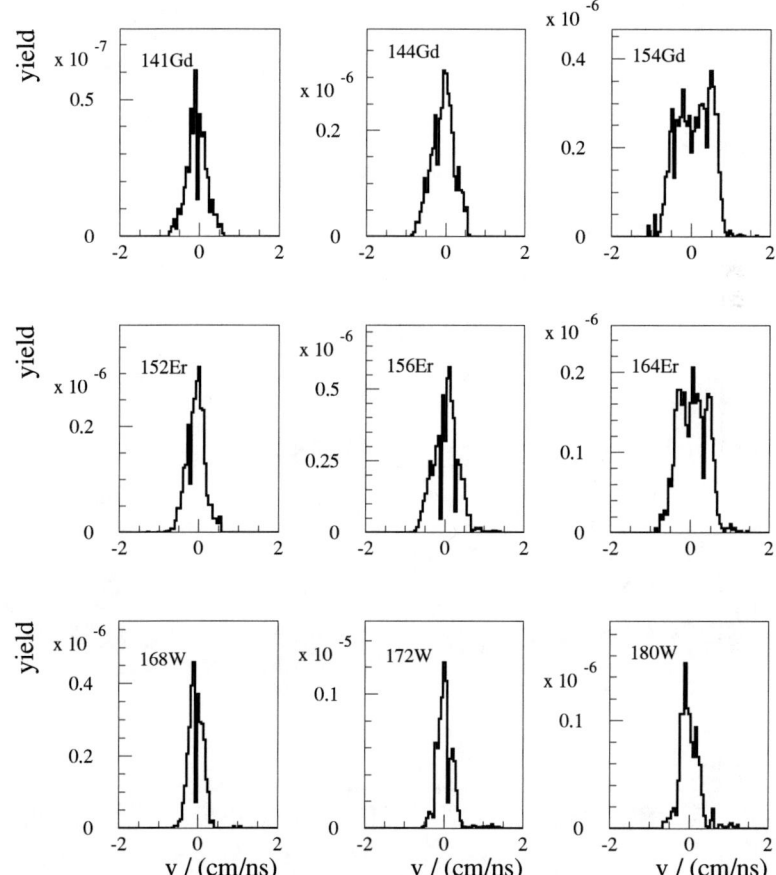

FIGURE 2. Fission fragment velocity spectra in the rest frame for selected isotopes of element Z = 64, 68, 74.

The velocity distribution of a fragment reveals the process from where it comes; either fission or evaporation from the excited spallation residue.

Two examples of FF spectrum are shown in Fig. 1. This basic experimental results are simulated using Monte-Carlo calculations including kinematics and FRS opening angle [2]. The fission velocity decreases with increasing mass A (or Z), as expected from momentum conservation in fission.

For heavier FF both wings tend to come close together and finally join in a somewhat rectangular plot as seen in Fig. 2 for ^{154}Gd, where velocity-distributions of isotopes of three elements of Z = 64, 68 and 74 are shown; The three lightest isotopes, in the left column, are EVR produced either in the Ti-windows of the H_2 target or to a smaller amount in secondary reactions of primary abundant heavier EVR [8, 9]. The heaviest isotopes in the right column are fission fragments. The isotopes shown in the central column are mixtures of EVR and FF with changing weights of both processes. The very different mass and charge dependencies of the velocity distributions for both types of fragment allow for a separation and unfolding of the two production mechanisms [3, 10].

RESULTS

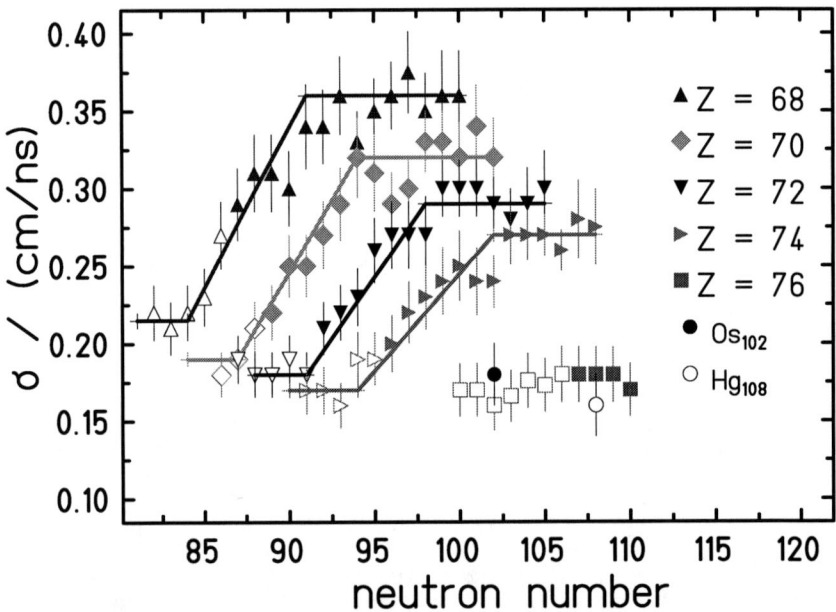

FIGURE 3. Variances of velocity distributions for nuclides produced by fission or as evaporation residues. The horizontal lines indicate the variances for fission products. When the variances drop down, the lines are used to interpolate the relative weight of fission. Empty symbols refer to isotopes where the fission contribution was not extracted. The circles shown on the Z = 76 curve correspond to measured widths from J. Taieb [3] for ΔA = 50 and 60.

The variance of the velocity distributions of the fragments are taken as the ingredients to evaluate the fission velocities and the share of FF among the transmitted isotopes. Fig. 3 reports the dependence of the variances as a function of the neutron number for the heaviest even elements.

For Z increasing from 68 to 74 all curves show a flat part for heavy isotopes, associated to a pure fission component, followed by a fall when the neutron number decreases. The fall indicates a mixing of fission and evaporation fragments. Towards the end, for the minimum width, a few points from each element correspond to EVR on the Ti windows. This last values increase with the corresponding mass losses as expected from Morrissey systematics [11].

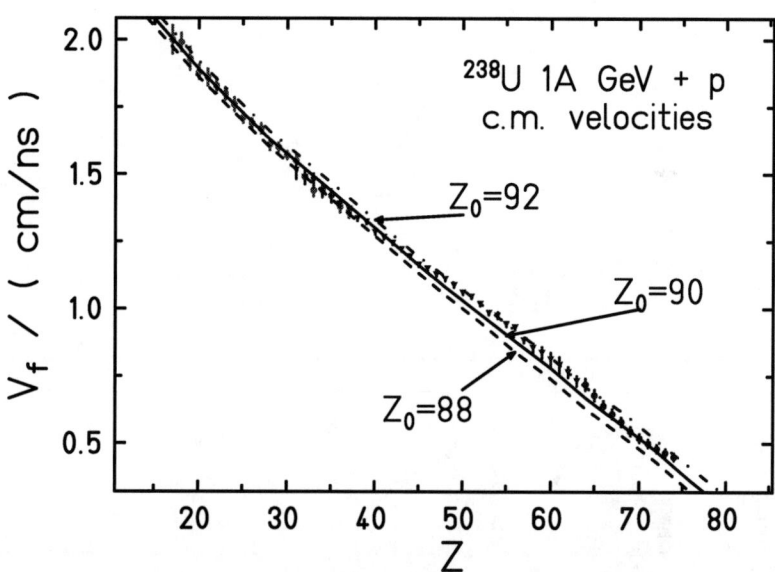

FIGURE 4. The c.m. velocity of fission fragments measured as a function of the atomic number. The three lines: dashed $Z = 88$, full line $Z = 90$, and dashed-dotted line $Z = 92$, are calculated assuming Coulomb repulsion with a radius r_0 being kept constant and fixed by the measured value of the velocity for symmetric fission of ^{220}Th taken as normalisation.

The longitudinal projection of the velocity vectors becomes a rectangle for heavy FF. The measured distribution is the convolution of a rectangle with a normal distribution due to fluctuations of the reduced momenta of the fissioning parent nuclei. These fluctuations are mainly due to a range of recoils by evaporation. In our previous work [2], the related width was evaluated to be $\sigma_r = 0.13$ cm/ns. The width (FWHM) of the rectangle is the sum of forward and backward velocities i.e. twice v_f. The variance of the measured peak σ_{meas} is related to the variance of fission σ_f by $\sigma^2_{meas} = \sigma_r^2 + \sigma_f^2$. As σ_f

is larger than σ_r, we set $\sigma_f^2 = v_f^2/3$, and it follows $v_f = \sqrt{3}\sigma_f$. The velocities v_f obtained for each Z, are shown in Fig. 4. The 3 lines are calculated assuming a coulomb potential acting between the two emitted fragments of mass $\overline{A}$. Three parent fissioning elements Z_0 are tested and the measured values are all included in a Z_0-window of 88 to 92.

Isotopic cross sections

Cross sections are obtained by integrating the c.m. velocity distributions of each isotope. A few other corrections are required [2]

1) For $64 \leq Z \leq 67$, the fission velocity varies between 0.71 to 0.6 cm/ns and the transmission increases from 90% to 100%. For $Z > 67$ the transmission is taken as 100%.

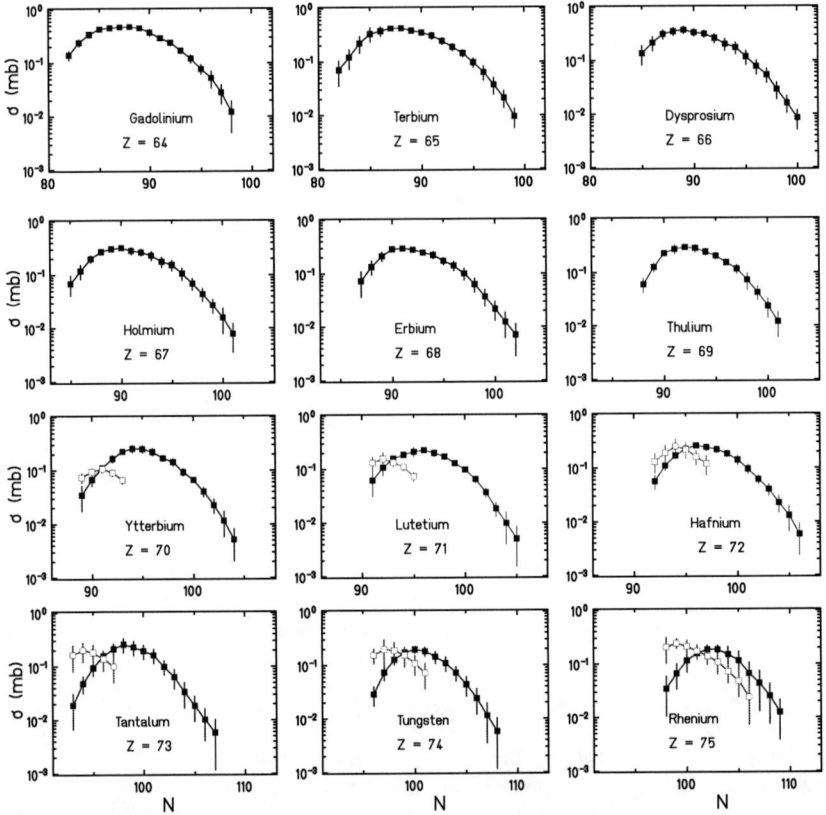

FIGURE 5. Isotopic distributions of cross sections for elements gadolinium to rhenium. fission fragments cross sections are represented by full symbols. The evaporation residue contributions are shown with empty symbols for the three heaviest elements. The large error bars reflect the uncertainties of the separation process.

2) Only fragments totally stripped (q = Z) are analyzed here. However, at increasing

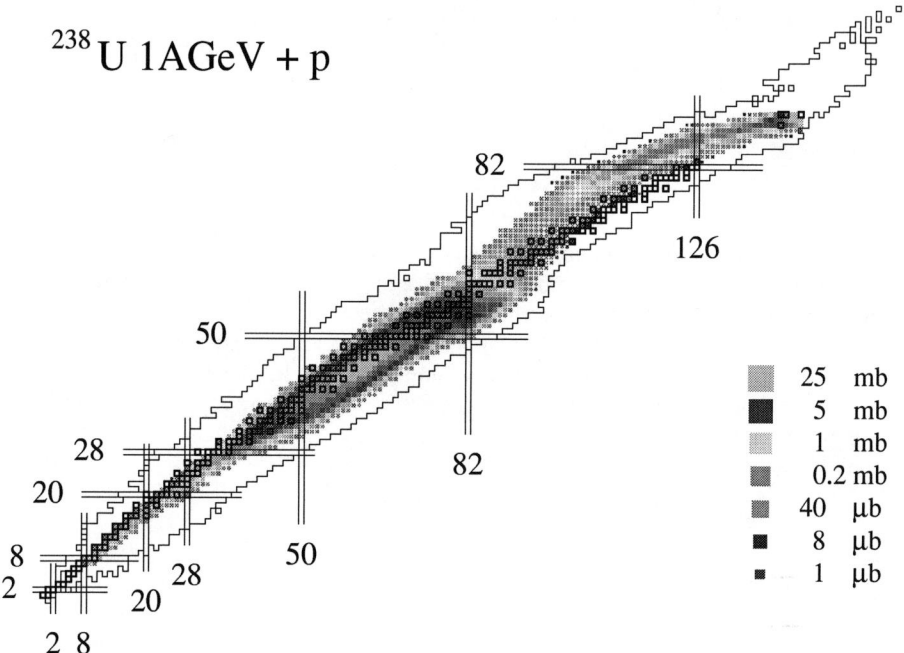

FIGURE 6. The identified isotopes are shown on a chart of nuclei. The logarithm of the experimental cross-sections are indicated by a color scale.

atomic numbers the proportion of the next ionic state ($q = Z - 1$) is no longer negligible and the correction grows from 1.09 for $Z = 64$ to 1.23 for $Z = 73$ [13].

3) Another loss of FF comes from secondary reactions in the scintillator SC2. This loss is calculated using the formulation of Karol [14] for the total cross section. The related correction factor increases from 1.17 for $Z = 64$ to 1.18 for $Z = 73$.

4) Cross sections of neutron-deficient isotopes are excluded when the contamination due to EVR becomes as large as the counting of FF.

The isotopic cross section distributions show similar bell shapes. The contribution of fission extends up to $Z = 75$. It is not possible to conclude about an increment of cross sections for heavy partners of the very light nuclides since the EVR produced on the Ti-windows -or on p- and secondary reaction fragments become predominant at $Z = 74$ and beyond.

On the neutron-rich side, the domain of investigation can be extended much further down to very small cross-sections (0.2 nb). This was demonstrated earlier with the identification of 117 new neutron-rich FF in U (0.75 GeV) + Be collisions. [6, 7].

The distribution of $\overline{N}$ as a function of Z indicates that many neutrons are emitted together with the formation of a very asymmetric pair of FF. Those rare fission processes are due to highly excited nuclei.

CONCLUSION

Fission fragments are separated and identified up to mass A = 184. Velocity measurements allow to deduce velocity distributions of fission fragments and to separate them from evaporation residues. Isotopic production cross sections are obtained for isotopes of the 66 elements produced by U + p fission, down to a threshold value of 0.1 mb covering a range of 0.1 to 14 mb, (Fig. 6), with a systematic uncertainty of 10%. The domain of very heavy FF-partners is investigated for the first time. They are attributed to fission of very excited intermediate parent nuclei in the range $88 < Z < 92$.

REFERENCES

1. P. Armbruster et al. *Phys. Rev. Lett.***93** (2004) 212701
2. M. Bernas et al. *Nucl. Phys.***725A** 213 (2003)
3. J. Taieb, *PhD Thesis* Univ. Paris XI (2000)
4. M. V. Ricciardi *Contribution to this meeting* M. V. Ricciardi, *PhD Thesis* (2005) Univ. Santiago de Compostela.
5. H. Geissel et al. *Nucl. Instrum. Methods***B70** 286 (1992).
6. M. Bernas et al. *Phys. Lett.***415B** 111 (1997)
7. Ch. Engelmann,*PhD Thesis* Univ. Tübingen (1998), GSI Rep. Diss. 1998-15, and *Zeit. Phys.***A352** 351 (1995)
8. T. Enqvist et al. *Nucl. Phys.***A686** 481 (2001)
9. P. Napolitani et al. *Nucl. Phys.***727A** 120 (2003)
10. E. Casajeros *Ph.D. Thesis* Santiago de Compostela (2002)**
11. D. J. Morrissey, *Phys. Rev.* **C39** (1989) 460
12. M. Bernas et al. Seminar on fission Pont d'Oye (2003)
13. C. Scheidenberger et al. *Nucl. Inst. and Methods***142** (1998) 441
14. P. J. Karol *Phys. Rev. C***11** (1975) 1203

Nuclide Cross-Sections Of Fission Fragments In The Reaction ^{208}Pb+p At 500 A MeV

C.Volant[1], B.Fernandez-Dominguez[1], P.Armbruster[2], L.Audouin[3], J.Benlliure[4], M.Bernas[3], A.Boudard[1], E.Casarejos[4], S.Czajkowski[5], J.E.Ducret[1], T.Enqvist[2], B.Jurado[2], R.Legrain[1], S.Leray[1], B.Mustapha[3], J.Pereira[4], M.Pravikoff[5], F.Rejmund[2,3], M.V.Ricciardi[2], K.-H.Schmidt[2], C.Stéphan[3], J.Taieb[2,3], L.Tassan-Got[3], W.Wlazlo[1]

1 DAPNIA/SPhN CEA/Saclay, F-91191 Gif sur Yvette Cedex, France
2 GSI, Planckstrasse 1, 64291 Darmstadt, Germany
3 IPN Orsay, F-91406 Orsay Cedex, France
4 Univ. de Santiago de Compostela, E-15706 Santiago de Compostela, Spain
5 CENBG, F-33175 Gradignan Cedex, France

Abstract. The isotopic distributions and recoil velocities of the fission fragments produced in the spallation reaction ^{208}Pb+p at 500 A MeV have been measured using the inverse kinematics technique. The shapes of the different distributions are found in good agreement with previously published data while the deduced total fission cross-section is higher than expected. From the experimental data, the characteristics of the average fissioning system can be reconstructed in charge, mass and excitation energy and the number of post-fission neutrons can be inferred. The results are also compared to different models describing the spallation reaction.

Keywords: spallation reaction, production of fission fragments, post-fission emissions
PACS: 24.10.-i;2475.+i; 25.40.Sc;25.85.Ge;25.85.-w;28.41.Kw

The understanding of the spallation mechanism is of great interest since it is out of the experimentalist possibilities to measure all data needed for the design of any technological applications. For that, simulations are required and the nuclear models that enter in the simulation codes need to be checked and/or tuned on a data base as large as reasonable. Experiments on various targets and bombarding energies are thus needed. Lead being one of the expected material for neutron spallation targets, an accurate measurement [1] of the production of the fragmentation and fission products at 1 GeV using the inverse kinematics techniques (lead beam on a liquid hydrogen target) with the FRS facility at GSI (Darmstadt, Germany) has been performed. This energy being somewhat the upper limit for the various projects, new measurements at lower energies appeared mandatory and the present work deals with measurements at 500 MeV, a near lower limit of the present technique which requires a large proportion of fully stripped ions to be tractable. This drawback is not effective for the fission fragments which are studied in the present communication and has been solved in an elegant way for the fragmentation residues by L.Audouin et al. [2]. Since its discovery, the fission process, remains a fascinating and still debated subject even for low excitation energies of the fissioning nuclei. A subtle interplay between static and dynamic properties is at work and the evolution with the excitation energy is also the object of controversies to describe the competition between evaporation and fission decay. Confrontations of the results of various models with accurate data taken at different energies are thus expected to improve our understanding of the fission process.

Experimental Results

The experimental procedure is given in further details in [3]. The Pb beam impinges on a liquid hydrogen target and the forward focused reaction products are analyzed by the FRagment Separator and its detection. Mass, charge and velocities are accurately measured. Several magnetic settings are needed to cover the full momentum and mass distributions. The velocity distributions are reconstructed to obtain the cross-section of the various nuclides. In the case of fission products, due to the finite angular acceptance (30 mrad), only the forward and backward components are measured and a justified assumption of isotropic distribution is done. With further corrections detailed in [3], the deduced cross-sections are accurate within 15-25% down to few 10 µbarns.

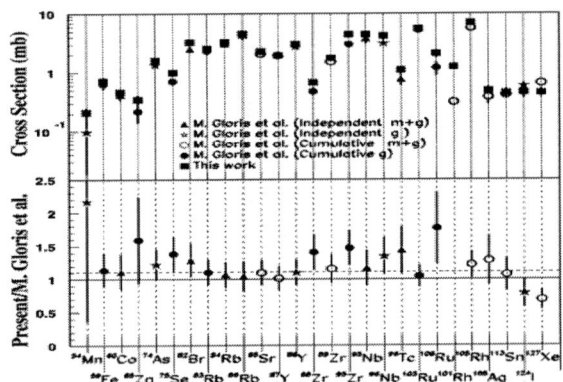

FIGURE 1. Upper panel: Experimental cross-sections on ^{208}Pb (this work) and on ^{nat}Pb [4]. Independent and cumulative yields as well as measurements of metastable (m) and ground (g) states are also indicated. Bottom panel: Ratio of the cross-sections.

The results on the production in the fission channel, of around 400 nuclides identified in mass and charge as well on their velocity distributions have been obtained. On Figure 1, comparisons are done with measurements using the direct kinematics and γ-spectrometry at 553 MeV [4]. Adequate summations of our data are done to be comparable to the data obtained after radioactive decays. Taking into account that natural lead was used for these data; the ratio between the two experiments is $R_{present}/R_{direct}$=1.22 ±0.04 which is compatible within the systematic uncertainties (~15%) of both experiments.

Summing up the production of the present identified isotopic channels and assuming that the unmeasured fission products have a negligible contribution, one obtains a fission cross-sections σ_f= 232 ±33 mb. Compared to previous measurements and systematics [5], the present data appears significantly larger than expected (almost a factor of 2) but despite checks, no experimental explanation has been found up to now.

Characteristics Of The Fissioning Nuclei

They are deduced from the average characteristics of the fission fragments. Neglecting the charge post-fission emission, using the measured mean masses and evaluating the number of post-fission neutrons (ν_{post}) thanks to the measured fragment velocities and assumptions on the Coulomb repulsion one obtains: ν_{post} = 8.0±3, Z_{fiss}=80.0±0.2, A_{fiss}=194±3.8 and the mean total excitation energy E^*= 107±22 MeV using the width of the Z-distribution. This last value is coherent with the obtained ν_{post} value assuming that in average 10 MeV is needed to evaporate one neutron. Furthermore the deduced total kinetic energy taking into account the ν_{post} values is in agreement (134±5 MeV) with the Viola systematic for fission.

Comparisons With Models

Commonly the spallation reaction is viewed as a two-step process, an intra-nuclear cascade stage (hard nucleon-nucleon collisions) leading to an excited nucleus, followed by evaporation and/or fission. Sometimes a pre-equilibrium stage is introduced between these two stages. Generally, the codes describing the reaction couple two different approaches, an INC model and a statistical de-excitation one. Different attempts have been done in [3], using different intra-nuclear cascade models (ISABEL [6], INCL4 [7]), followed by the same evaporation-fission approach,
the ABLA model [8], for observables more related to the first step of the reaction or with INCL4 followed by two different de-excitation models, ABLA and GEM [9], for quantities more dependent on this stage. Since none of these combinations of models is able to reproduce the total fission cross-section measured at 500 MeV, mainly comparisons to the shapes of the different distributions were done.

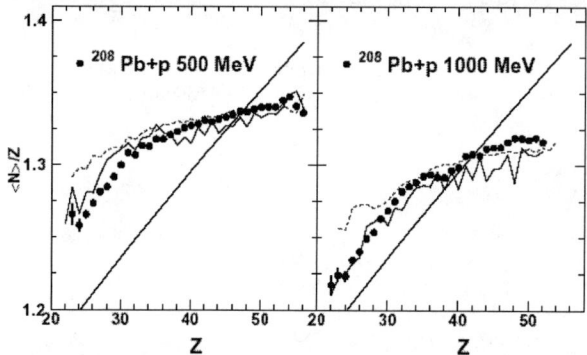

FIGURE 2. Ratio between the mean neutron number and the charge as a function of the nuclear charge, points are data at two energies. Dashed curves are results of INCL4/ABLA and full curves of INCL4/GEM calculations. The full line corresponds to the stable isotopes.

Attempts, using the deduced fissioning system characteristics, tried to evaluate the merit of the INC codes to describe the interplay between the excitation energy, the angular momentum and the fissility parameter at the end of the INC stage. Although the INCL4-ABLA combination seemed better, no clear statement can be achieved with the present single data. On Figure 2, is illustrated the influence of the evaporation/fission models following the INCL4 code at two incident energies: ABLA, taking into account angular momenta and dissipation effects in fission, and GEM. Both account reasonably well for the data for the heavier nuclei. For the lightest nuclei, GEM appears to better reproduce the centroids but ABLA is in better agreement with the production cross-sections [3], GEM predicting a too strong fall-off with decreasing fission-fragment charge.

A last comparison is done on Figure 3 for the total mass distributions at two energies. At 1 GeV, the fission cross-section is well reproduced, but the light fragmentation residues are largely underestimated. On the contrary, at 500 MeV, the fission cross-section is underestimated by about a factor of 2 but the light evaporation residues are well accounted for. One can wonder if really the theory has all the needed ingredients to modelize this spallation reaction or if other processes enter into play, like for instance a more complex fragment production, which could be expected at high energy but which could also have a role at lower energies. More complete experiments measuring multifragment coincidences are foreseen in a near future.

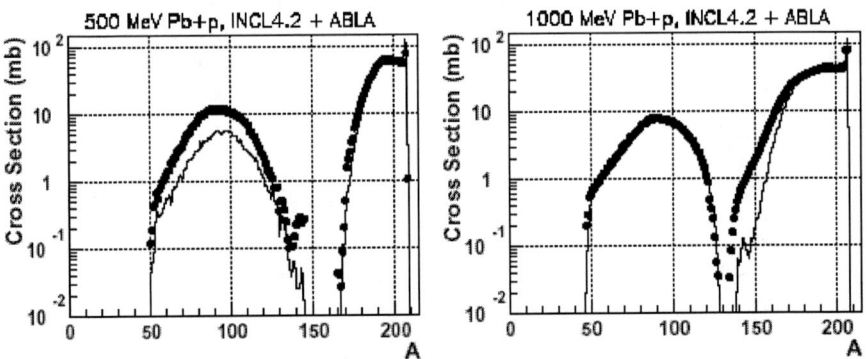

FIGURE 3. Comparison between the experimental mass distributions at two energies and the predictions of the combination INCL4/ABLA. Full lines are the calculated distributions.

REFERENCES

1. Enqvist, T. et al., *Nucl. Phys.* **A686**, 481-524 (2001).
2. Audouin, L. et al., submitted to *Nucl. Phys.* **A**. and PhD Orsay (2003).
3. Fernandez-Dominguez, B. et al., *Nucl. Phys.* **A747**, 227-267 (2005).
4. Gloris, M. et al., *Nucl. Inst. Methods* **A463**, 593-633 (2001).
5. Prokofiev, A. V., *Nucl. Inst. Methods* **A463**, 557-575 (2001).
6. Yariv, Y. and Fraenkel, Z., *Phys. Rev.* **C20**, 2227-2243 (1979).
7. Boudard, A. et al., *Phys. Rev.* **C66**, 044615-1,-28 (2002).
8. Junghans, A. R. et al., *Nucl. Phys.* **A629**, 635-655 (1998).
9. Furihata, S., *Nucl. Inst. Methods* **B171**, 251-258 (2000).

Experimental Cross Sections And Velocities Of Light Nuclides Produced In The Proton-Induced Fission Of ^{238}U At 1 GeV

M. V. Ricciardi[a], K. -H. Schmidt[a], F. Rejmund[a], T. Enqvist[a],
P. Armbruster[a], J. Benlliure[b], M. Bernas[c], B. Mustapha[c], L. Tassan-Got[c],
C. Stephan[d], A. Boudard[d], S. Leray[d], C. Volant[d], S. Czajkowski[e],
M. Pravikoff[e]

[a]*GSI, Planckstr. 1, D-64291 Darmstadt, Germany*
[b]*Universidad de Santiago de Compostela, Dpto. de Fisica, E-15706, Spain*
[c]*IPN Orsay, IN2P3, F-91406 Orsay, France*
[d]*CEA Saclay, F-91191 Gif sur Yvette, France*
[e]*CEN Bordeaux-Gradignan, F-33175 Gradignan, France*

Abstract. Light nuclides produced in collisions of 1 A GeV ^{238}U with hydrogen have been observed with a high-resolution forward magnetic spectrometer, the fragment separator (FRS), at GSI. Fragments were identified in A and Z and their production cross-sections measured. For each nuclide the velocity was precisely determined from the measured magnetic rigidity. This insight into the kinematics of the relativistic nuclear collisions allowed disentangling different reaction mechanisms. Thanks to the combined results on A, Z, and velocity of the fragments it was found out that all the observed isotopes, from $Z=37$ down the last element measured ($Z=7$), were formed in a binary decay process, interpreted as fission. A qualitative analysis of the cross sections revealed that the charge distribution of these light fragments, which forms a plateau around $Z=15$ and increases below $Z=13$, is in agreement with the theoretical expectations of the statistical model.

Keywords: Nuclear fission; light fission fragments; IMF emission; statistical model.
PACS: 25.85.-w; 25.85.Ge; 24.60.Dr; 21.10.Ft; 21.10.Gv.

INTRODUCTION

In 1996, at GSI, Darmstadt, a European collaboration started a dedicated experimental program, devoted to reaching a full comprehension of the proton-induced spallation reactions in the energy range 0.3-1.5 GeV per nucleon. Within this project, the systems ^{56}Fe + ^{1,2}H, ^{136}Xe + ^{1}H, ^{197}Au + ^{1}H, ^{208}Pb + ^{1,2}H and ^{238}U + ^{1,2}H have been studied. The study of light-residue production (from $Z=7$ to $Z=37$) in hydrogen-induced reactions of ^{238}U presented here belongs to this systematic study. Together with three other investigations which proceeded in parallel and which are dedicated to the formation of heavier residues by fission[1] (from $Z=28$ to $Z=73$) and by fragmentation[2] (from $Z=74$ to $Z=92$) in the system ^{238}U + hydrogen, the whole chart of the nuclides from $Z=7$ on is covered. An overview of all the results for the reaction ^{238}U on hydrogen at 1 A GeV are given in a dedicated letter[3]. In the reaction ^{238}U +

hydrogen at 1 A GeV, apart from spallation reactions, which end up in rather heavy fragments (see ref. 2 and also the contribution of M. Bernas to this workshop), most part of the cross sections of the medium-mass residues results from fission reactions. One of the most important signatures of fission is the binary nature of the decay process. The light residues, investigated in this work, also show a binary nature. However, binary decay can occur also via multifragmentation, which is the process responsible for the production of light nuclei in most high-energy nuclear reactions. The possible scenarios behind fission and multifragmentation are indeed strongly different, because the first presupposes the slow decay of a compound nucleus, while the second one the passage through a fast break-up phase. In this work, we will discuss whether the light residues that we observed are consistent with scenario of a binary decay of a compound nucleus, namely, fission.

THE EXPERIMENT

The experiment was performed with a high-resolution, forward, two-stage magnetic spectrometer, the fragment separator (FRS), at GSI. The primary beam of ^{238}U, at the energy of 1·A GeV, impinged on a liquid-hydrogen target, placed inside a titanium container at the entrance of the FRS. The primary-beam intensity was constantly monitored. The fragments produced in the interaction of the beam with the target were detected at the exit of the FRS. The equipment along the FRS consisted mainly of two scintillators, placed at the intermediate image plane and at the exit, and of an ionisation chamber, placed behind the FRS. The ionisation chamber recorded the energy loss of the produced ions, and from that, the nuclear charge of the reaction products was deduced. The two scintillation detectors were used to detect the horizontal positions as well as the time-of-flight between the mid-plane and the exit. The combined information on positions at mid- and final-image-plane and velocity (measured by the time-of-flight) gave a measurement of the magnetic rigidity ($B\rho$) of the ion passing through the FRS. According to the equation:

$$B\rho = \frac{A \cdot \gamma \cdot v \cdot m_0}{Z \cdot e} \qquad (1)$$

(where v is the velocity of the ion, γ is the relativistic parameter, m_0 the nuclear mass unit, and e the charge of an electron), the mass number A of any reaction product could be determined once the nuclear charge Z was determined. The full identification was possible for every produced nuclide, thanks to high resolving power in Z and A ($Z/\Delta Z$ = 150, $A/\Delta A$ = 400). Once every produced nuclide was identified (thus its mass A and charge Z were integer numbers), its velocity could easily be calculated from equation (1) with an accuracy that depends only on $B\rho$. This method gives an absolute measurement of the velocity, which does not suffer from calibration problems and produces a very accurate result, being the resolution in magnetic rigidity about $3 \cdot 10^{-4}$.

Due to the limited momentum acceptance of the FRS it was necessary to combine several measurements with fixed $B\rho$ in order to cover all A/Z and velocities. Once this was done, it was possible to reconstruct the two-dimensional cluster-plot of the velocity distribution as a function of the neutron number for every element. In Figure 1-right, the velocity of iron isotopes produced in fission and fragmentation reactions is presented in the beam frame. Every vertical line represents the velocity distribution of one isotope. The pictures must be observed keeping in mind the limited angular acceptance of the spectrometer, represented by the cone of Figure 1-left, which produced the characteristic triple-humped velocity distributions: fragments distributed in the central part of the spectrum come from the fragmentation reactions, while those with higher positive and negative velocities are produced in fission events. The spectrum in the Figure 1-right collects the counts from the hydrogen target, including a contribution from the titanium windows. From the comparison with data taken with a thin titanium target of the same thickness of the windows of the container, we deduced that the nuclei with the extreme velocity values (fission fragments) are mostly due to the interaction of uranium with protons, while the central part (fragmentation products) is mostly due to the interaction of uranium with titanium.

The recoil velocity of the fissioning nuclei could be determined from the difference of the mean positions of the two external peaks of the velocity spectrum. From this knowledge, the mean velocity of the fission fragments in the fissioning-nucleus frame could be determined too.

From the integral of the three different components of the velocity spectrum, knowing the beam intensities, the target properties and the ratios of the transmitted reaction residues (which can be calculated), the production cross sections were determined both for fission and for fragmentation products.

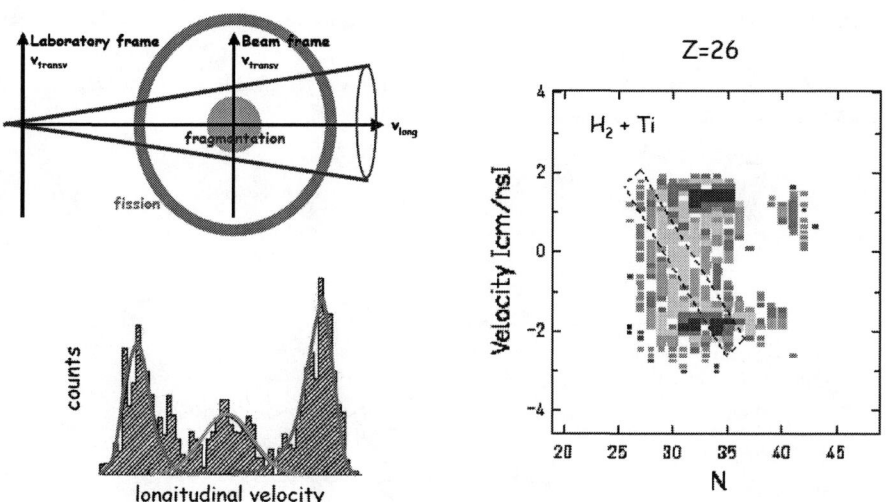

FIGURE 1. Left: Schematic representation of the limited angular acceptance of the FRS, which produces the characteristic triple-humped velocity distribution of the fission and fragmentation products. Right: Longitudinal velocity of iron isotopes produced in fission and fragmentation reactions, presented in the beam frame (the zero of the y axis represents the velocity of the beam).

INTERPRETATION OF THE EXPERIMENTAL RESULTS

Velocity of the fission fragments

The mean velocity of fission fragments can be estimated by the following empirical liquid-drop description of the total kinetic energy:

$$TKE = \frac{Z_1 Z_2 e^2}{D} \quad \text{with} \quad D = r_0 A_1^{1/3}\left(1 + \frac{2\beta_1}{3}\right) + r_0 A_2^{1/3}\left(1 + \frac{2\beta_2}{3}\right) + d \quad (2)$$

where A_1, A_2, Z_1, Z_2 denote the mass and charge numbers of a pair of fission fragments prior to neutron evaporation. D repesents the distance between the two charges and is given by the fragment radii ($r_0 A^{1/3}$), corrected for the deformation (β), plus the neck (d). The parameters (r_0=1.16 fm, d=2.0 fm, β_1=β_2=0.625) were deduced from experimental data in ref. 4. The formula (2) is valid for sufficiently excited nuclei, where shell effects are negligible. When the momentum conservation is imposed to the reaction, the velocities of the two fission fragments are determined.

We have estimated the mean velocities of the fission fragments for two compound nuclei: ^{238}U and ^{185}Au. They are compared with the experimental data in Figure 2 (for technical reasons, the mean velocity could no be determined of fragments below Z=17). While for the heavier fragments, the experimental data fall in between these two estimates, for fragments below $Z = 25$, the mean velocity tends to be higher than the estimation for the ^{238}U compound nucleus. This is due to the fact that the experimental parameters of equation (2) that were obtained in symmetric fission are not applicable to very asymmetric mass splits. In very asymmetric fission, both the neck (parameter d) and the deformation (parameter β) might be smaller, with a consequent increase of the kinetic energy.

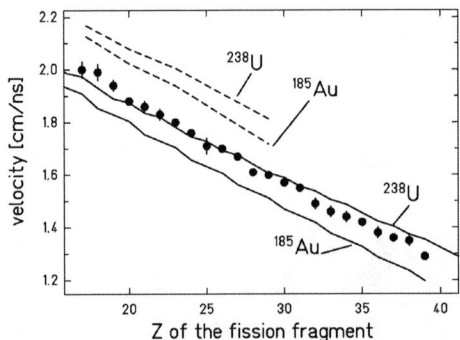

FIGURE 2. Measured mean values of the velocities of fission fragments in the frame of the fissioning nucleus (●). The lines represent expected values of the velocity of fragments originating from the compound nuclei ^{238}U and ^{185}Au. The solid lines represent the expected velocities for the scission-point model (deformed nuclei) and the dashed lines the values obtained by the nucleus-nucleus fusion approach (undeformed nuclei).

To verify the validity of this assumption, we have performed a calculation of the mean velocity in case of asymmetric binary decay from undeformed nuclei. We assumed that the binary decay can be described as the inverse process of fusion. The shape of the potential is given in terms of the nuclear, Coulomb and centrifugal contributions. In our calculations, the empirical nuclear potential of R.Bass[5] was used. The total kinetic energy of the two nuclei is assumed to be equal to the height of the potential barrier. Imposing momentum conservation, the velocity of the two fragments was determined. The result of this calculation for the compound nuclei ^{238}U and ^{185}Au is represented in Figure 2 by the dashed lines. The experimental data fall in between the cases of a split into highly deformed nuclei and undeformed nuclei.

This result gives an indication that the lightest fragments are produced in configurations which are more compact than predicted by the systematics of equation (2) that is based on more symmetric fission. We take this fact as an indication of the transition from fission to the evaporation of intermediate-mass fragments.

Cross sections

In Figure 3-right the production cross sections of the fission fragments are presented in form of charge distribution. The Figure includes the heavier fragments obtained in the parallel analysis[1,2]. Please note that our measurement was technically limited to $Z \geq 7$, but the production of light nuclides would extend even farther down to $Z=1$. One can notice that, as expected, the cross sections are very high in the main fission region and decrease rapidly from $Z=30$ to $Z=20$. But then they stay constant and finally slightly increase again below $Z=10$.

The observed change of slope of the charge distribution can be explained by means of the statistical model. Figure 3 includes also the prediction obtained with a statistical code obtained combining the output results of the intranuclear cascade stage predicted by INCL3[6] with the deexcitation stage ABLA[7,8]. ABLA is a statistical model where the compound nucleus at every step of its evolution has two possible decay channels: evaporation and fission. Evaporation is treated as described in ref. 7 and fission as described in ref. 8. In the statistical model of fission for a given excitation energy the yield of a certain fission fragment is determined by the statistical weight of the transition states above the potential barrier, i.e. at the saddle point. This weight is in turn correlated to the density of nuclear levels. In ABLA, the latter are calculated using the thermodynamic Fermi-gas picture, i.e. assuming the nucleus as a system of non-interacting fermionic particles. The potential energy at the saddle-point depends on mass-asymmetric deformations, which lead to the formation of two fragments of different sizes. In the fission model of ABLA, the barrier as a function of mass asymmetry is defined by three components. The first is the symmetric component defined by the liquid-drop potential by means of a parabolic function with a curvature obtained from experimental data[9]. This parabola is modulated by two neutron shells, represented by Gaussian functions. Shells are supposed to wash out with excitation energy[10]. The heights and the widths of the Gaussians representing the shell effects and additional fluctuations in mass asymmetry acquired from saddle to scission are derived from experimental data. The above representation of the barrier as a function of mass asymmetry is valid only for the main fission region (fron $Z=20$ to $Z=65$

approximately), while for very asymmetric mass splits the potential energy is expected to inverse the slope and start to decrease. This aspect has been widely discussed in the past in terms of "conditional saddle"[11]. From the physical point of view an extremely asummetric binary split corresponds to an evaporation of a light nucleus. Up to now the evaporation part of ABLA considered only the emission of light particles, specifically: neutrons, protons, tritons, deuterons, ^{3}He and alphas. We implemented in the code the evaporation of intermediate-mass fragments (IMF), i.e. light nuclei with Z>2. The statistical weigth for the emission of these fragments is calculated on the basis of the detailed-balance principle, and depends mostly on the ratio of the level density of the nuclear system at the barrier and the levele density of the compound nucleus (at ground state). The barrier is calculated using the fusion nuclear potential of Bass[5].

In Figure 3 the result of the code is presented in form of chart of the nuclides (left) and of charge distribution (right). The latter result is compared with the experimental data. The full line is obtained by the sum of the three components: the evaporated IMF, the fission fragments and the evaporation residues.

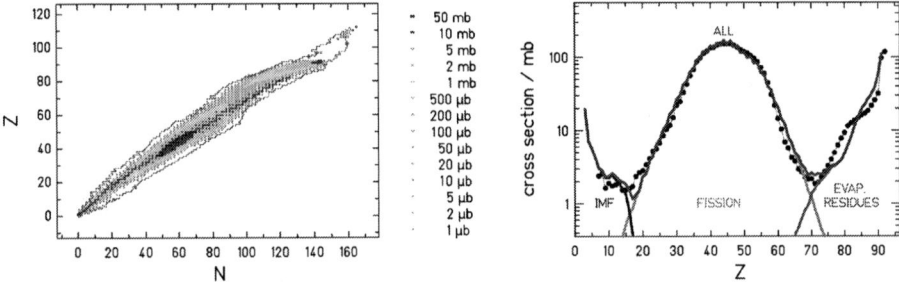

FIGURE 3. Production cross-sections for the reaction ^{238}U (1·A GeV) + p. Left: Prediction of INCL+ABLA presented on the chart of the nuclides. Right: Experimental data (full dots) are compared with the results of INCL+ABLA.

In Figure 4, the isotopic distributions of the first eight measured elements are compared with the prediction of INCL+ABLA. The agreement is extremely good. Results for the mean values of the isotopic distributions of every element are summarised in Figure 5. There, the mean N/Z-ratio is shown as a function of the proton number. The solid line is the result of the INCL+ABLA prediction. In the fission model inside ABLA, the population of the fission channels is assumed to be basically determined by the statistical weight of transition states above the potential-energy landscape at the fission barrier, as described previously. Several properties, however, are finally determined at scission, among them the mean value and the fluctuations in the neutron-to-proton ratio, which are responsible for the so-called "charge polarisation". It can be noticed that the calculation reproduces rather well the mean values (the <N>/Z-ratio) of the isotopic distributions. This is an indication that both the charge polarisation in the fission process and the competition with the evaporation of neutrons in the statistical model are rather well described in the code.

The lightest products are neutron rich, as expected to be in fission. Compared to electromagnetic-induced fission, where the mean N/Z is close to that one of ^{238}U, here the neutron excess is lower, demonstrating that the process occurred at higher excitation energies. The neutron enrichment decreases slightly with the decreasing mass. The reason for this tendency is connected to the fact that the valley of stability becomes quickly narrow and steep. Large fluctuations in N/Z become more and more unlikely.

To conclude this section, we like to point out that the result of the ABLA code is remarkable, because the theoretical model behind it could never be compared before with experimental results on fully identified nuclide distributions in the region of light fission fragments from proton-induced fission.

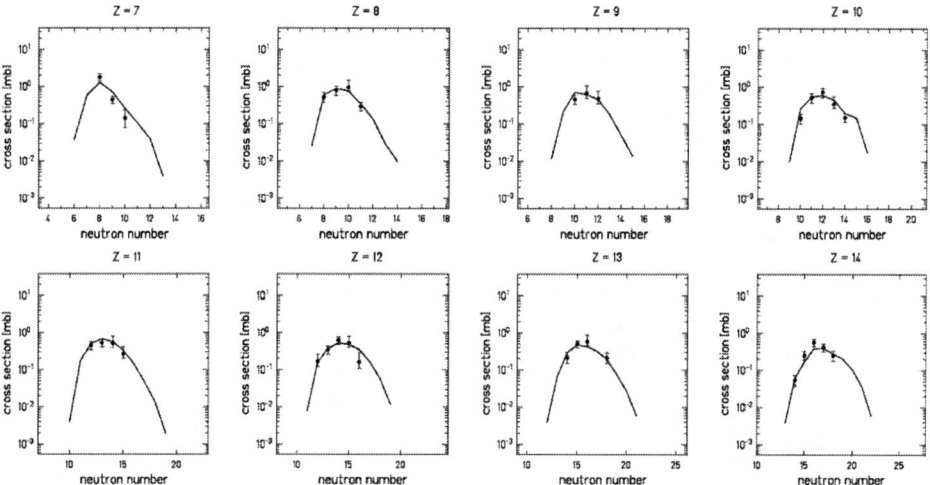

FIGURE 4. Cross sections for the isotopes of the eigth lightest elements measured in the reaction ^{238}U (1·A GeV) + p. The solid line represents the prediction of INCL+ABLA.

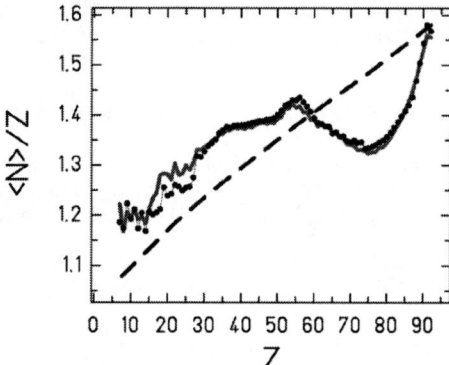

FIGURE 5. Mean neutron-to-proton ratio of isotopic distributions as a function of the atomic number. The experimental data (dots) are compared to the the prediction of the INCL+ABLA code (full line) and to the stability line (dashed line).

CONCLUSIONS

Despite the long study of fission, there is still very little experimental information available on the light residues produced in very asymmetric fission of actinides. Here, 254 light residues in the element range $7 \leq Z \leq 37$ formed in the proton-induced reactions of ^{238}U at 1 A GeV were presented. This experimental work belongs to a systematic study of the reaction ^{238}U + ^{1}H at 1 A GeV, where the production of nuclides with $7 \leq Z \leq 92$ was measured and analysed. The other experimental data, complementing those presented here, can be found in refs. 1, 2, 3.

The light fragments presented here, which populate the chart of the nuclides far down, could be qualified as binary-decay products thanks to the available kinematic information. The study of all the experimental observables, the production cross sections and the velocities, showed that these signatures are consistent with the binary decay of a fully equilibrated compound nucleus. As discussed in ref. 11, the binary decay of a compound nucleus includes fission and evaporation with a natural transition in-between, and it might be called fission in a generalized sense. Thus, very asymmetric fission of the system ^{238}U + ^{1}H at 1 A GeV seems to reach down to rather light nuclei, extending below $Z = 7$. In the spallation-fission reaction of ^{238}U this feature is unambiguously identified for the first time. The production cross sections could be satisfactorily reproduced by a statistical model.

ACKNOWLEDGMENTS

The work has been supported by the European Community with the programmes under the contracts: ERBCHB CT 94 0717, HPRI 1999 CT 00001, FIKW CT 2000 00031, and HPRI-1999-CT-50001.

REFERENCES

1 M. Bernas, P. Armbruster, J. Benlliure, A. Boudard, E. Casarejos, S. Czajkowski, T. Enqvist, R. Legrain, S. Leray, B. Mustapha, P. Napolitani, J. Pereira, F. Rejmund, M.-V. Ricciardi, K.-H. Schmidt, C. Stéphan, J. Taieb, L. Tassan-Got, and C. Volant, *Nucl. Phys. A* **725**, 213-253 (2003).
2 J. Taïeb, K.-H. Schmidt, L. Tassan-Got, P. Armbruster, J. Benlliure, M. Bernas, A. Boudard, E. Casarejos, S. Czajkowski, T. Enqvist, R. Legrain, S. Leray, B. Mustapha, M. Pravikoff, F. Rejmund, C. Stéphan, C. Volant and W. Wlazło, *Nucl. Phys. A* **724**, 413-430 (2003).
3 P. Armbruster, J. Benlliure, M. Bernas, A. Boudard, E. Casarejos, S. Czajkowski, T. Enqvist, S. Leray, P. Napolitani, J. Pereira, F. Rejmund, M. V. Ricciardi, K.-H. Schmidt, C. Stéphan, J. Taïeb, L. Tassan-Got and C. Volant, *Phys. Rev. Lett.* **93**, 212701 (2004).
4 C. Böckstiegel, S. Steinhäuser, J. Benlliure, H.-G. Clerc, A. Grewe, A. Heinz, M. de Jong, A.R. Junghans, J. Müller and K.-H. Schmidt, *Phys. Lett.* B **398** 259-267 (1997).
5 R. Bass, Proceeding of the Symposium on Deep-Inelastic and Fusion Reactions with Heavy Ions, Berlin 1979, Springer Verlag, Berlin.
6 J. Cugnon, C. Volant, S. Vuillier, *Nucl. Phys. A* **620** 475-496 (1997).
7 A.R. Junghans, M. de Jong, H.-G. Clerc, A.V. Ignatyuk, G.A. Kudyaev, K.-H. Schmidt, *Nucl. Phys. A* **629** 635-654 (1998).
8 J. Benlliure, A. Grewe, M. de Jong , K.-H. Schmidt, S. Zhdanov, *Nucl. Phys. A* **628** 458-469 (1998)
9 S.I. Mulgina, K.-H. Schmidt, A. Grewe, S.V. Zhdanov, *Nucl. Phys. A* **640** 375-387 (1998).
10 A.V. Ignatyuk, K. K. Istekov, G. N. Smirenkin, *Yad. Fiz.* **29** (1979) 875-883 (*Sov. J. Nucl. Phys.* **29** 450-454 (1979))
11 L. G. Moretto, G. Wozniak, "The Categorial Space of Fission", LBL-25744 (1988)

Transmutation and Neutron Flux Studies with Fission Chambers in the MEGAPIE Target

CHABOD S.[1], BLANDIN CH.[2], CHARTIER F.[3], FIONI G.[4], FOUCHER Y.[1], LETOURNEAU A.[1], MARIE F.[1], TOUSSAINT J.C.[1]

1 CEA/DSM/DAPNIA Saclay, Gif-sur-Yvette - France
2 CEA/DEN/DER Cadarache, Saint-Paul les Durance – France
3 CEA/DEN/DER Saclay, Gif-sur-Yvette – France
4 CEA/DSM Saclay, Gif-sur-Yvette - France

Abstract. 8 fission micro chambers will be inserted inside the central rod of the 1 MW liquid Pb-Bi MEGAPIE target in order to study the transmutation of two major actinides and to measure the neutron flux at a level of 5%. These chambers were developed for high neutron fluxes and tested at Laue Langevin Institute.

Keywords: fission chambers, neutron flux, neutron sources, spallation, transmutation
PACS: 29.25.Dz

Introduction

The MEGAwatt PIlot Experiment (MEGAPIE) [1] will be the first demonstrator of a liquid Pb-Bi spallation target running with a 1MW proton beam. Operations will begin in summer 2006 at the SINQ installation of Paul Scherrer Institute. MEGAPIE experiment will provide the scientific community with unique information on the behaviour of a liquid spallation target under "realistic" and long term irradiation conditions (proton beam energy = 575 MeV, beam intensity = 1.4 mA, irradiation time = 6-9 months).

In the framework of the Mini-INCA project [2,3] and in collaboration with PHOTONIS Company, new designed micro fission chambers [4] were developed for studies under intense neutron fluxes. These prototypes will be used to monitor on line the neutron flux inside the MEGAPIE target and to study the transmutation of two major actinides: ^{241}Am and ^{237}Np. The arrangement of the neutron detectors and the precise electronic that will be used will provide a significant characterisation of the neutron flux and of its spatial and temporal variations in order to validate transport codes and to provide information on safety in ADS. The on line monitoring of the fission rates for the minor actinides irradiated will answer experimentally on the incineration potential of various systems presenting similar neutron spectra (Molten salt reactors, GTMHR, etc.).

The experimental conditions of irradiation are very stringent inside the target (high neutron flux, high temperature frequently varying, electromagnetic constraints, etc.). Consequently, a specific design of fission chambers have been proposed and tested inside the ILL reactor.

Neutron Detectors for MEGAPIE

For the on line measurement of the neutron flux and the transmutation potentials of ^{241}Am and ^{237}Np, 8 fission chambers will be deployed inside the MEGAPIE central rod. The inner diameter of the central rod is 1.3 cm in its closest part from the target window. Consequently, small chambers (Ø 4.7 mm for 8 cm length) will be used. These chambers will be positioned by pairs all along the central rod for an optimal background correction and spatial flux variation study (Fig 1).

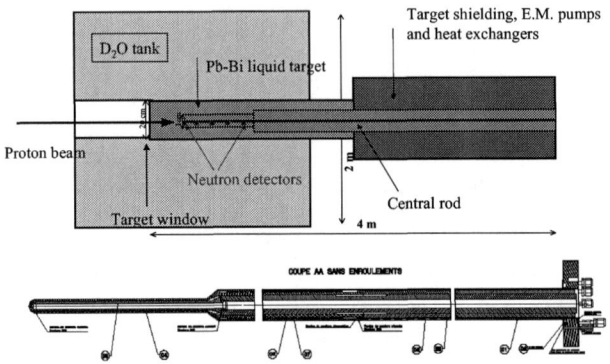

FIGURE 1. MEGAPIE target and its central rod with the neutron detectors

Two 40 and 150 µg ^{235}U deposit fission chambers will monitor the thermal component of the flux. A 150 µg ^{235}U deposit chamber shielded with 200 µm Gadolinium will perform the measurement of the fast component of the flux. 3 chambers without deposit will be used for the gamma background correction. Two ^{237}Np and ^{241}Am deposit chambers will be irradiated for transmutation purposes. The integration over time of their fission rates during a long term irradiation period will provide incineration potentials data in a "realistic" spallation flux.

In addition to mechanical constraints due to the limited amount of space inside the central rod, the fission chambers will work under extreme conditions of neutron fluency, thermal constraints and electromagnetic pollution. The temperature will fluctuate inside the Pb-Bi liquid from 250°C to 420°C within 20 seconds time periods. Thermal calculations using CASTEM code were performed and showed that the static working temperature of the detectors is near 400°C. 3 thermocouples will constantly monitor on line the temperature inside the central rod. A code allowing calculation of the current delivered by a fission chamber has been developed. It demonstrates that the drifts induced by the thermal fluctuations can easily be compensated by adjusting the voltage.

Despite of these unfavourable working conditions, the current intensities delivered by the chambers will be measured with an absolute uncertainty of 5%. For ^{235}U chambers, this current obeys to the classical formula:

$$I = \frac{N_a \frac{m}{M} <\sigma_F> \Phi}{\Gamma}$$

The current I is function of the Avogadro constant N_a, the mass m of the ^{235}U deposit, the molar mass M of ^{235}U, the fission cross section σ_F, a calibration coefficient Γ measured prior to the irradiation, and finally φ, the neutron flux. In MEGAPIE, this current will range from 1 to 100 µA, as the deposit mass and the distance from the neutron production spot differ.

In addition, 9 threshold reaction flux monitors will be used for the measurement of each components of the neutron flux (Table 1).

TABLE (I). Characteristics of some flux monitors that will be used for the MEGAPIE experiment. Monitors are Ø 6mm ultra pure metal discs. The thickness and reactions of interest are also indicated with the neutron threshold energy

Monitor	Thickness [µm]	Reactions of interest	Half-life [day]	Gamma lines [keV]	Threshold Energy [MeV]
Gd	200	$^{154}Gd(n,2n)^{153}Gd$ $^{154}Gd(n,p)^{154}Eu$	241.5 3137.5	97 & 103 123 & 723 & 1274	9 0.5
Fe	50	$^{54}Fe(n,p)^{54}Mn$	312.1	834	0.7
Mn	powder	$^{55}Mn(n,2n)^{54}Mn$	312.1	834	10
Ni	50	$^{58}Ni(n,p)^{58}Co$ $^{60}Ni(n,p)^{60}Co$	70.88 1925.6	511 & 810 1173 & 1332	0.5 3
Rh	25	$^{103}Rh(n,2n)^{102}Rh$ $^{103}Rh(n,p)^{103}Ru$	207 39.25	475 & 511 497	10 1
Ti	125	$^{46}Ti(n,p)^{46}Sc$	83.81	889 & 1120	2.5
Y	200	$^{89}Y(n,2n)^{88}Y$ $^{89}Y(n,p)^{89}Sr$	106.61 50.55	898 & 1836 909	10 2

Monte Carlo simulations of the central rod using MCNPX code have been performed [5] to estimate the neutron and gamma rays fluxes. The neutron flux inside the chambers ranges from $5.57 \cdot 10^{13}$ (4 % with an energy inferior to 1 eV) to $8.15 \cdot 10^{12}$ n.cm^{-2}.s^{-1} (74 % with an energy inferior to 1 eV) between the closest and the most distant position from the target window (Fig. 2).

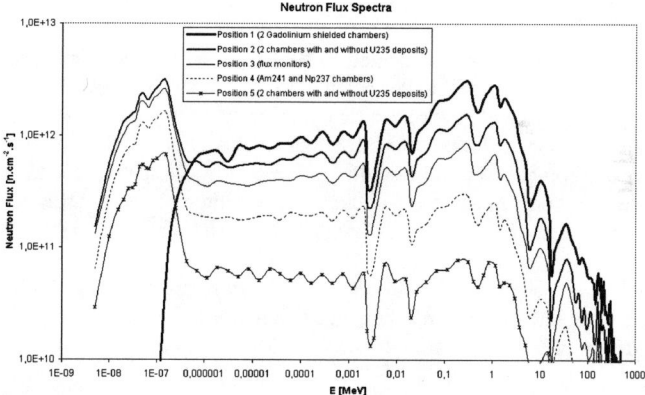

FIGURE 2. Neutron flux spectra inside the detectors at different distances from the target window.

Results of Test Campaigns at Laue Langevin Institute

At the end of year 2003, 4 new designed fission chambers have been tested at the High Flux Reactor of Laue Langevin Institute under severe neutron flux conditions. In terms of fluency, 1 day of irradiation at ILL is equivalent to 1 month in MEGAPIE. Thus, the 43 days of tests at ILL are largely above the expected MEGAPIE conditions. The main objectives of these integral tests were to demonstrate the capability of the prototype chambers to measure high thermal fluxes (until $1.07\ 10^{15}$ n.cm^{-2}.s^{-1}) at an absolute level of 5 %. We wanted also to test their resistance to high temperature, thermal fluctuations and electromagnetic pollution. Using a new precise electronics, cabling and acquisition system, the monitoring of the flux fluctuations has been performed with accuracy better than 1 %. Fig 3 shows an example of fast and slow monitoring of the neutron flux using ^{235}U deposit chambers.

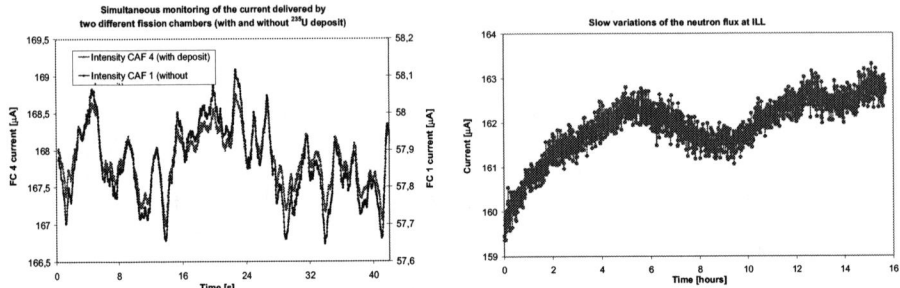

FIGURE 3. Fast (left) and slow (right panel) monitoring of neutron flux using ^{235}U deposit chamber

The gamma-rays induced background has been successfully subtracted. The use of triaxial shielded cables and specific connectors minimizes the electromagnetic perturbations. A study of the behaviour of the detectors in transitory modes is in progress.

Conclusions

New types of fission chambers dedicated to neutron flux measurements and transmutation studies under severe conditions have been developed for MEGAPIE experiment. This challenging project will provide the scientific community with unique data on the neutron flux levels and transmutation potentials inside high intensity spallation targets. In addition, constraints over neutron generation by spallation process and transport codes will be fixed.

The prototype chambers have been validated during integral tests at ILL over the past two years. With an adapted prior heat treatment, they resist high temperatures (600°C) and thermal ageing during a long term irradiation. The chambers perform an absolute monitoring of the neutron flux at a level better than 5% and follow the fluctuations at a level of 1%. The assembly of the complete detector has begun, its delivery to PSI is foreseen by end 2005 and irradiation at MEGAPIE will start mid 2006.

References

1. M. Salvatores, G.S. Bauer and G. Heusener, *The MEGAPIE Initiative*, Report MPO-1-GB-6/0_GB, Paul Scherrer Institute, Zurich (1999).
2. G. Fioni, F. Cribier, F. Marie & al, Nucl. Phys. A693 (2001) 546.
3. F. Marie & al, ND2004, International Conference on Nuclear Data for Science and Technology, Sept 26 – Oct 01 2004, Santa Fe, USA.
4. M. Fadil, *thesis INPG*, Grenoble, DAPNIA-03-01-T (2003).
5. Y. Foucher, internal report (2004).

NEA NUCLEAR DATA SERVICES: EXFOR, JANIS AND THE JEFF PROJECT

Yolanda Rugama, Hans Henriksson

*OECD/NEA Data Bank, 12 Bld des Iles,
92130 ISSY-LES-MOULINEAUX, FRANCE*

Abstract. The OECD Nuclear Energy Agency (NEA) Data Bank is part of an international network of data centres in charge of the compilation and dissemination of basic nuclear data. Through its activities in the nuclear data field, the NEA participates in the production of data and their distribution to nuclear data users. The high priority request list is an example of such a project. The NEA thus provides an essential link between producers and users of nuclear data. The NEA Data Bank distributes the main computer codes and nuclear databases with bibliographical information, evaluated libraries, e.g. JEFF, and experimental data in the data base EXFOR comprising published neutron induced as well as charged particle induced nuclear reaction data. The new data library JEFF-3.1 will be presented here, as well as the data display tool JANIS. The NEA is also involved in the work in the Generation IV International Forum (GIF) technical working groups that are developing research programs for advanced reactor concepts.

1. INTRODUCTION

The Nuclear Energy Agency (NEA) is a specialized agency within the Organisation for Economic Co-operation and Development (OECD), an intergovernmental Organization of industrialized countries, based in Paris, France.

NEA Data Bank works internationally in a network of several data centres in charge of the compilation and dissemination of basic nuclear data. Through its activities in the nuclear data field, the NEA participates in the production of data and their distribution to nuclear data users. The NEA thus provides an essential link between producers and users of nuclear data. The NEA web site (www.nea.fr) offers interfaces to the main nuclear databases with bibliographical information, evaluated libraries, e.g. JEFF, ENDF/B and JENDL, and experimental data in the data base EXFOR comprising published neutron induced as well as charged particle induced nuclear reaction data.

The display program JANIS (JAva-based Nuclear Information Software) has been developed at the NEA, and its latest version (JANIS-2.2) was released in June 2005. JANIS is designed to facilitate the visualisation and manipulation of nuclear data, and to allow the user access to numerical and graphical representations without prior knowledge of the storage format. In this paper an overview will be given of EXFOR as well as the JEFF evaluation library project including the contents of the new JEFF-3.1 nuclear data library released in May 2005. The JANIS display program will also be presented with examples.

2. THE NEA

The mission of the NEA is to assist its member countries in maintaining and further developing the scientific, technological and legal bases for safe, environmentally friendly and economical use of nuclear energy for peaceful purposes. The NEA works as a forum for international co-operation and as a centre of excellence which helps member countries with technical expertise. The NEA's current membership consists of 28 countries, in Europe, North America and the Asia-Pacific region. Together these countries account for approximately 85% of the world's installed nuclear capacity. NEA work areas can be divided into nuclear safety and regulation, nuclear energy development, radioactive waste management, radiation protection and public health, nuclear law and liability, information and communication, nuclear science and the nuclear data bank activities.

The nuclear data evaluation co-operation activities involve the evaluation projects in the following regions: Japan (JENDL), United States (ENDF), Western Europe (JEFF), and non-OECD countries (BROND, CENDL, and FENDL). The participation of the evaluation projects in non-OECD Member countries will be channeled through the Nuclear Data Section of the International Atomic Energy Agency (IAEA).

3. HIGH PRIORITY REQUEST LIST OF NUCLEAR DATA NEEDS

The Nuclear Energy Agency's HPRL project started in the early 1980s by the NEA committees on Reactor Physics and Nuclear Data. The aim was to provide targets for the improvement of nuclear data, primarily for application in the nuclear industry. The users were asked to define their most important requirements, taking into account the limited resources available for measurement and evaluation of data.

The request list is collected to provide targets for the improvement of nuclear data, primarily for application in the nuclear industry through the evaluated data projects, and is a compilation of the highest priority nuclear data requirements. The purpose of the list is to provide a guide for those planning measurement, nuclear theory and evaluation programmes. The HPRL is a place where data users meet data producers.

The High Priority Request List (HPRL) is in a stage of renewal, [1]. A totally new list is going to be presented in 2005, and each year there will be a review of the requests by external referees coordinated by the subgroup C of the OECD NEA Nuclear Science Committee's Working Parties on International Evaluation Co-operation (WPEC). This group consists of both data users and producers from industry, representing Europe, Japan, Russia and USA.

The NEA is at the moment collecting new requests for experimental nuclear data. The requests are divided in high priority ones, where a quantitative justification is needed, and general requests where a more qualitative justification is sufficient. All requests need to be tied to a certain project including a project life span, that is to be stated. The list will be maintained by the NEA Data Bank and is presented on the NEA home page.

4. The NEA Data Bank Services

The Data Bank primary role is to provide scientists in member countries with reliable nuclear data and computer programs for use in different nuclear applications. The services include also thermochemical data for radioactive waste management applications.

The 22 member countries of the NEA Data Bank are: Austria, Belgium, Czech Republic, Denmark, Finland, France, Germany, Greece, Hungary, Italy, Japan, R.o. Korea, Mexico, Netherlands, Norway, Portugal, Slovak Republic, Spain, Sweden, Switzerland, Turkey, United Kingdom. By arrangement with the IAEA, the Data Bank computer program services cover both Data Bank countries and member states of IAEA, except USA and Canada. A separate agreement covers nuclear data and computer program exchanges with the USA and Canada. Users of the Data Bank services include governmental research institutes, industry and universities.

The NEA Data Bank administrates and distributes the main evaluated nuclear data libraries on the NEA web site (www.nea.fr) as well as on CD/DVD on request. The NEA has been using relational databases since 1993 to provide a centralised repository of data and has used web-based technology to allow interactive retrieval of the data. The NEA home page offers interfaces to the main nuclear databases: EVA for evaluated data, CINDA for bibliographical information and EXFOR for experimental data. This latter also includes on-line plotting capabilities.

The nuclear data services are also in charge of the collection and validation as well as the distribution of the Joint Evaluated Fusion and Fission (JEFF) library. The JEFF project has evolved from the two separate EFF (Fusion) and JEF (Fission) projects to a joint collaboration in 1995, with a first library, JEFF-3.0 [2], released in 2002. A preliminary version of JEFF-3.1 was tested and evaluated in November 2004 and a final test version was presented in March 2005. The official release of JEFF-3.1 was in May 2005 as scheduled. A summary report of JEFF-3.1 is planned to be published in the end of 2005.

The computer program services (CPS) group provides more than 2000 documented packages and group cross-section data sets related to nuclear energy applications. The CPS group publishes news letters regularly (see www.nea.fr/html/dbprog), describing the acquisition of basic nuclear data, computer codes and experimental system data needed over a wide range of nuclear and radiation applications. Independent verification and validation of these data is offered using quality assurance methods, adding value through international benchmark exercises, workshops and meetings and by issuing relevant reports with conclusions and recommendations. The CPS disseminate the different products to authorized establishments in member countries and integrates user feedback (more than 600 establishments are served in member countries and about 80 from other countries through agreement with the IAEA). The services include collection of programs, compilation and verification in an appropriate computer environment, and that the computer program package is complete and adequately documented.

The NEA Data Bank organises seminars and workshops to present information on computer programs or groups of programs that are considered to be of special interest to users, such as the NJOY workshop in May 2005 at the NEA. Training courses on widely used computer programs are organised a few times a year to ensure a correct and effective use of them.

Below follows three examples of services from the nuclear data services of the Data Bank, namely international collaboration, collection and maintenance of experimental data in EXFOR, the data manipulation and display tool JANIS, and finally the new evaluated data library JEFF-3.1.

5. EXFOR

A comprehensive set of experimental reaction data are stored in a database called EXFOR [3], that was initiated already 1969. EXFOR has always been coordinated through an international network [4], and the other three main nuclear data centres are, besides the OECD/NEA Data Bank, the National Nuclear Data Center (NNDC) at Brookhaven National Laboratory (USA), the Nuclear Data Services (NDS) at IAEA and the Russian Nuclear Data Center (CJD) at the Institute of Physics and Power Engineering (IPPE) in Russia.

In addition to storing the experimental data points and their bibliographic information, experimental information including source of uncertainties is also compiled. EXFOR is complete with respect to neutron reaction data, and is intended to also cover all charged particle data up to ^{12}C with incident energies up to about 1 GeV. Selected heavy-ion-induced and photon-induced reaction data are also included. EXFOR contains at present about 15,000 experiments from 1935, divided in 114,000 different reactions with a total of about 8.4 Million data points.

The bibliographic database CINDA (Computer Index of microscopic Neutron Data) is closely linked to EXFOR, and contains a complete bibliography of all neutron data published since 1932, as well as an index to corresponding EXFOR entries and evaluated data. Besides neutron data, CINDA also covers photo-neutron, photo-fission and spontaneous fission data. CINDA is available on web retrieval, through the JANIS program, and as a book that can be requested from the NEA.

6. JANIS

JANIS (JAva-based Nuclear Information Software) is a display program designed to facilitate the visualization and manipulation of nuclear data [5]. Its objective is to allow the user of nuclear data to access numerical and graphical representations without prior knowledge of the storage format. It offers maximum flexibility for the comparison of different nuclear data sets. Features included in the latest release are described such as direct access to centralized databases through JAVA Servlet technology.

One of the main missions of the OECD/NEA Data Bank is to provide nuclear data services to research laboratories, universities and industry in member countries. In the last decade, these services have developed along two parallel paths, namely, the use of the Internet and the distribution of dedicated software. Each approach has its advantages. For instance, the Internet option enables the user to access centralised (and thus up-to-date) databases. However, the display of the data is limited by the capabilities of web-pages. Likewise, software running on the user's personal computers can implement advanced, user-friendly interfaces enabling the display of complicated structures.

JANIS was implemented as application-like software with direct access to large databases. The software is free of charge and can be downloaded or launched from the JANIS home page: http://www.nea.fr/janis, where the complete manual can be found as well. JANIS accesses data contained in comprehensive databases. The formats supported are ENDF-6 (along with the linearised pointwise option PENDF and the group-wise option GENDF) and the computational format derived from EXFOR.

JANIS comprises a number of functionalities. The main browser window shows the nuclide chart where basic isotope data can be shown, from NUBASE or from evaluated data libraries. This gives overall information of the isotope and of the evaluated data libraries, e.g. JEFF, JENDL, CENDL, BROND and ENDF/B. Searches can also be performed in the EXFOR and CINDA databases.

In Fig 1, an example is shown on how JANIS displays data on the left side the fission yields of ^{235}U and ^{238}U from JEFF-3.1 and on the right side the user has compared the ^{241}Am fission cross-section from JEFF-3.1 with a set of data form EXFOR.

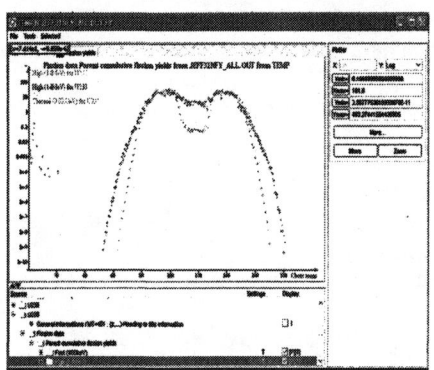
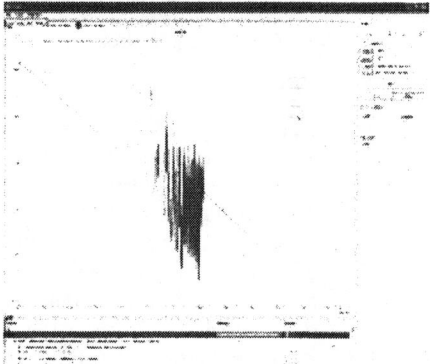

A variety of output formats exist in JANIS. For the graphical display, the PS/EPS and PNG formats are possible, and tabular data can be stored in CSV format (Comma Separated Values) for further use in other software (e.g. MS Excel). Updates of the software can be automatically downloaded through the live-update feature. Version 2.2 will be available soon on the JANIS web page. Feedback is appreciated and can be posted at janisinfo@nea.fr.

7. JEFF LIBRARY

The Joint Evaluated Fission and Fusion (JEFF) project is a collaboration between the countries participating in the NEA Data Bank. The JEFF library comprises of sets of evaluated nuclear data, mainly for fission and fusion applications; it contains a number of different data types, including neutron and proton interaction data, radioactive decay data, fission yield data, thermal scattering law data and photo-atomic interaction data.

The topics covered are benchmarking, testing and evaluations, radioactive decay and fission yield data, experimental data needs, fission product data, and fusion relevant data.

The JEFF project also consists of the European Fusion File (EFF) and European Activation File (EAF) projects (funded by the EC Fusion Programme). These projects are directed by the EFF/EAF monitor group.

The JEFF-3.1 Nuclear Data Library is the latest version of the Joint Evaluated Fission and Fusion Library. The complete suite of data was released in May 2005, and contains general purpose nuclear data evaluations compiled at the OECD Nuclear Energy Agency (NEA) Data Bank in co-operation with several laboratories in NEA Data Bank member countries. Within the framework of the JEFF-3 project, the JEFF Working Group on Radioactive Data and Fission Yields decided to produce improved versions of the decay-data and fission-yield libraries with a release in conjunction with the JEFF library. Activation data has also been included in the latest version.

JEFF-3.1 combines the efforts of the JEFF and EFF/EAF Working Groups who have contributed to this combined fission and fusion file.

- The *neutron data library* covers 381 isotopes or elements, which is an increase from 340 in JEFF-3.0.

A great achievement was to include covariance data for many isotopes in the neutron data library.

All actinides have now extended information on delayed neutron data in that they all are presented in eight-group formalism.

- There are 26 isotopes in the *proton data library*
- 9 materials are covered in the *thermal scattering* law file.
- The special purpose library on *activation data* contains 774 target nuclei with over 12600 neutron induced reactions.
- *Radioactive decay data* with about 3852 isotopes and spontaneous and neutron induced fission yield data.

Processed data for Monte Carlo applications will be made available during 2005 as well as full documentation of JEFF-3.1.

The data can be downloaded from the NEA web site, www.nea.fr/html/dbdata/JEFF, or CDs can be sent on request.

Acknowledgments

The authors wish to thank the entire JEFF collaboration.

References

1. H. Henriksson, Y. Rugama 'Nuclear Data Services from the NEA' Nuclear Data Needs for Generation IV (2005)
2. NEA Data Bank, '*The JEFF-3.0 Nuclear Data Library*', JEFF Report 19, (2005), ISBN 92-64-01046-7
3. V. McLane, "EXFOR Basics. A Short Guide to the Nuclear Reaction Data Format", report BNL-NCS-63380 (IAEA-NDS-206) (May 2000)
4. O. Schwerer, V. McLane, H. Henriksson and S. Maev, AIP Conf. Proc. **769**, 83 (2005), Int. Conf. on Nuclear Data for Science and Industry 2004.

FISSION: MASS AND CHARGE YIELDS

Low energy fission: dynamics and scission configurations

H. Goutte*, J.-F. Berger*, D. Gogny[†] and W. Younes[†]

*CEA/DAM Ile de France, DPTA/Service de Physique Nucléaire
BP 12, 91680 Bruyères le Châtel, France
[†]L-414 Lawrence Livermore National Laboratory, Livermore, California 94551, USA

Abstract. In the first part of this paper we recall a recent study concerning low energy fission dynamics. Propagation is made by use of the Time Dependent Generator Coordinate Method, where the basis states are taken from self-consistent Hartree-Fock-Bogoliubov calculations with the Gogny force. Theoretical fragment mass distributions are presented and compared with the evaluation made by Wahl. In the second part of this paper, new results concerning scission configurations are shown. Deviations of the fission fragment proton numbers from the Unchanged Charge Distribution prescription and fission fragment deformations are discussed.

Keywords: Hartree-Fock-Bogoliubov approach, time-dependent generator coordinate method, fission fragment distributions
PACS: 21.60.Jz,21.60.Ev,24.75.+i,25.85.-w

INTRODUCTION

Fission has been intensively studied for many years (see Ref. [1] and references therein) and fundamental informations on both the fissioning system and the fragments have been extracted from experiments: resonances in class I and class II states from fission cross sections, structure effects from fragment mass distributions [2], intrinsic excitations from odd-even effects [2, 3], and fragment deformations from total kinetic energy distributions or neutron number distributions. In the present work, a theoretical analysis of low-energy fission is made by using a completely microscopic approach, where the sole input is the effective interaction proposed by D. Gogny [4, 5]. The fission phenomenon is described by evolving a time-dependent wave function of the Hill-Wheeler type, which is taken as a linear combination of Hartree-Fock-Bogoliubov solutions characterized by their elongation (quadrupole moment) and asymmetry (octupole moment). The principle of this method and the first results obtained with it have been presented in a recent publication [6]. The advantage of such an approach is to take into account on the same footing collective and intrinsic degrees of freedom along the fission channel. It also includes explicitely the influence of the dynamics of the phenomenon on fission observables, in particular fragment distributions.

In the first part of this report the main results derived [6] in the case of ^{238}U are recalled: i) the broadening of the fragment mass distributions due to dynamical effects, and ii) the influence of the initial conditions on the symmetric fragmentation yields. In the second part, new results concerning scission configurations are presented. Deviations of the fission fragment proton numbers from the Unchanged Charge Distribution prescription and fission fragment deformations are discussed.

FORMALISM

In the constrained Hartree-Fock-Bogoliubov (HFB) theory the deformed states of a nucleus are obtained from a minimization principle of the energy functional:

$$\delta <\Phi(\{q_i\})|\hat{H} - \lambda_N \hat{N} - \lambda_Z \hat{Z} - \sum_i \lambda_i \hat{Q}_i|\Phi(\{q_i\})> = 0, \quad (1)$$

where $\hat{H}$ is the nuclear microscopic Hamiltonian, $\hat{Q}_i$ are multipole operators and λ_N, λ_Z, and λ_i are the Lagrange parameters associated with the constraints on nucleon numbers N, Z and average deformations q_i:

$$\begin{aligned} <\Phi(\{q_i\})|\hat{N}|\Phi(\{q_i\})> &= N, \\ <\Phi(\{q_i\})|\hat{Z}|\Phi(\{q_i\})> &= Z, \\ <\Phi(\{q_i\})|\hat{Q}_i|\Phi(\{q_i\})> &= q_i. \end{aligned} \quad (2)$$

In the present study, the Hamiltonian $\hat{H}$ is built using the D1S effective nucleon-nucleon Gogny force [4, 5] and constraints $\hat{Q}_i$ include both the axial quadrupole and octupole moments, Q_{20} and Q_{30}. Let us mention that the dipole moment has been constrained to a value of zero to avoid spurious translational motion.

Time evolution in the fission channel has been described in terms of a wave function of Hill-Wheeler type based on the HFB intrinsic configurations $|\Phi(\{q_i\})\rangle$ of Eq. (1):

$$|\Psi(t)\rangle = \int d\{q_i\}\, f(\{q_i\},t)\, |\phi(\{q_i\})\rangle, \quad (3)$$

where $f(\{q_i\},t)$ is a time-dependent weight function which is obtained by applying the variational principle:

$$\frac{\delta}{\delta f^*(\{q_i\},t)} \int_{t_1}^{t_2} \langle\Psi(t)|\hat{H} - i\hbar\frac{\partial}{\partial t}|\Psi(t)\rangle dt = 0. \quad (4)$$

The result of Eq. (4) is the well-known Hill-Wheeler equation which reduces to a time-dependent Schrödinger equation when the Generator Coordinate Method (GCM) problem is solved using the Gaussian Overlap Approximation (GOA):

$$\hat{H}_{coll}\, g(\{q_i\},t) = i\hbar \frac{\partial g(\{q_i\},t)}{\partial t}. \quad (5)$$

In equation (5) the collective Hamiltonian is:

$$H_{coll} = -\frac{\hbar^2}{2}\sum_{i,j}\frac{\partial}{\partial q_i} B_{ij}(\{q_i\}) \frac{\partial}{\partial q_j} + V(\{q_i\}) + \sum_{i,j} ZPE_{ij}(\{q_i\}), \quad (6)$$

with $B_{ij}(\{q_i\})$ the inverse of the inertia tensor, $V(\{q_i\})$ the HFB potential and $ZPE_{ij}(\{q_i\})$ the vibrational plus rotational zero-point energy corrections. The wave functions $g(\{q_i\},t)$ can be related back to the $f(\{q_i\},t)$ weight functions through a weighted integral.

Eq. (5) has been solved by discretizing $\{q_i\}$ on a mesh. In order to avoid reflections of the wave function g on the edges of the box, the rectangular $\{q_i\}$ domain includes an extra-region at large elongation where the wave function is absorbed. Time evolution has been performed from Eq. (5) using discrete time steps and by applying the Crank-Nicholson method [7].

Fragment mass distributions $Y(A_H)$ have been derived by a time-integration of the flux $\vec{J}(\{q_i\},t).\vec{n}ds$ of the wave function through scission at a given fragmentation:

$$Y(A_H) = \int_0^T dt\, \vec{J}(\{q_i\},t).\vec{n}ds, \tag{7}$$

where T is the time for which the time-dependent flux along the scission line becomes stabilized, $\vec{n}$ is a vector normal to the scission line in the (q_{20},q_{30}) plane, and $\vec{J}$ is the current associated with the wave function g.

RESULTS

Scission line

Constrained Hartree-Fock-Bogoliubov calculations have been performed in ^{238}U from the first well to very elongated shapes by increasing step by step the quadrupole moment. We consider that scission occurs when the system falls from the fission valley into the fusion valley: we say that we have post-scission configurations when the density in the neck is lower than 0.01 nucleon/fm^3. With this definition it appears that when scission occurs, a 15 MeV drop in the energy of the total system and a 30% decrease of the hexadecapole moment are observed, in good agreement with the results of Ref. [4]. The quadrupole moment of the fissioning system at scission is displayed in figure 1 as a function of the octupole moment. Very elongated shapes are found for symmetric fragments whereas a minimum is observed for $q_{30} = 44\ b^{3/2}$ corresponding to the most probable fragmentation, $A_H \simeq 134$ and $A_L \simeq 104$. The curve of figure 1 is considered as the scission line.

Dynamical effects

In order to define the initial condition of the time evolution, we imagine that the nucleus is a compound system described in terms of complicated quasi-stationnary states which decay into neutron and γ-ray emissions and fission. These initial states can be chosen as eigenstates of the parity operator with eigenvalues $\pi = \pm 1$. However, no symmetric fission is found when the initial state has a negative intrinsic parity. This is due to the fact that the collective wave-function in that case is an odd function of q_{30}. As a consequence, the flux of the wave function through the scission line at $q_{30} = 0$ vanishes. Therefore, symmetric fission can be obtained only if the initial wave-function

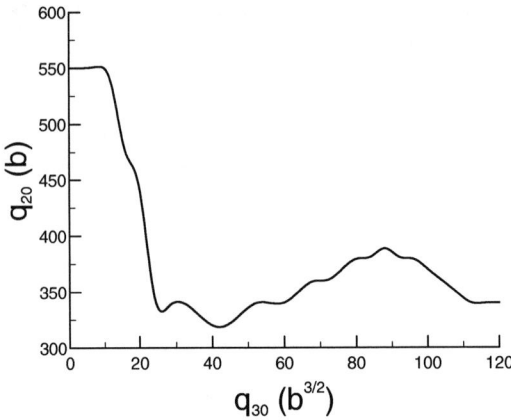

FIGURE 1. Quadrupole deformation of the scission configurations of ^{238}U as a function of the octupole deformation.

possesses a non-zero positive intrinsic parity component. From this, it follows that our predictions for symmetric fission yields will depend on the percentage of the collective states with negative and positive intrinsic parities which are populated in the compound nucleus at the initial stage of the fission process. Since we are only interested in intrinsic states decaying to fission, we assume that the percentages of intrinsic parities (+) or (-) in the initial states with excitation energy E are:

$$p^-(E) = \frac{\sigma(\pi=-1,E)}{\sigma(\pi=-1,E)+\sigma(\pi=+1,E)},$$

$$p^+(E) = \frac{\sigma(\pi=+1,E)}{\sigma(\pi=-1,E)+\sigma(\pi=+1,E)},$$
(8)

where the fission cross sections $\sigma(\pi = \pm 1, E)$ are determined from the formation cross section and the fission probability of the compound nucleus. We assume that these probabilities $p^+(E)$ and $p^-(E)$ represent the mixing of parities in the initial state, which is an eigenstate of the parity quantum number P related to the intrinsic parity π by $P = \pi(-)^I$, where I is the angular momentum. Using the Hauser-Feschbach theory for the reaction n+^{237}U with the optical potential model of Ref. [8] and fission probabilities deduced from a statistical model calculation we find $p^+(E) = 77\%$ and $p^-(E) = 23\%$ for an excitation energy of 1.1 MeV above the first barrier and $p^+(E) = 54\%$ and $p^-(E) = 46\%$ for an excitation energy of 2.4 MeV above the first barrier. These results indicate that structure effects are important: positive parity states decaying to fission are favoured at low energy whereas parity equipartition takes place at higher energy. The mass distributions obtained from these values of p^+ and p^- are displayed in figure 2: the solid and dashed lines correspond to the two energies, 2.4 MeV and 1.1

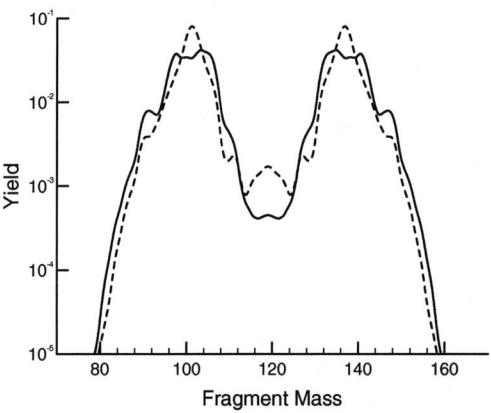

FIGURE 2. Fragment mass distributions obtained for two initial states. Solid line: E = 2.4 MeV with p^+ =54 % and p^- =46 %. Dashed line: E = 1.1 MeV with p^+ = 77 % and p^- = 23 %.

MeV respectively. One observes that the symmetric fragmentation yield is slightly higher at excitation energy 1.1 MeV than at 2.4 MeV, as expected from the parity composition of the initial state. However, this approach does not reproduce an essential feature of measured or evaluated mass fragment distributions namely, the increase of the symmetric fission yield with increasing neutron energy. Nevertheless, the main features of Wahl's distribution, position and height of the maxima, peak to valley ratio and broadening of the distribution as well, are satisfactorily reproduced by the present approach.

The role of the collective dynamics is made manifest in Fig. 3, where the fragment mass distribution obtained from a fully q_{20}-q_{30} dynamical calculation with an initial state located 2.4 MeV above the barrier (solid line), is compared with a "one-dimensional" distribution (dotted line) and with the Wahl systematics (dashed line) [9]. Let us recall that in the "one-dimensional" model, the probability of occurence of a given mass asymmetry is taken as the squared amplitude of the positive parity eigenstate with lowest energy of a one-dimension collective Hamiltonian defined along the scission line [6]. Therefore, in this approach, only collective vibrations along the asymmetry degree of freedom are considered, and the evolution of the system from first well to scission is ignored. We clearly see that the widths of the peaks obtained from the full dynamical calculation are much larger- about twice - than those of the "one-dimensional" one, and appear in much better agreement with the Wahl evaluated data. As a consequence, dynamical effects prior to scission are found to play a major role in the broadening of the fragment mass distribution.

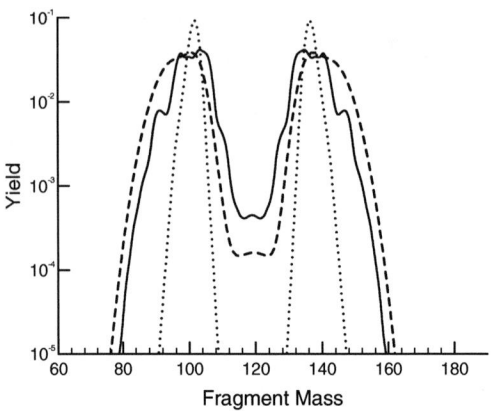

FIGURE 3. Comparison between the "one-dimensional" mass distribution (dotted line), the mass distribution resulting from the dynamical calculation (solid line) with the initial state state located 2.4 MeV above the barrier, and the Wahl evaluation (dashed line) [9].

Fission fragment properties

Along the scission line displayed in Fig. 1, many properties of the fission fragments can be determined from integrations in the left and right half-spaces on either sides of the $z = z_{neck}$ plane. Let us note that the location z_{neck} of the neck is determined as the z-value for which the nucleon density integrated on x and y is minimum. We will concentrate here on the exoticity of the nuclei and on their deformations. Total kinetic energy distributions can be found in Ref. [6].

For each imposed q_{30} value along the scission line, a pair of a Heavy (H) and Light (L) fragments is obtained in our calculations. As no constraints on $A_{H(L)}$, $Z_{H(L)}$, $N_{H(L)}$ are introduced in our calculations, the calculated values represent the most probable $Z_{H(L)}$ and $N_{H(L)}$ fragmentation for a given fragment mass $A_{H(L)}$.

The theoretical ratio between the neutron and proton numbers of the fission fragments is displayed in the left panel of Fig. 4. We see that this ratio evolves on the average between 1.5 and 1.7 as a function of fragment mass. These values for the nascent fragments are closed to the N/Z ratio of the compound nucleus, which is around 1.6 in the case of ^{238}U. Nevertheless, microscopic effects appears and light fragments are found to be less neutron rich nuclei than the heavy ones.

In the right panel of Fig. 4 the deviation of the theoretical most probable charges from the expected values following the "Unchanged Charge Distribution" (UCD) hypothesis is shown as a function of the fragment mass. Experimental data are taken from photofission experiments at excitation energy in the barrier region [3]. A general charge polarization is found for all nuclei: the charge of the heavy fragment is always less than the one given by the UCD model, whereas the mean charge of the light fragment is found to be greater than Z_{ucd} corresponding to nuclei close to the stability valley. These features were already discussed in Ref. [1]. The order of magnitude and the large structure around A $\simeq$ 132 and 140 are well reproduced. Nevertheless, the experimental fact that the charge deviation changes sign near A $\simeq$ 128-130 nuclei is not reproduced by our results. This point should be studied in more detail in the future.

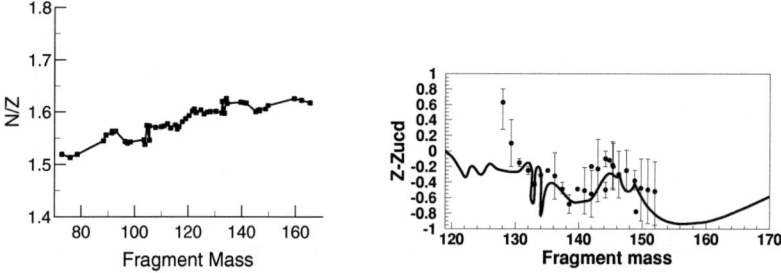

FIGURE 4. Left panel: N over Z ratio of the fragments as a function of the fragment mass. Right panel: Deviation of the fission fragment proton number from the Unchanged Charge Distribution as a function of the fragment mass. Dots indicate experimental data [3] and the continuous line shows the theoretical prediction.

The quadrupole deformation parameter β of the fission fragments is plotted on the left panel of figure 5 as a function of the fragment mass A. The parameter β is related to the mass quadrupole moment Q_{20} by: $\beta = Q_{20}A^{-\frac{5}{3}}$.

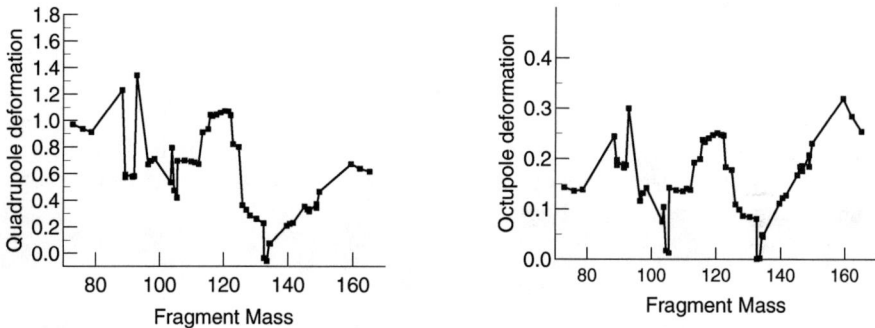

FIGURE 5. Left panel: Quadrupole deformation parameter β as a function of the fragment mass. Right panel: Octupole deformation parameter β_3 as a function of the fragment mass.

From figure 5 it appears that the quadrupole deformation curve has a saw-tooth structure as a function of the fission fragment mass:

- very asymmetric light and heavy fragments have a large deformation $\beta \simeq 0.9$,
- the most probable heavy fragment A $\simeq$ 134 is found to be spherical,
- the corresponding most probable light fragment A $\simeq$ 104 is less deformed than its neighbours $\beta \simeq 0.5$,
- symmetric-fission fragments also have a large deformation $\beta \simeq 1$.

These results are in good agreement with the fact that, due to shell effects, the most probable fission corresponds to a heavy spherical fragment associated with a deformed light one.

The octupole deformation parameter $\beta_3 = Q_{30}A^{-2}$ of the fragments is displayed on the right panel of figure 5 as a function of the fragment mass A. We can see that the octupole deformation is always lower than 0.4 and that the most probable heavy and light fragments do not exhibit any octupole deformation. These results may depend slightly on the definition adopted for scission configurations, and sensitivity tests will have to be performed. Other calculations related to fragment properties, such as excitation energies and the number of evaporated neutrons are under way.

CONCLUSION

In the first part of this paper, we have presented fragment mass distributions for ^{238}U fission calculated in the framework of a microscopic time-dependent Generator Coordinate approach. Results show that the large spread of mass yields cannot be reproduced without including the dynamics from the first well to scission. Also the parity content of the initial state chosen has been found to strongly influence symmetric fragmentation yields. In the second part, first results related to scission configurations as a function of the mass number have been shown. In particular we present: i) the deviation of the mean charge of the fragments from the Unchanged Charge Distribution values and ii) fission fragment deformations. The promising results reported here encourage us to pursue this study of fission fragment properties, where the sole input of the calculations is the effective Gogny D1S force.

ACKNOWLEDGMENTS

The work of D. G. and W.Y. was performed under the the auspices of the U.S. Department of Energy by the University of California, Lawrence Livermore National Laboratory under contract No. W-7405-Eng-48.

REFERENCES

1. "The nuclear fission process" edited by C. Wagemans, CRC Press (1991).
2. K.H. Schmidt, *et al.*, Nucl. Phys. A665 (2000) 221.
3. S. Pommé, E. Jacobs, K. Persyn, D. De Frenne, K. Govaert and M.-L. Yoneama, Nucl. Phys. A560 (1993) 689.
4. J.-F. Berger, M. Girod, and D. Gogny, Nucl. Phys. A428 (1984) 23c.
5. J.-F. Berger, M. Girod, and D. Gogny, Comput. Phys. Commun. 63 (1991) 365.
6. H. Goutte, J.F. Berger, P. Casoli and D. Gogny, Phys. Rev C71 (2005) 024316.
7. W. H. Press, B. P. Flannery, S. A. Teukolsky,and W. T. Vetterling, in Numerical recipes: the art of scientific computing, (Cambridge University Press, Cambridge, 1986)
8. W.Younes and H.C. Britt, Phys.Rev. **C67** (2003) 024610
9. A. C. Wahl, Systematics of fission-product yields, bf LA-13928, Los Alamos National Laboratory, (2002)

Microscopic calculations of potential energy surfaces: fission and fusion properties

L. Bonneau* and P. Quentin[†,*]

*Los Alamos National Laboratory, Theoretical Division, MS B283, Los Alamos, New Mexico 87545 USA
[†]Centre d'Etudes Nucléaires de Bordeaux-Gradignan, CNRS-IN2P3 and Université Bordeaux I, BP 120, 33175 Gradignan, France

Abstract.
Various valleys of the deformation potential energy surface relevant to fission and fusion processes have been investigated within the same Skyrme–Hartree–Fock plus BCS microscopic model in the $A = 70$ and $A > 220$ mass regions. The available experimental fission barrier heights of actinides are reproduced within a rms error of 1.5 MeV whereas the conditional barriers of the considered light nucleus are overestimated by about 10 MeV. The fission paths describing the descent from saddle to scission have been found consistent with the results obtained with the Gogny force used in the Hartree–Fock–Bogolyubov approach and in agreement with the experimental mass distributions. In general, the valleys corresponding to very asymmetric separated shapes (close to the Pb plus light partner configuration) are in agreement with the most favorable target-projectile combinations in cold fusion reactions experimentally used to form such compound nuclei. The deduced fusion barrier heights have been found about 10 MeV lower than those obtained within the Extended Thomas Fermi model.

Keywords: self-consistent microscopic approach, potential energy surface, fission barrier, most probable fragmentation, fusion barrier
PACS: 21.60.Jz, 24.75.+i, 25.70.-z

1. INTRODUCTION

The nuclear fission process constitutes a remarkable challenge for any theory, especially for a microscopic one. A large amplitude collective motion as well as nuclear configurations far from the ground state (GS) equilibrium solution are involved: both dynamic and static properties thus play significant roles. The same can also be said about the reverse process, namely the fusion of colliding heavy ions. From the static standpoint, these processes are both governed by the topology of the potential energy surface (PES) of the considered compound nucleus (CN), especially when the excitation energy of the latter with respect to its GS is low.

In this respect, we aim at investigating the PES of various nuclei with a view to extracting low energy fission and fusion static properties at once. In this perspective, we have applied earlier the self-consistent mean field Skyrme–Hartree–Fock plus BCS model to the study of fission barriers in the actinide region [1] and very recently, we have extended it to the $A = 70$ mass region (below the Businaro–Gallone point), taking the example of the ^{70}Se isotope [2]. The compilation of the obtained barrier heights is presented in Sect. 2 together with new results for three additional actinides. Then we have investigated the various fission paths leading from the outer saddle point to separated fragments configurations. The results and their consequences in terms of fragment mass and kinetic energy distributions are discussed in Sect. 3. Finally we have shown in various mass regions that the potential energy surface possesses several fusion valleys with high mass asymmetry, from which we have deduced fusion barrier heights. This is discussed and compared to other calculations and experimental data in Sect. 4.

2. FISSION BARRIERS

In a previous paper [1], the above mentioned model has been decribed and applied to the calculation of fission barriers for twenty-six actinides among which six could be compared with available experimental data, namely the ^{230,232}Th, ^{234,236}U, ^{240}Pu and ^{252}Cf isotopes. Their barrier heights and those of three additional actinides (^{228}Ra, ^{226}Th and ^{238}U) are reported in Tab. 1. They differ on average by about 1.5 MeV (rms error) from the experimental values. It is worth reminding here that the latter are extracted from

TABLE 1. Inner (E_A) and outer (E_B) fission barrier heights for nine actinides (left values) compared with experimental data (right values) from Ref. [3] (from Ref. [4] for ^{252}Cf).

Nucleus	E_A (MeV)		E_B (MeV)	
^{228}Ra	5.3	8.0±0.5	6.0	8.5±0.5
^{226}Th*	6.4	5.9±0.3	6.5	6.6±0.3
^{230}Th	4.9	6.1±0.2	4.4	6.5±0.3
^{232}Th	5.5	5.8±0.2	4.1	6.2±0.2
^{234}U	5.3	5.6±0.2	5.1	5.5±0.2
^{236}U	6.2	5.6±0.2	4.6	5.5±0.2
^{238}U	7.3	5.7±0.2	4.4	5.7±0.2
^{240}Pu	7.1	5.6±0.2	4.1	5.1±0.2
^{252}Cf	7.1	5.3	2.9	3.5

* experimental values for ^{227}Th

fission cross-section data assuming a double-humped, parabolic shape of the barriers. They include many-dimension effects in the effective, one-dimension barrier parameters. Instead, the theoretical values are obtained by exploring a restricted deformation space, the other degrees of freedom taking their values giving a local minimum (not necessarily the lowest a priori) in the full potential energy surface. As can be seen on the left panel of Fig. 1, the barrier profiles can be far from looking parabolic. Finally, one has also to bear in mind that the pairing strength entering the present model has been kept fixed for all the studied nuclei and has not been adjusted to reproduce such deformation properties.

Very recently, we have applied the same model (with a different pairing strength, though) to the lightest nucleus for which experimental data are currently available, namely the ^{70}Se isotope [2]. This nucleus lies below the Businaro–Gallone point and is slightly neutron-deficient, which makes challenging the task of calculating conditional fission barriers in this mass region. We have found two shallow local minima close to the spherical point, one being prolate, the other one oblate (the lower among both). Starting from the spherical point we have determined the path leading to fission, called ground state ascending valley (full line in the right panel of Fig. 1), and we have shown

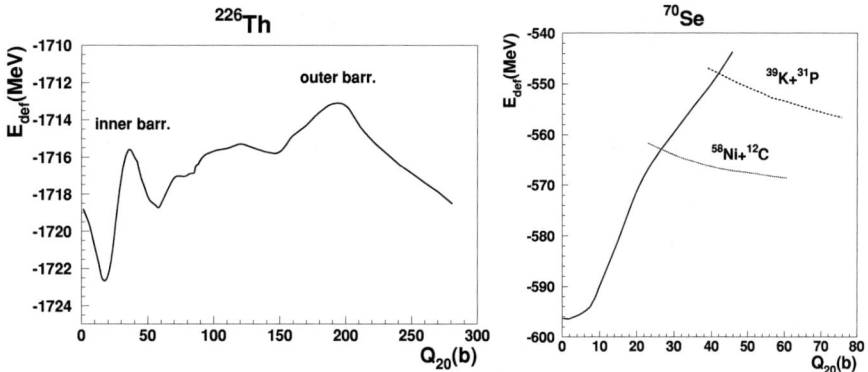

FIGURE 1. Left panel: deformation energy (in MeV) along the fission barrier of the ^{226}Th isotope as a function of the total axial quadrupole moment Q_{20} (in barns). Right panel: deformation energy along the fission and fusion valleys of ^{70}Se as a function of Q_{20}.

that it is stable against left-right asymmetric as well as triaxial distortions. We have also found two fusion valleys corresponding to two different fragmentations, namely ^{58}Ni+^{12}C and ^{39}K+^{31}P (dotted and dashed line in Fig. 1, respectively). We have then sought for a continuous path connecting the GS ascending valley to each of these exit channels. For a given exit channel, the highest point along the corresponding path gives a priori only an upper limit of the conditional fission barrier height. To obtain the actual value of the latter, it is necessary to explore the entire PES restricted to the most relevant shape coordinates, for example in a similar way as the calculations by P. Möller and collaborators in five dimensions (see, e.g., Ref. [5]). This is a very demanding task which is nowadays out of reach of any Hartree–Fock type calculations, essentially for computation time reasons. In Tab. 2 we have reported our theoretical values, compared to the experimental values of T. S. Fan and collaborators [6]. As can be noticed, our

TABLE 2. Conditional fission barriers for the ^{70}Se nucleus compared with experimental data taken from Ref. [6].

Z_{light}	SHF+BCS [2]	exp.
6	34.7	25.3±0.8
15	44.9	35.1±0.8

barrier heights overestimate by almost 10 MeV the experimental ones. To put this result in the proper perspective, in addition to the upper limit character of our values, it is also worth recalling that it was, before this work, commonly thought that our overestimation should be much larger on the basis of a semi-classical estimate of the curvature liquid drop energy, which relies on the leptodermous approximation for the nuclear density [7]. The latter might no longer be valid for saddle point shapes close to scission point shapes, as it is the case for such light nuclei as ^{70}Se.

3. FISSION PATHS AND MOST PROBABLE FRAGMENTATIONS

Beyond the outer saddle point we have determined several fission paths leading to separated fragments configurations. To do so we have let the degrees of freedom other than the elongation Q_{20} (essentially the mass asymmetry Q_{30} and the neck coordinate Q_N) take the values minimizing the deformation energy of the CN. We have searched for several possible such local minima by varying the starting point of the iterative Hartree–Fock process. This has led to the various valleys displayed in Figs. 2, 3 and 4 for six

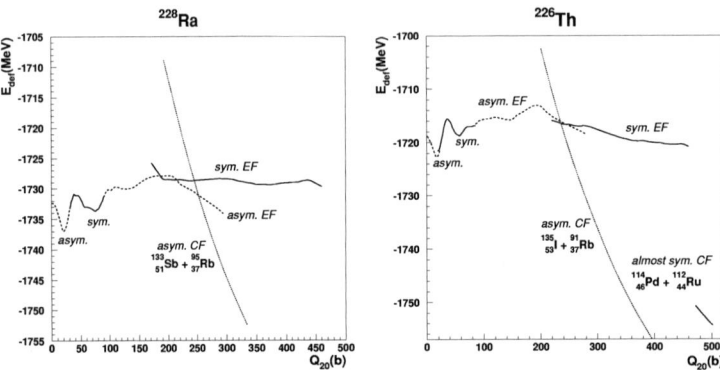

FIGURE 2. Fission and fusion valleys of the ^{228}Ra and ^{226}Th isotopes. Left-right symmetric solutions are displayed in full line, whereas the asymmetric ones appear as dashed lines. The dotted lines represent two-fragments solutions, regardless of their symmetric or asymmetric character.

actinides: ^{228}Ra, ^{226}Th, ^{238}U, ^{252}Cf, ^{256}Fm and ^{258}Fm.

The first two of these isotopes exhibit similar PES patterns. First their GS shape is found to be left-right asymmetric. Moreover two valleys corresponding to one-body shaped, elongated configurations develop around and beyond the outer saddle point: one for left-right asymmetric solutions called asymmetric elongated fission path, the other one for symmetric solutions called symmetric elongated fission path. They are labeled "asym. EF" and "sym. EF" in Fig. 2, respectively. In the same elongation range, we have found a valley called asymmetric compact fission path and labeled "asym. CF" in Fig. 2 describing two-fragments solutions. Whereas the asymmetric EF valley is continuously connected to the GS, the symmetric EF valley is separated from the former by a ridge. From this topological features we can infer that the very low energy fission (up to about 10 MeV) of these two isotopes would preferably proceed through the asymmetric path, leading to an asymmetric fragmentation with a mass ratio $A_H/A_L \approx 133/95$ (heavy fragment mass over light fragment mass) for ^{228}Ra and a charge ratio $Z_H/Z_L \approx 53/37$ for ^{226}Th. These values are consistent with the experimental peak-to-peak ratios for the mass distribution of ^{228}Ra (of about 136/92) [8] and for the element distribution of ^{226}Th (amounting to 54/36) [9]. At higher energy, in addition to the previous asymmetric path, the symmetric EF one becomes energetically accessible, which opens the symmetric division channel. This is in qualitative agreement with the experimental observation of three-peak fragment mass and element distributions for the neutron- and electromagnetic-induced fission of ^{228}Ra [8] and ^{226}Th [9], respectively.

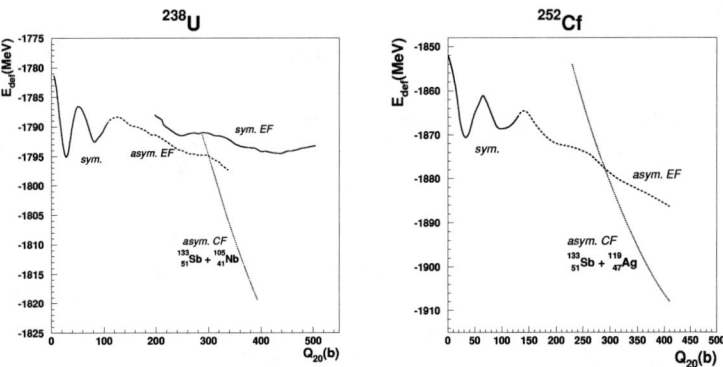

FIGURE 3. Same as Fig. 2 for the ^{238}U and ^{252}Cf isotopes.

As for the ^{238}U isotope (left panel of Fig. 3), we have obtained similar valleys. However the symmetric EF valley is found to lie between 2.5 MeV and 5 MeV above the asymmetric EF one all along the Q_{20} range where they coexist, which makes it virtually inaccessible in low energy fission. As a result, the most favorable path towards fission is the asymmetric one, ending with a fragmentation whose calculated mass ratio ($\approx 133/105$) is somewhat lower than the peak-to-peak ratio (about 142/96) for the experimental mass distribution of ^{238}U spontaneous fission [10]. In the case of ^{252}Cf, only an asymmetric path has been found, leading to a mass ratio $A_H/A_L \approx 133/119$ also underestimating the experimental value (143/109) [11]. It is interesting to note that similar valleys have been obtained for ^{252}Cf by M. Warda and collaborators in the Hartree–Fock–Bogolyubov approach using the D1S Gogny interaction [12]. In particular, these authors have found almost the same mass ratio $A_H/A_L \approx 134/118$.

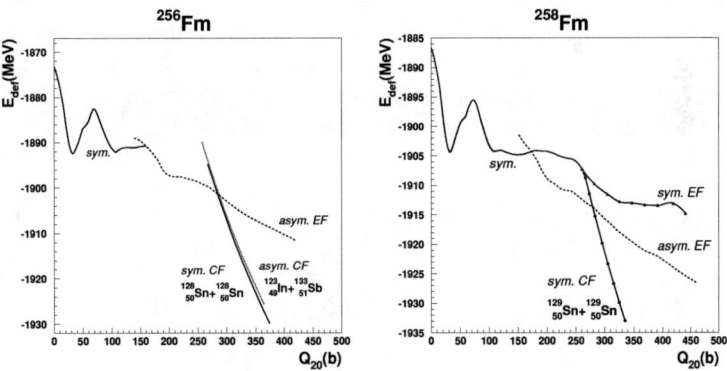

FIGURE 4. Same as Fig. 2 for the ^{256}Fm and ^{258}Fm isotopes.

Finally, the case of the ^{256,258}Fm isotopes is of particular interest since they exhibit totally different fragment mass distribution patterns (for spontaneous fission): whereas the ^{256}Fm nucleus has a broad, asymmetric mass distribution [13], that of the ^{258}Fm is symmetric, sharply peaked at $A = 129$ [14]. On the one hand, only an asymmetric path has been found for the ^{256}Fm isotope (dashed line in the left panel of Fig. 4), eventually falling down into an asymmetric fusion valley (dotted line). It is interesting to note that a symmetric fusion valley has been also obtained, separated from the latter by a ridge (of about 1-2 MeV in the upper part of the valleys). On the other hand, the predominant path that has been found in the ^{258}Fm case turns out to be symmetric. It remarkably connects the GS to the (symmetric) fusion valley (solid line with full circles) in a continuous way. Correlatively, the asymmetric fission valley also present in the PES of ^{258}Fm is separated from the symmetric paths by a ridge, making it much less accessible. Another interesting feature is the existence of a second symmetric valley (EF path represented as a solid line with full triangles in Fig. 4). Whereas the two fragments are formed very close to each other at the early stages of the CF path, leading to a high TKE mode, the much larger center of mass distance between fragments at the very end of the EF path is responsible for a lower TKE mode. This could explain the experimentally well known bimodal fission of ^{258}Fm [15], in contrast to the explanation of M. Warda and co-authors in Ref. [16] invoking the asymmetric EF path to account for the low TKE mode.

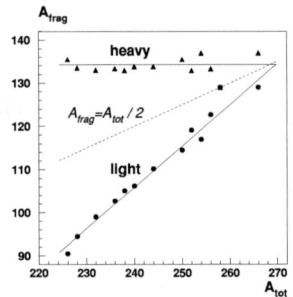

FIGURE 5. Most probable heavy and light fragments masses as a function of compound system mass.

In view of the preceding discussion, the obtained fusion valleys (two-fragments configurations) correspond to the most probable fragmentations. They are due to shell effects in the heavy fragment, close to the doubly-magic ^{132}Sn. As shown in Fig. 5 for the above presented actinides and several additional ones (as well as for the ^{266}Hs super-heavy nucleus whose main valleys are displayed in Fig. 6 and discussed in Sect. 4), the heavy fragment mass is rather constant as the CN mass increases. This behaviour is consistent with microscopic-macroscopic calculations [17] and with the similar trend experimentally observed for the mean heavy fragment mass (see Fig. 4 of Ref. [13]).

4. FUSION BARRIERS

In addition to the above discussed fusion valleys corresponding to the $N = 82$ and $Z = 50$ shell effects in the heavy fragment, we have also found one valley corresponding to the $N = 50$ shell effect in the light fragment and another one where the heavy partner is close to the doubly-magic ^{208}Pb (refered here as to the hyper-asymmetric (HA) fusion valley). In Fig. 6 we have displayed all the fusion valleys found for three very heavy nuclei

(^{256}Fm, ^{258}Fm and ^{266}Hs). In fact the HA fusion valley can be interpreted as the most energetically favored target+projectile combination in a cold fusion reaction forming the considered CN. The deduced fusion barrier height $B_{fus}^{(HF)}$, defined as the difference

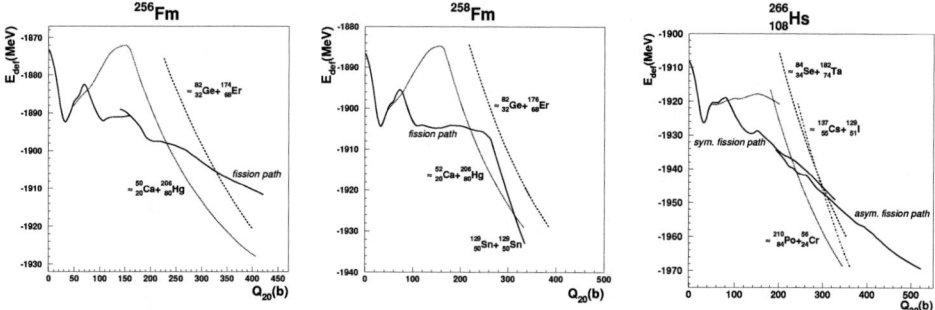

FIGURE 6. Fusion and fission valleys of the compound nuclei ^{256}Fm, ^{258}Fm and ^{266}Hs.

between the energy at the top of the fusion valley and the sum of the projectile and target GS energies, is reported in Tab. 3 for each of these 3 nuclei together with the theoretical values obtained in three other models: the microscopic-macroscopic model [18], the Extended Thomas Fermi model [19] and the Bass model with the parameters of Ref. [20]. In the present work, $B_{fus}^{(HF)}$ is actually calculated as

$$B_{fus}^{(HF)} = E_{CN}^{(min)} + Q_{fis},$$

where $E_{CN}^{(min)}$ is the height of the fusion barrier with respect to the GS of the CN (i.e, the minimal excitation energy at which the CN is formed just at the top of the fusion barrier) and Q_{fis} is the Q-value corresponding to the fission of the considered CN into the target+projectile fragmentation (for which we have used the experimental value). On the

TABLE 3. Fusion barrier heights and minimal CN excitation energies (with respect to the GS of the CN) compared with other theoretical results and with the experimental value [21], respectively.

CN	reaction	$E_{CN}^{(min)}(exp)$	$E_{CN}^{(min)}(HF)$	$B_{fus}^{(HF)}$	$B_{fus}^{(mic-mac)}$	$B_{fus}^{(ETF)}$	$B_{fus}^{(Bass)}$
^{256}Fm	^{206}Hg+^{50}Ca		20.0	166.3		175.5	
^{258}Fm	^{206}Hg+^{52}Ca		19.5	163.3		174.7	
^{266}Hs	^{212}Po+^{56}Cr		9.7	202.1		219.8	
^{266}Hs	^{208}Pb+^{58}Fe	~10			221.96	232.5	226.8

experimental side, the ^{266}Hs CN has been formed by the ^{208}Pb(^{58}Fe,1n)^{265}Hs reaction and the corresponding excitation function has been measured by S. Hoffman and collaborators [21]. Our most favorable target+projectile combination ^{212}Po+^{56}Cr being very close to the one experimentally used, we have also included the experimental $E_{CN}^{(min)}$-value in Tab. 3. The latter is the minimal energy above which the excitation function takes appreciable values. Whereas we underestimate the other theoretical barrier heights by 10 to 20 MeV, our $E_{CN}^{(min)}$-value is close to the experimental one, which might indicate that the deformations of *both* fragments have to be taken into account.

83

5. CONCLUSION AND PERSPECTIVES

We have shown that we can learn more about the fission and fusion properties from the static study of the PES in the Skyrme–Hartree–Fock plus BCS approach. The obtained results generally agree with the experimental data and are consistent in most cases with the calculations in the Hartree–Fock–Bogolyubov and microscopic-macroscopic models. However, the constant pairing interaction (so-called seniority force) does not seem to be appropriate for two-fragments shapes. A δ-interaction for example would better describe the pairing correlations separately in both fragments. Moreover, a particle number conserving approach should rather be used instead of the BCS or Bogolyubov approximations, especially when one is interested in properties varying rapidly with the nucleon number (like the shape transition of the mass distribution in the Fm isotopic chain). Finally, the center of mass correction (performed here upon using a one-body operator approximation) should require a better treatment around and beyond the scission point.

ACKNOWLEDGMENTS

We are grateful to A. J. Sierk and P. Möller for fruitful discussions. One of the authors (Ph. Q.) thanks the Theoretical Division at LANL for the excellent working conditions extended to him during numerous visits. This work has been supported by the U.S. Department of Energy under contract W-7405-ENG-36.

REFERENCES

1. L. Bonneau, P. Quentin and D. Samsœn, *Eur. Phys. J. A* 21, 391 (2004)
2. L. Bonneau and P. Quentin, submitted to *Phys. Rev. C*
3. S. Bjørnholm and J. E. Lynn, *Rev. Mod. Phys.* 52, 725 (1980)
4. G. N. Smirenkin, *IAEA Report*, INDC(CCP)-359 (1993)
5. P. Möller, A. J. Sierk and A. Iwamoto, *Phys. Rev. Lett.* 92, 072501 (2004).
6. T. S. Fan et al., *Nucl. Phys. A* 679, 121 (2000).
7. W. Stocker, J. Bartel, J. R. Nix and A. J. Sierk, *Nucl. Phys. A* 489, 252 (1988).
8. J. Weber, H. C. Britt, A. Gavron, E. Konecky and J. B. Wilhelmy, *Phys. Rev. C* 13, 2413 (1976)
9. K.-H. Schmidt et al., *Nucl. Phys. A* 685, 60c (2001) *Phys. Rev. C* 13, 2413 (1976)
10. T. R. England and B. F. Rider, LA-UR 94-3106, ENDF-349 (1994)
11. F. J. Hambsch and S. Oberstedt, *Nucl. Phys. A* 617, 347 (1997)
12. M. Warda, K. Pomorski, J. L. Egido and L. M. Robledo, submitted to *J. Phys. G*
13. K. F. Flynn et al., *Phys. Rev. C* 5, 1725 (1972)
14. D. C. Hoffman et al., *Phys. Rev. C* 21, 972 (1980)
15. E. K. Hulet et al., *Phys. Rev. C* 40, 770 (1989)
16. M. Warda, J. L. Egido, L. M. Robledo and K. Pomorski, *Phys. Rev. C* 66, 014310 (2002)
17. P. Möller, D. G. Madland, A. J. Sierk and A. Iwamoto, *Nature (London)* 409, 785 (2001).
18. P. Möller, A. J. Sierk, T. Ichikawa and A. Iwamoto, *Prog. Th. Phys.* Suppl. 154, 21 (2004)
19. A. Dobrowolski, K. Pomorski and J. Bartel, *Nucl. Phys.* A729, 713 (2003)
20. R. Bass, *Nuclear Reactions with Heavy Ions* (Springer Verlag, Berlin, 1980)
21. S. Hofmann et al., *Z. Phys. A* 358, 377 (1997)

Microscopic description of fission in odd nuclei

S. Pérez* and L.M. Robledo*

*Departamento de Física Teórica C-XI, Universidad Autónoma de Madrid, 28049 Madrid, Spain

Abstract. We present preliminary results for the fission properties of odd mass nuclei obtained in the framework of the mean field approximation with the equal filling approximation to handle the unpaired odd nucleon. In the calculations the Gogny force with the D1S parameterization has been used. The results for the nucleus ^{235}U are discussed and the hindrance factor for the spontaneous fission half life is partially attributed to the reduction of pairing correlations.

Keywords: Odd nuclei, Mean field Approximation, Equal Filling Approximation, Gogny force
PACS: 21.60.Jz, 21.10.Hw, 25.85.Ca

INTRODUCTION

The fission phenomenon is one of the most fascinating aspects of nuclear dynamics (see [1] for a review). Fission is governed by a delicate balance between bulk properties of the nucleus (the surface and Coulomb repulsion energies) and the quantum mechanical "shell effects". The delicate interplay between both determines the number of barriers separating the nuclear ground state from the scission point as well as their height. The properties of the barriers are very important since the spontaneous fission decay of the parent nucleus involves the penetration through the barriers. In even-even nuclei the fission phenomenon has been studied with phenomenological effective interactions like the Skyrme [2, 3, 4] or Gogny [5, 6, 7, 8] ones in the framework of the mean field approximation of Hartree-Fock-Bogoliubov (HFB). In those studies the nucleus was driven to fission by constraining in an adequate quantity like the mass quadrupole moment. The evolution of the system from its ground state up to the scission point involves a wide range of quadrupole moments and an accurate description of the phenomenon demands a fine mesh on this quantity making the calculations extremely time consuming. In addition, it is rather common that the final fission configuration is mass asymmetric and therefore reflection asymmetric (octupole deformed) shapes have to be considered as possible HFB solutions increasing again the computational cost. Due to these considerations most of the fission calculations have been restricted to axially symmetric (but reflection asymmetric) configurations. On the other hand, and as far as we can tell, there are no attempts to describe fission in odd mass nuclei in the above framework. The most likely reason is that, in order to study odd nuclei in the HFB method, "blocked" HFB wave functions have to be considered. Such "blocked" wave functions break time reversal invariance making the solution of the HFB equation much more time consuming. In addition, the selfconsistent character of the HFB equations does not grant that blocking the quasiparticle with lowest excitation energy will yield the lowest energy solution. As a consequence, several blocking possibilities have to be considered. An alternative to reduce the computational cost of the HFB method for odd nuclei is to consider the so called Equal Filling Approximation (EFA) which is an approximation that deals, in an approximate way, with the odd nucleon but preserving time reversal invariance. The problem with this prescription was that it lacked a theoretical justification and therefore it was not possible to justify the expression needed to compute collective masses needed to estimate spontaneous fission half lives. Recently, we have found that the EFA can be described in solid theoretical grounds by resorting to the concepts of "statistical quantum mechanics". Using these concepts it is possible to justify the whole procedure of the EFA (the EFA-HFB equations, the EFA-HFB energy, etc) and extend it to compute collective masses.

In this paper we will use the framework developed for the EFA to study the fission properties of the odd nucleus ^{235}U with the effective interaction of Gogny [9].

THEORETICAL FORMULATION

In order to study odd nuclei in a mean field framework with pairing correlations it is necessary to consider "blocked" one quasiparticle wave functions $|\Psi_{ODD}^{\mu_B}\rangle = \alpha_{\mu_B}^+|\Phi\rangle$ [10] where $|\Phi\rangle$ is the wave function of the fully paired core and $\alpha_{\mu_B}^+$ is the quasiparticle creation operator for the state to be blocked μ_B. This wave function as well as the associated density matrix

$$\rho_{kk'}^{(\mu_B)} = \langle \phi|\alpha_{\mu_B} c_{k'}^+ c_k \alpha_{\mu_B}^+|\phi\rangle = \left(V^*V^T\right)_{kk'} + \left(U_{k'\mu_B}^* U_{k\mu_B} - V_{k'\mu_B} V_{k\mu_B}^*\right)$$

and the pairing tensor

$$\kappa_{kk'}^{(\mu_B)} = \langle \phi|\alpha_{\mu_B} c_{k'} c_k \alpha_{\mu_B}^+|\phi\rangle = \left(V^*U^T\right)_{kk'} + \left(U_{k\mu_B} V_{k'\mu_B}^* - U_{k'\mu_B} V_{k\mu_B}^*\right)$$

are not invariant under time reversal. This property also translates to the Hartree-Fock and pairing fields making the calculations more complicated and therefore time consuming. A way to restore time reversal invariance in a phenomenological way is to substitute the density matrix and pairing tensor by the averages

$$\rho_{kk'}^{EFA} = \left(V^*V^T\right)_{kk'} + \frac{1}{2}\left(U_{k'\mu_B} U_{k\mu_B}^* - V_{k'\mu_B}^* V_{k\mu_B} + U_{k'\bar{\mu}_B} U_{k\bar{\mu}_B}^* - V_{k'\bar{\mu}_B}^* V_{k\bar{\mu}_B}\right)$$

and

$$\kappa_{kk'}^{EFA} = \left(V^*U^T\right)_{kk'} + \frac{1}{2}\left(U_{k\mu_B} V_{k'\mu_B}^* - U_{k'\mu_B} V_{k\mu_B}^* + U_{k\bar{\mu}_B} V_{k'\bar{\mu}_B}^* - U_{k'\bar{\mu}_B} V_{k\bar{\mu}_B}^*\right)$$

that now preserve time reversal invariance as they involve and average with equal weights of the blocked level μ_B and its time reversed partner $\bar{\mu}_B$. Intuitively, the above densities should correspond to an occupancy of $1/2$ for the quasiparticle states μ_B and $\bar{\mu}_B$. This procedure is called the "Equal Filling Approximation" (EFA) (see [9] for an example of application with the Gogny force) and to our knowledge it was lacking a theoretical justification up to recently when we have explored the similarities of the above expressions for the EFA density and pairing tensor with the corresponding quantities defined in the framework of the Finite Temperature mean field theory. These similarities allowed us [11] to find a theoretical justification of the EFA in the framework of "statistical quantum mechanics" and based on the variational principle over an statistically averaged energy. With this justification we have been able to derive consistent expressions for the collective masses to be discussed below.

We have performed calculations in the framework of the EFA with the finite range and effective interaction of Gogny [9]. As it is customary in all the mean field calculations with the Gogny force, we have subtracted the kinetic energy of the center of mass motion from the Routhian to be minimized in order to ensure that the center of mass is kept at rest. We have also dealt with the exchange Coulomb energy in the Slater approximation and neglected the contribution of the Coulomb interaction to the pairing field. For the Gogny force we have used the parameter set known as D1S that was adjusted almost twenty years ago [5, 12] in order to reproduce basic nuclear matter parameters, the binding energies of several magic nuclei and the fission barrier of ^{240}Pu. In the present context, one of the advantages of the Gogny interaction is its good performance in describing the fission phenomena [5, 6, 7, 8]. The quasi-particle operators have been expanded in an axially symmetric harmonic oscillator (HO) basis and special attention has been paid to the convergence of the results with the number of basis states included.

DISCUSSION OF THE RESULTS

We have carried out calculations for the nucleus ^{235}U as a representative example in the actinide region. As a first step we have performed a calculation for the "seed" nucleus ^{234}U by constraining on the quadrupole moment from sphericity up to a very elongated configuration near scission. In a second step we have performed calculations for ^{235}U by using the HFB code for even-even nuclei but constraining the number of particles to be the corresponding ones, that is $\langle N \rangle = 143$. From now on we will refer to this calculation as the "average" one. The corresponding result will be used as a reference to compare the more fundamental results obtained by "blocking" the corresponding nucleons in the Equal Filling Approximation (EFA). The HFB energies for the two nuclei are depicted in the lower panel of Fig. 1. In this plot we observe that, apart from an overall displacement of the HFB energies of ^{235}U with respect to the one of ^{234}U, the behavior with the quadrupole moment is essentially the same. This is more clearly seen by comparing the ^{235}U curve with the dashed one that corresponds to the ^{234}U result shifted as to make the ground state minima of both nuclei coincide. In the upper panel of Fig. 1 we have plotted the "particle-particle" correlation energies (the HFB pairing energies but with the opposite sign) for both protons and neutrons and the nucleus considered. As expected the proton pairing energies are essentially the same for the two nuclei and the neutron ones follow the same trend with quadrupole deformation but are slightly shifted.

As it was mentioned previously, the selfconsistent character of the EFA-HFB method does not grant that the lowest energy solution for a given value of the quadrupole moment is obtained by blocking the lowest lying quasiparticle configuration in the neighbor even-even nucleus. This nice property from a physical point of view has the disadvantageous side effect of forcing us to explore all the reasonable possibilities of blocking those configurations with the lowest lying one quasiparticle energies. For the present

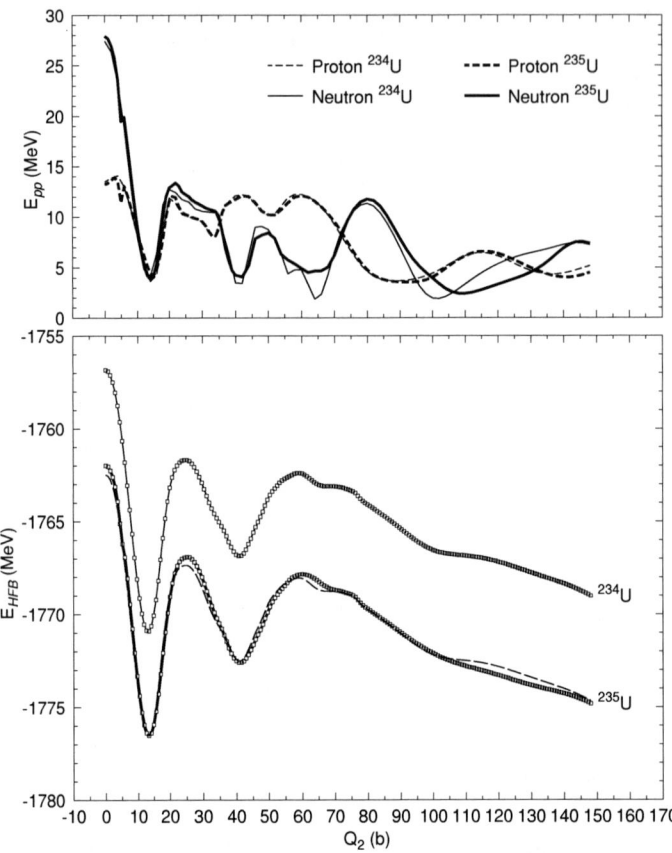

FIGURE 1. In the lower panel the HFB energy as a function of the quadrupole moment Q_2 if depicted by means of full curves with boxes for both ^{234}U and ^{235}U (in the "average" approximation, see text for details). The dashed curve corresponds to the ^{234}U energy but shifted as to make its ground state energy coincide with the one of ^{235}U. In the upper panel, the "particle-particle" (pairing) correlation energies are shown as a function of the quadrupole moment.

calculations we have followed the recipe of blocking the three lowest quasiparticle configurations for each value of J_z from $1/2$ up to $11/2$. This amounts to blocking 18 configurations for each value of the quadrupole moment considered. We have checked in some selected points that we are not missing any relevant configuration by considering the four lowest quasiparticle configurations and by going up to J_z values of $13/2$. In Fig. 2 we present the results obtained for the energies of the odd nucleus ^{235}U computed in the EFA-HFB framework for each of the J_z values blocked. Only the results corresponding to blocking the configuration with the lowest one quasiparticle energy are presented as they lead, in this case, to the lowest selfconsistent energies. In addition, we have also plotted (dashed line) in all panels the energy curve corresponding to the "average" calculation for ^{235}U of Fig 1 in order to use it as a reference curve. We observe that for all

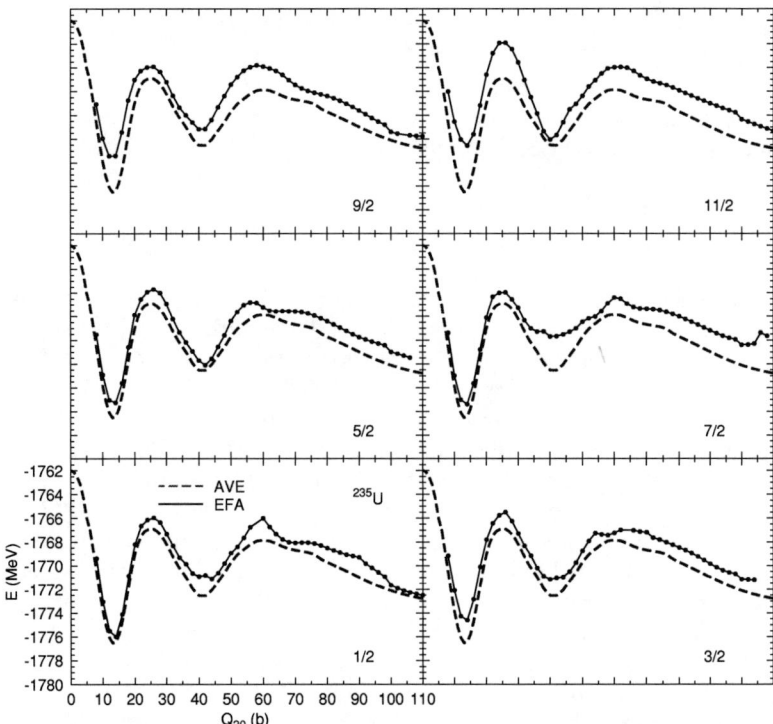

FIGURE 2. The EFA-HFB energies for the nucleus ^{235}U corresponding to the blocking of the lowest energy quasiparticle neutron configuration with J_z values from 1/2 to 11/2 (full curves with symbols). Additionally, the result for the "average" calculation (see text for details) is also plotted as a dashed line

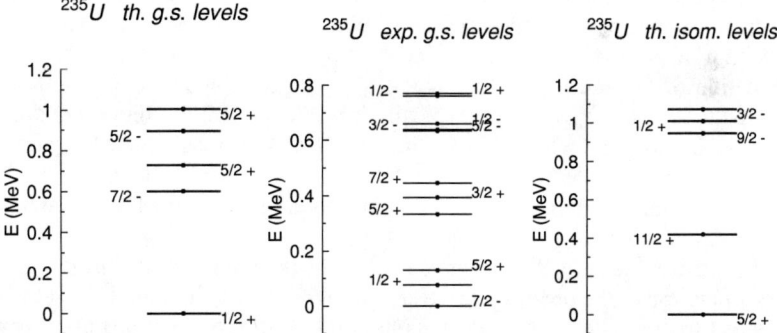

FIGURE 3. On the left panel, the theoretical prediction in the framework of the EFA-HFB for the low lying energy spectrum of ^{235}U. All the label in this panel have almost the same quadrupole deformation $Q_2 \approx 13$ b. On the center panel, the experimental spectrum is shown. Finally, on the right panel the theoretical spectrum corresponding to the fission isomer is shown

J_z values the curves show two minima at around $Q_2 = 13$ b (ground state) and $Q_2 = 42$b (fission isomer) as in the "average" result. Clearly, the ground state corresponds to the J_z value of 1/2 and the fission isomer to the value 5/2. Additionally, the "blocked" results show a very similar behavior as the one observed in the "average" calculation but the potential energy surfaces (PES) tend to lie higher in energy specially for higher deformations. This shift upwards of the blocked PES as compared to the "average" calculation corresponds to the so called "specialization" energy [13] that was introduced to explain the hindrance factors in the fission half lives of odd nuclei as compared to the values for the neighboring even-even nuclei.

In Fig. 3 we show the theoretical predictions of the low lying spectrum (left panel) and the fission isomer spectrum (right panel) and in the middle panel we show the experimental data. Concerning the low lying spectrum we notice that the ground state spin and parity is not reproduced as we predict a $1/2^+$ and experimentally a $7/2^-$ is found. However, the first excited state observed experimentally is a $1/2^+$ and our predicted first excited state is a $7/2^-$. A closer look at both plots reveals that the same kind of levels appear in both the experiment and theory but ordered in a different way. In fact, if we artificially shift up the $1/2^+$ level as to position it in between the $7/2^-$ and the $5/2^+$ levels we will obtain a much nicer agreement with experiment. At this point it is difficult to attribute this discrepancy to the interaction or to other correlation effects not considered at the mean field level like the rotational energy correction. Finally, let us mention that the fission isomer is predicted to be a $5/2^+$ and the first excited state is a $11/2^+$. This prediction, to our knowledge, has still to be confirmed experimentally. Another important piece of information corresponds to the proton and neutron pairing energies for each "blocked" configuration. A look at these quantities reveals that the proton pairing energies are always very similar to the ones of the "average" calculation (see upper panel of 1) whereas the neutron pairing energies are in most of the cases substantially smaller than the corresponding quantities obtained for the "average" calculation of ^{235}U. The quenching of the neutron pairing correlations for the "blocked" calculations is the expected behavior: the pairing correlations for "blocked" configurations are always smaller than the ones of the reference "even" configuration [10]. As a consequence of the quenching of pairing correlations the quadrupole collective mass, depicted in Fig. 4 as a function of Q_2 and for all the "blocked" configurations, increases substantially as compared to the "average" result (typically it is a factor 1.2-1.4 bigger) and in some regions it can be up to a factor 2 bigger than the "average" result. This is again an expected behavior as the collective mass depends on some power of the inverse of the two quasiparticle excitation energies and this quantity gets smaller for those systems with less pairing correlations.

Once the quadrupole collective mass as well as the potential energy surface (PES) are known as a function of the quadrupole deformation Q_2 one might think that all the ingredients needed to compute the spontaneous fission half life in the WKB approximation are at hand. However, the EFA-HFB potential energy surface has still to be corrected for zero point energy corrections associated to the collective quadrupole motion as well as to the restoration of the rotational invariance (rotational energy correction). The rotational energy correction is known [5, 8] to modify the topology of the PES since it is an increasing function with the quadrupole moment. For the zero point energy correction of the quadrupole collective motion one just expects an overall shift downwards of the PES

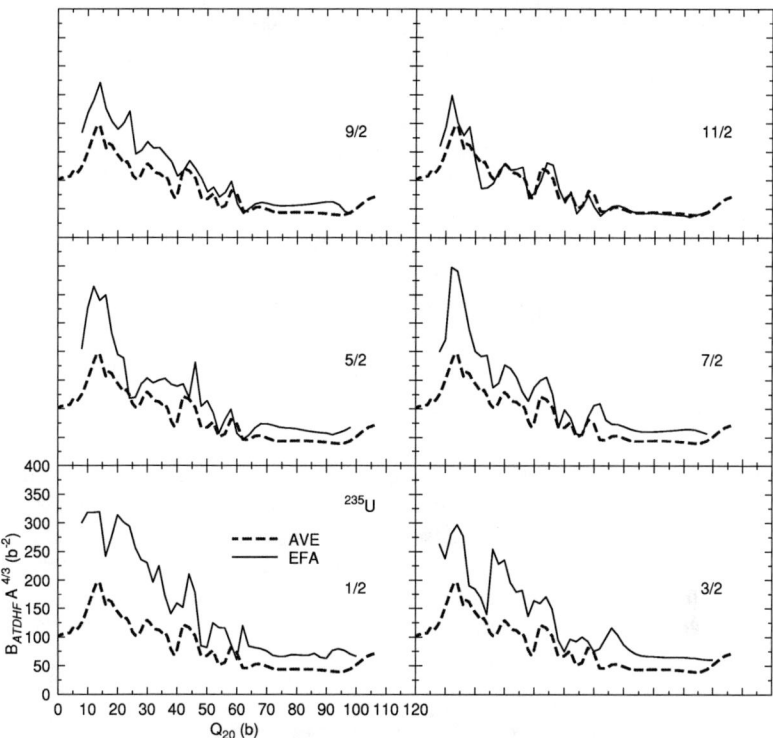

FIGURE 4. The collective quadrupole mass parameter computed using the approximations detailed in the text, for the nucleus ^{235}U and corresponding to the blocking of the lowest energy quasiparticle neutron configuration with J_z values from 1/2 to 11/2 (full curves with symbols). Additionally, the result for the "average" calculation (see text for details) is also plotted as dashed lines

but local modifications of the topology might eventually appear at some specific points. Given the relevance of these two quantities for the fission half live and taking into account that we are still in the process of evaluating them consistently in the framework of the EFA, we are forced to defer the evaluation of $t^{sf}_{1/2}$ for spontaneous fission for a future publication. However, we can make a quick and dirty estimate of the expected values of the relevant quantities in order to give a hint on the fission hindrance factor for odd nuclei. In the framework of the WKB method, $t^{sf}_{1/2}$ is given as the product of the exponential of the action times a quantity of the order of $3 \, 10^{-21}$s. The experimental value of $t^{sf}_{1/2}$ for ^{234}U is $t^{sf}_{1/2} = 4.48 \, 10^{23}$s (see [14] for this and other Uranium isotopes) and therefore one would need an exponential of the action of the order of 10^{44} or an action of the order of 100 for ^{234}U in order to reproduce the experimental datum. The action is proportional to the square root of the collective mass and therefore one expects an increase of this quantity for ^{235}U as compared to the ^{234}U values as a consequence of the "specialization energy" and the enhanced quadrupole collective mass parameter. If one assumes that the collective mass for ^{235}U is on the average a factor 1.2 bigger than the collective mass of ^{234}U, the ^{235}U action can be expected to be of the order of $\sqrt{1.2} \times 100 = 109.5$ and the hindrance factor would be then of the order of 10^4 as observed experimentally [14]. In our calculations for the blocked configuration with $J_z = 7/2$ (the one corresponding to the experimental ground state) the collective mass is roughly a factor 1.3 bigger than the reference one for ^{234}U in reasonable agreement with the previous estimations. We can conclude from this estimate that the hindrance factor in the spontaneous fission half lives of odd nuclei is not only due to the "specialization" energy but it is also strongly influenced by the increase of the quadrupole collective mass due to the quenching of pairing correlations.

CONCLUSION

We have performed preliminary calculations for the fission path of the odd nucleus ^{235}U using the Equal Filling Approximation to the HFB method and the Gogny force as interaction. We get a reasonable description of the lowest lying excited states and make predictions for spin and parity of the fission isomer states. We have also discussed the effect of the specialization energy and the quenching of pairing correlations in the fission half life in order to explain the hindrance factor for this quantity. We find that both effects are relevant.

ACKNOWLEDGEMENTS

This work is partly sponsored by the Ministry of Education and Science (Spain) under Project FIS2004-06697. S.Perez acknowledges the Spanish Ministry of Education and Science for financial support by a FPU fellowship (Ref. AP2001-0182).

REFERENCES

1. S. Bjornholm and J.E. Lynn, Rev. of Mod. Phys. **52** (1980) 725
2. H. Flocard, P.H. Heenen, S.J. Krieger and M.S. Weiss, Prog. Theor. Phys. 72 (1984) 1000.
3. M. Bender, K. Rutz, P.-G. Reinhard, J.A. Maruhn and W. Greiner Phys. Rev. **C58** (1998) 2126
4. T. Buervenich, M. Bender, J.A. Maruhn and P.-G. Reinhard, Phys. Rev. **C69** (2004) 014307
5. J. F. Berger, M. Girod and D. Gogny, Nucl. Phys. **A428** (1984) 23c.
6. J.F. Berger and K. Pomorski, Phys. Rev. Lett. **85** (2000) 30
7. J.L. Egido and L.M. Robledo, Phys. Rev. Lett. **85** (2000) 1198.
8. M. Warda, J.L. Egido, L.M. Robledo and K. Pomorski Phys. Rev. **C66** (2002) 014310; Intl. J. of Mod. Phys. **E13** (2004) 169.
9. J. Dechargé and D. Gogny, Phys. Rev. **C21** (1980) 1568.
10. P. Ring and P. Shuck, *The Nuclear Many Body Problem* (1980), Springer–Verlag Edt. Berlin.
11. S. Pérez and L.M. Robledo, to be published
12. J.F. Berger, M. Girod and D. Gogny, Comp. Phys. Comm. **63** (1991) 365.
13. P. Fong, Phys. Rev. 122 (1961) 1545.
14. H. R. von Gunten, A. Grütter, H. W. Reist, and M. Baggenstos, Phys. Rev. C 23 (1981) 1110.

Self-Consistent Study of Fission Barriers of Even-Even Superheavy Nuclei

A. Staszczak*, J. Dobaczewski† and W. Nazarewicz**,†

*Institute of Physics, Maria Curie-Skłodowska University,
pl. M. Curie-Skłodowskiej 1, 20-031 Lublin, Poland
†Institute of Theoretical Physics, Warsaw University, ul. Hoża 69, 00-681 Warsaw, Poland
**Department of Physics and Astronomy, University of Tennessee, Knoxville, TN 37996, USA,
Physics Division, Oak Ridge National Laboratory, P.O. Box 2008, Oak Ridge, TN 37831, USA

Abstract. Static fission barriers of even-even nuclei with $100 \leq Z \leq 110$ are investigated using the Skyrme-Hartree-Fock model with particular attention paid to symmetry-breaking effects along the fission path. Effects of reflection-asymmetric and triaxial degrees of freedom on the fission barriers are discussed.

Keywords: Skyrme-Hartree-Fock, Fission barriers, Superheavy nuclei
PACS: 21.60.Jz, 25.85.Ca, 27.90.+b

INTRODUCTION

Since the middle of the 1960s, studies of superheavy nuclei have provided rich and unique information on the structure of atomic nuclei; for recent reviews see, e.g., Refs. [1, 2, 3, 4, 5, 6]. Recent experimental works [7, 8, 9, 10] illustrate continuous progress in this area.

Spontaneous fission, together with α-decay, is a major decay channel of superheavy nuclei. Microscopically, the phenomenon of fission can be viewed as many-body tunneling through a potential barrier. Studies of fission barriers allow for a determination of stability of the heaviest nuclei (see, e.g., Refs. [11, 12, 13, 14, 15, 16]).

Recently, a number of theoretical calculations of static fission barriers of nuclei in the actinide and trans-actinide regions have been carried out. These include calculations based on the microscopic-macroscopic treatment [17], the self-consistent approach with the Gogny [18] and Skyrme [19, 20, 21] forces, and also within the relativistic mean-field model [20, 22].

The present paper is a continuation of our previous work [21] with the objective to study static fission barriers of even-even nuclei with $100 \leq Z \leq 110$ within the Skyrme-Hartree-Fock (SHF) approach. Nuclei from this region are well deformed in their ground states. It is worth noting that their stability is specified by two 'magic' neutron numbers: $N = 152$ for the isotopes of fermium ($Z = 100$) and nobelium ($Z = 102$), and $N = 162$ for the isotopes of hassium ($Z = 108$) (see, e.g., Refs. [23, 24]). In this study, we investigate the influence of reflection-asymmetric (non-zero octupole mass moment, $Q_{30} \equiv \langle \hat{Q}_{30} \rangle \neq 0$) and triaxial ($Q_{22} \neq 0$) degrees of freedom on static fission paths.

FISSION BARRIERS IN THE SHF+BCS MODEL

Description of the model

Calculations have been carried out using the Hartree-Fock+BCS code that solves self-consistent HF equations by using the Cartesian (3D) harmonic oscillator (HO) finite basis [25, 26, 27]. This code makes it possible to break all self-consistent symmetries of the nuclear mean field, including simultaneous breaking of axial and reflection symmetries, which are of particular interest to our study.

We use the energy density functional defined by the Skyrme interaction SLy4 [28] and a seniority pairing force treated in the BCS approximation. The pairing strengths have been adjusted to reproduce the experimental proton (0.803 MeV) and neutron (0.696 MeV) pairing gaps in ^{252}Fm.

In Ref. [21] we investigated the stability of SHF results to the number of deformed HO states used in the basis. It has been found that reliable calculations can be carried out with 1140 HO states. At the spherical shape, this number corresponds to 17 oscillator shells.

For all the nuclei considered, the self-consistent binding energy (E^{tot}) is computed by using a quadratic [29] constraint on the mass quadrupole moment Q_{20}. Our study covers the prolate shapes in the range of Q_{20} =0–300 b (barns) that have been binned with the step of 10 b. To fix the position of the center of mass, an additional constraint on the mass dipole moment Q_{10} =0 has been imposed.

Static fission barriers

In Figs. 1 and 2 the total binding energies (E^{tot}) and the mass octupole moments (Q_{30}) calculated for the even-even fermium and nobelium isotopes, respectively, are plotted as functions of the mass quadrupole moment Q_{20}. All calculated static barriers exhibit a similar, two-humped pattern with the inner barrier higher than the outer one. The calculated barrier widths and heights are well correlated with the 'magic' neutron number N=152. One can see that the barriers in ^{252}Fm and ^{254}No reach their local maxima. For the heavier isotopes with N >152, the outer barrier collapses and practically disappears for N=162 (Fm isotopes) and N=160 (No isotopes).

As one can see, the disappearance of the outer barrier is related to the transition from the reflection-symmetric fission path ($Q_{30} = 0$) to the reflection-asymmetric fission path ($Q_{30} \neq 0$). For the heavier Fm and No isotopes, such a transition occurs at greater values of Q_{20}. In the extreme cases of ^{264}Fm (when a fission into the two doubly magic ^{132}Sn nuclei is expected) and ^{264}No, the static fission paths are predicted to be totally reflection-symmetric.

The reduction of the outer barrier plays a crucial role in the standard interpretation of the experimentally known rapid decrease of the spontaneous fission half-lives in the heavy Fm and No isotopes.

In Figs. 1 and 2, the influence of the triaxial asymmetry (for $Q_{22} \neq 0$) on the height of the first fission barrier is shown as a difference between the open and solid circles. The

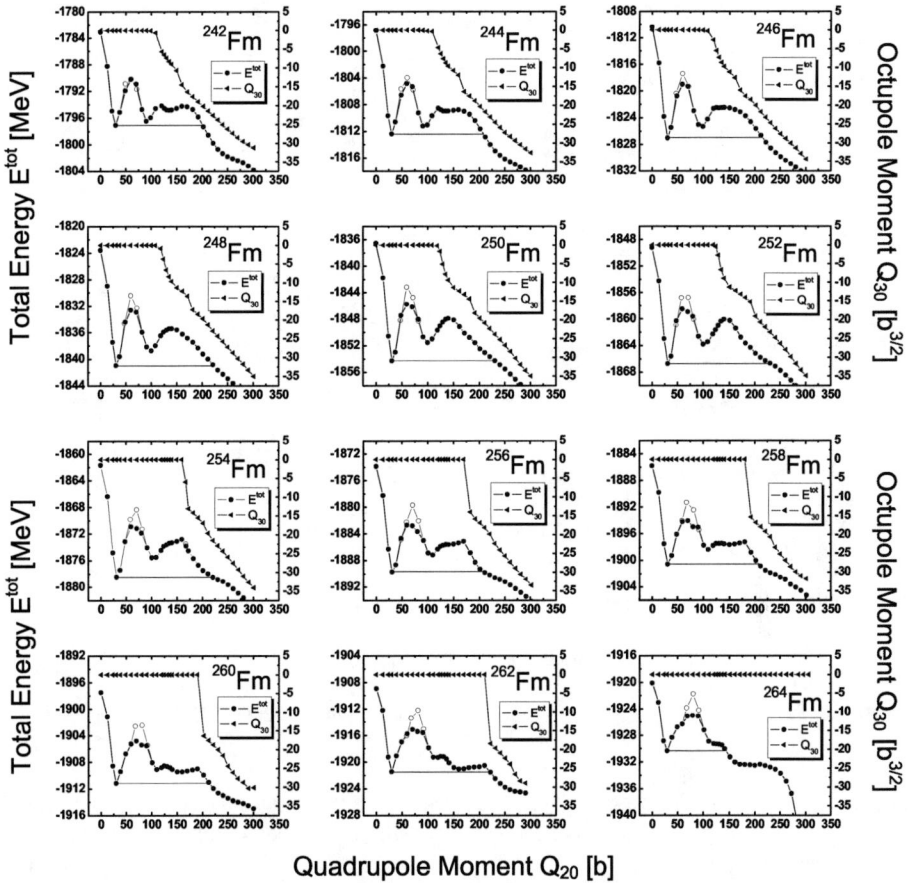

FIGURE 1. Total binding energies E^{tot} (solid circles) and mass octupole moments Q_{30} (solid triangles) calculated along the lowest static fission paths for the even-even fermium ($Z=100$) isotopes with $N=142$–164. The results corresponding to axial shapes are marked by open circles.

effect of triaxiality increases with the neutron number, reaching peak values of about 3 MeV.

The total binding energies and mass hexadecapole moments (Q_{40}) calculated along the static fission paths for the even-even rutherfordium ($Z=104$), seaborgium ($Z=106$), hassium ($Z=108$), and darmstadtium ($Z=110$) isotopes are shown in Figs. 3 and 4. We have found that almost all of these nuclei have purely reflection-symmetric paths with $Q_{30}=0$. Only in the case of ^{254}Rf, ^{256}Rf, and ^{258}Sg have the reflection-asymmetric paths been predicted (results not shown). The effects of the triaxial degree of freedom are again represented by the differences between open and solid circles.

One can see that the hexadecapole moments increase along the static paths. It is worth noting that a similar behavior of Q_{40} is also seen in the Fm and No isotopes discussed

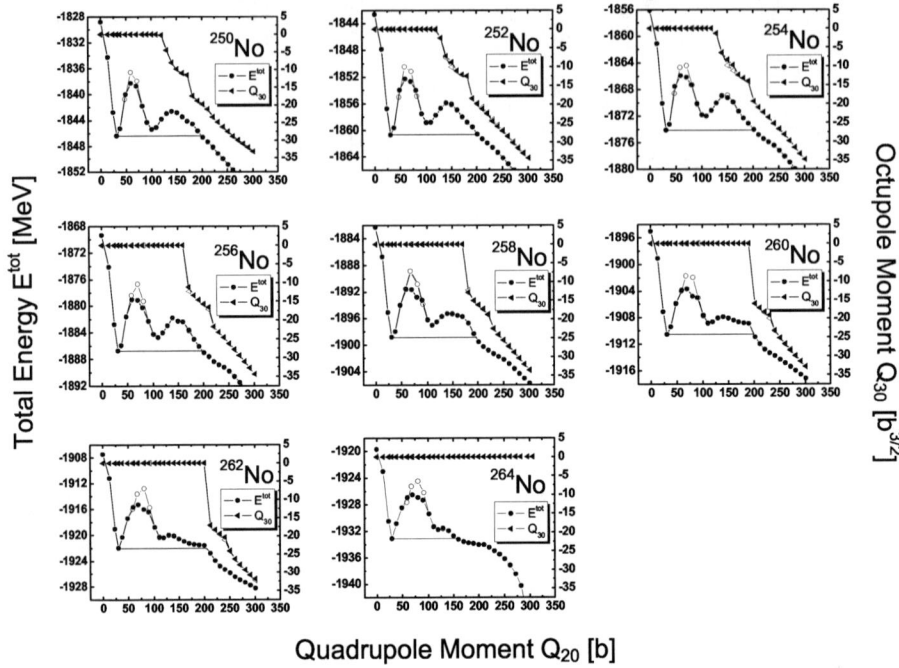

FIGURE 2. Similar to Fig. 1 except for the even-even nobelium (Z=102) isotopes with N=148–162.

above. However, in contrast to smooth changes of Q_{40} as a function of Q_{20} seen in Figs. 3 and 4, sudden jumps of Q_{30} along the lowest static fission paths appear for the heavier Fm and No isotopes.

When the proton number increases from Z=100 to 110, one can notice two effects: (i) the disappearance of the outer fission barrier, and (ii) the decrease of the inner barrier height. The first effect can be seen in Figs. 3B and 4 which display one-humped, narrow barriers obtained for Sg, Hs, and Ds. The second effect is particularly evident for the heaviest ^{276,278}Ds isotopes, where the barriers are twice as low as that in ^{252}Fm.

The even-even superheavy nuclei considered in this paper are schematically indicated in Fig. 5. Nuclei exhibiting the mass-asymmetry along the fission path are represented by gray, while those having symmetric fission paths are shown as dark gray. In the same figure, the experimentally known even-even nuclei are shown as black squares and diamonds. (The diamonds correspond to superheavy elements produced in hot-fusion reactions.) The deformed and postulated (spherical) 'magic' numbers are indicated by light gray.

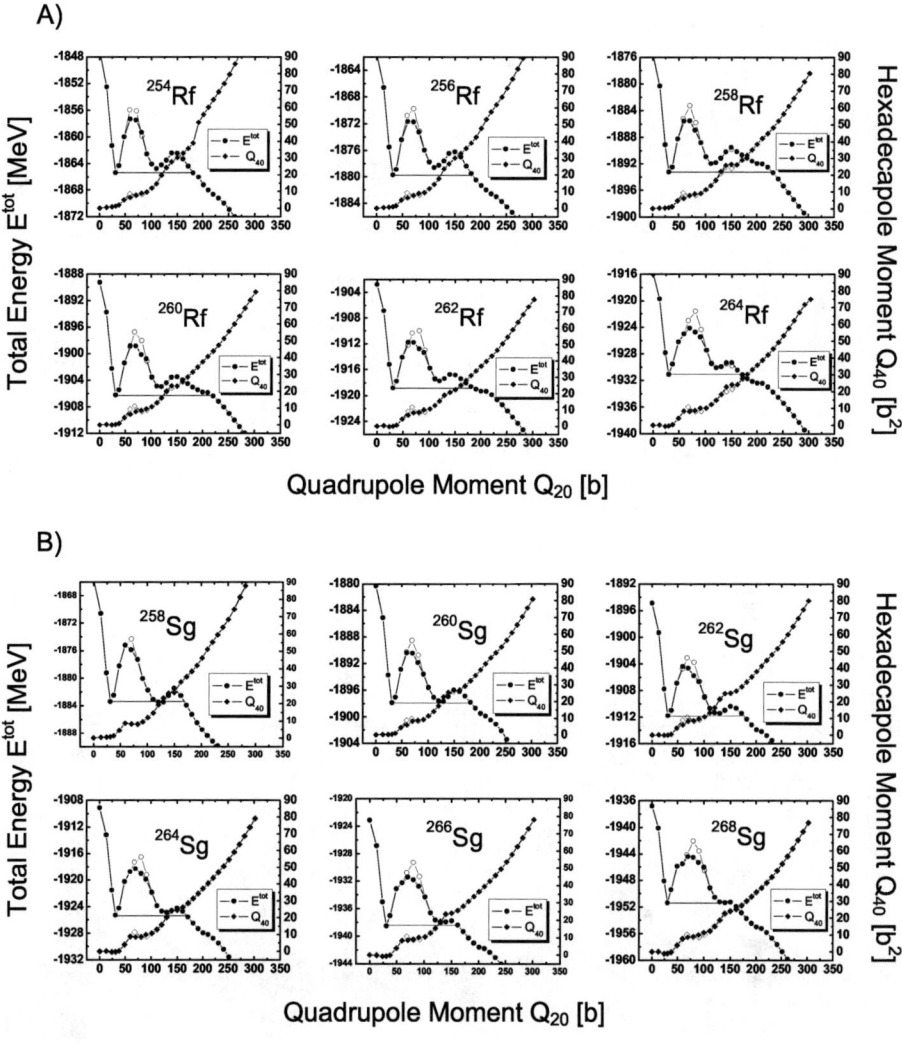

FIGURE 3. Total binding energies E^{tot} (solid circles) and mass hexadecapole moments Q_{40} (solid diamonds) as functions of Q_{20} for even-even rutherfordium (Z=104) isotopes with N=150–160 (A) and seaborgium (Z=106) isotopes with N=152–162 (B). Open circles show barriers for axial shapes.

CONCLUSIONS

We have applied the self-consistent SHF+BCS method with the Skyrme parametrization SLy4 to study static fission barriers of the even-even superheavy elements with $100 \leq Z \leq 110$. All calculations have been done taking into account the non-axial and reflection-symmetry-breaking shapes along the fission paths.

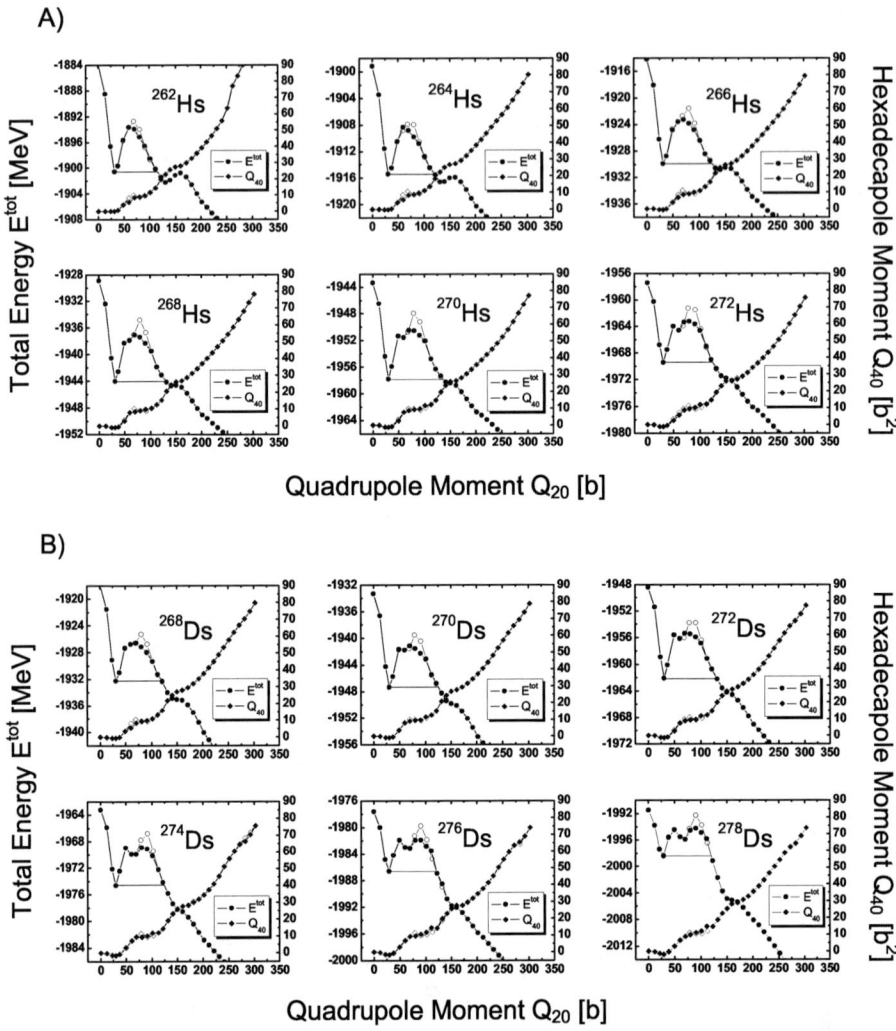

FIGURE 4. Similar to Fig. 3 except for the even-even hassium (Z=108) isotopes with N=154–164 (A) and darmstadtium (Z=110) isotopes with N=158–168 (B).

For all the investigated superheavy nuclei, the appearance of triaxial distortion reduces the height of the first barrier by up to 3 MeV. It is also found that the breaking of intrinsic reflection-symmetry appears for ^{258}Sg, ^{254,256}Rf, and for the Fm and No isotopes (excluding ^{264}Fm and ^{264}No). In all other cases purely reflection-symmetric fission paths are obtained.

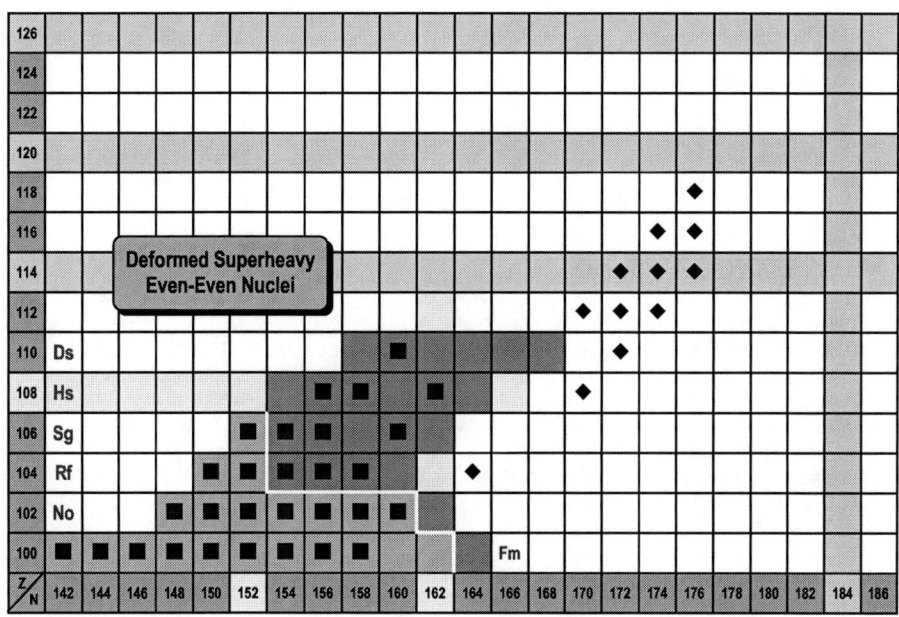

FIGURE 5. The portion of the nuclear chart showing even-even superheavy nuclei with $Z \geq 100$ considered in the present study.

ACKNOWLEDGMENTS

This work was supported in part by the National Nuclear Security Administration under the Stewardship Science Academic Alliances program through the U.S. Department of Energy Research Grant DE-FG03-03NA00083; by the U.S. Department of Energy under Contract Nos. DE-FG02-96ER40963 (University of Tennessee), DE-AC05-00OR22725 with UT-Battelle, LLC (Oak Ridge National Laboratory), and DE-FG05-87ER40361 (Joint Institute for Heavy Ion Research); by the Polish Committee for Scientific Research (KBN) under Contract No. 1 P03B 059 27; and by the Foundation for Polish Science (FNP).

REFERENCES

1. S. Hofmann and G. Münzenberg, *Rev. Mod. Phys.* **72**, 733–767 (2000).
2. P.-H. Heenen and W. Nazarewicz, *Europhys. News* **33**, 1–9, (2002).
3. P. Armbruster, *Acta Phys. Polon. B* **34**, 1825–1866 (2003).
4. S. Hofmann, G. Münzenberg, and M. Schädel, *Nucl. Phys. News* **14**, 5–13 (2004).
5. M. Leino, *Nucl. Phys.* **A751**, 248c–263c (2005).
6. S. Ćwiok, P.-H. Heenen, and W. Nazarewicz, *Nature* **433**, 705–709 (2005).
7. Yu. Ts. Oganessian *et al.*, *Nucl. Phys.* **A734**, 109–123 (2004).
8. Yu. Ts. Oganessian *et al.*, *Phys. Rev. C* **69**, 054607 (2004).
9. Yu. Ts. Oganessian *et al.*, *Phys. Rev. C* **70**, 064609 (2004).

10. K. Morita et al., *J. Phys. Soc. Jpn.* **73**, 2593–2596 (2004).
11. S. Bjørnholm and J. E. Lynn, *Rev. Mod. Phys.* **52**, 725 (1980).
12. B. S. Bhandari and Y. B. Bendardaf, *Phys. Rev. C* **45**, 2803–2818 (1992).
13. S. Ćwiok and A. Sobiczewski, *Z. Phys A* **342**, 203–213 1992.
14. K. Rutz, J. A. Maruhn, P.-G. Reinhard, and W. Greiner, *Nucl. Phys.* **A590**, 680–702 (1995).
15. A. Mamdouh, J. M. Pearson, M. Rayet, and F. Tondeur, *Nucl. Phys.* **A644**, 389–414 (1998); **A648**, 282(E) (1999).
16. M. G. Itkis, Yu. Ts. Oganessian, and V. I. Zagrebaev, *Phys. Rev. C* **65**, 044602 (2002).
17. A. Baran, Z. Łojewski, and K. Sieja, *Int. J. Mod. Phys.* **E 13**, 353–356 (2004).
18. M. Warda, J. L. Egido, L. M. Robledo, and K. Pomorski, *Phys. Rev. C* **66**, 014310 (2002).
19. L. Bonneau, P. Quentin, and D. Samsœn, *Eur. Phys. J.* **A 21**, 391–406 (2004).
20. T. Bürvenich, M. Bender, J. A. Maruhn, and P.-G. Reinhard, *Phys. Rev. C* **69**, 014307 (2004).
21. A. Staszczak, J. Dobaczewski, and W. Nazarewicz, *Int. J. Mod. Phys.* **E 14**, No. 3 (2005) - in print.
22. W. Zhang, J. Meng, S. Q. Zhang, L. S. Geng, and H. Toki, *Nucl. Phys.* **A753**, 106–135 (2005).
23. R. Smolańczuk, J. Skalski, and A. Sobiczewski, *Phys. Rev. C* **52**, 1871–1880 (1995).
24. A. Staszczak, Z. Łojewski, A. Baran, B. Nerlo-Pomorska, and K. Pomorski, in *Proc. Third International Conference on Dynamical Aspects of Nuclear Fission, Častá-Papernička, Slovak Republik, 1996*, edited by J. Kliman, B. I. Pustylnik, (JINR Dubna, 1996), p. 22–35, (nucl-th/9609065).
25. J. Dobaczewski and J. Dudek, *Comput. Phys. Commun.* **102**, 166–182 (1997); *ibid.* **102**, 183–209 (1997); *ibid.* **131**, 164–186 (2000).
26. J. Dobaczewski and P. Olbratowski, *Comput. Phys. Commun.* **158**, 158–191 (2004); *ibid.* **167**, 214–216 (2005).
27. http://www.fuw.edu.pl/~dobaczew/hfodd/hfodd.html
28. E. Chabanat, P. Bonche, P. Haensel, J. Meyer, and F. Schaeffer, *Nucl. Phys.* **A635**, 231 (1998); **A643**, 441(E) (1998).
29. H. Flocard, P. Quentin, A.K. Kerman, and D. Vautherin, *Nucl. Phys.* **A203**, 433–472 (1973).

LIGHT PARTICLES AND CLUSTER EMISSION

Microscopic description of cluster emission with the Gogny force

L.M. Robledo* and J.L. Egido*

Departamento de Física Teorica C-XI, Universidad Autonoma de Madrid, 28049 Madrid, Spain

Abstract. Cluster radioactivity is microscopically described using the mean field approach with the effective phenomenological Gogny interaction. As driving coordinate the axially symmetric octupole moment is used. Owing to the microscopic approach used, it is possible to compute collective masses consistently and therefore it is also possible to compute spontaneous emission half lives. The present approach has been applied to the cluster radioactivity emission in ^{226}Ra and the results for the half lives and other quantities agree well with experiment.

Keywords: Cluster radioactivity, Mean field description, Gogny force
PACS: 21.60.Jz, 21.60.Ev, 23.70.+j

INTRODUCTION

The phenomenon of cluster radioactivity (the emission of particles a few mass units heavier than the α particle) was theoretically predicted in the early eighties by Sandulescu and collaborators [1] in the framework of a model of very asymmetric fission. Four years later the prediction was experimentally confirmed by Rose et al. [2] with the discovery of the ^{14}C emission by the nucleus ^{223}Ra. Since then, the emission of ^{14}C, ^{20}O, ^{23}F, ^{24,26}Ne, ^{28,30}Mg and ^{32}Si light fragments by several isotopes of the actinides Fr, Ra, Ac, Th, U and Pu has been observed -see [3] for a review. A hint on the physics of the process is given by the fact that the heavier daughter nucleus is always in the vicinity of the double magic ^{208}Pb isotope.

From a theoretical point of view, cluster radioactivity has been mainly studied in the context of super asymmetric fission model (SAFM) [4] in which cluster radioactivity is viewed as a fission process and the fission barrier is introduced in a phenomenological fashion by the use of phenomenological potentials like the folded Yukawa one. In these models, phenomenological prescriptions have been used for the calculation of the collective masses. For a rather complete account of the theoretical status see Ref. [5]. However, and to our knowledge, no attempt has been made in order to understand the process from a more microscopic point of view like the ones used to describe standard fission with effective interactions (see, for instance [6, 7]).

The purpose of our study is to show that cluster radioactivity can be fully described microscopically with the Gogny interaction (D1S parameter set) in the usual framework used to describe fission but using the octupole degree of freedom as driving coordinate. We have performed HFB calculations constraining the octupole moment from the reflection symmetric configuration up to those values where a two fragments solution arises. Owing to the microscopic character of the calculations we can also evaluate collective masses and spurious energy corrections [8] (stemming from the restoration of broken symmetries) in a fully consistent way. With all these microscopic parameters we can evaluate, in the framework of the WKB approximation, the corresponding cluster radioactivity half lives as well as the mass distribution of the fragments.

The results presented here for the isotope ^{226}Ra agree well with experimental findings. This is an encouraging result as our calculations are parameter free and therefore the present agreement with experiment imply that the procedure has predictive power. Our findings can also serve as a guide to constraint the parameters used in more phenomenological approaches.

THEORETICAL FORMULATION

Our theoretical framework uses the Hartree- Fock- Bogoliubov (HFB) approximation [9] with the phenomenological density dependent Gogny interaction[10]. This is a finite range interaction which was devised to have the capability of describing at the same time the nuclear mean field and pairing correlations. This force has been successfully applied to many studies of nuclear structure including those related to fission and clustering effects.

As we are interested in studying the nucleus as a function of a driving coordinate (in our case the octupole moment) the equation to be solved corresponds to the constrained HFB method where the minimum of the HFB energy is found with the restriction that the mean value of the constraining operators take the desired value. The solution of the constrained HFB equation is best achieved by means of the gradient method where the Thouless theorem [9] is used to parameterize the multidimensional energy surface. The constraints are imposed by requiring the energy's gradient to be orthogonal to the constraints' ones.

We have solved the constrained HFB equation for the Gogny force and we have imposed axial symmetry (but allowing the breaking of the reflection symmetry) in the solution. As constraining operators we have used, depending on the case, the multipole operators $\hat{Q}_{\lambda 0} = r^\lambda Y_{\lambda 0}(\theta, \varphi)$ with $\lambda = 1$ (the center of mass operator) and $\lambda = 3$ or the "slice" operator

$$\hat{S}(z_0, z_1) = \begin{cases} 0 & z < z_0 \\ 1 & z_0 < z < z_1 \\ 0 & z > z_1 \end{cases}$$

that measures the number of particles in a slice of matter located between z_0 and z_1.

As it is customary in HFB calculations with the Gogny force, we have subtracted the kinetic energy of the center of mass motion to the Routhian to be minimized in order to ensure that the center of mass is kept at rest. We have also dealt with the exchange Coulomb energy in the Slater approximation and neglected the contribution of the Coulomb interaction to the pairing field. For the Gogny force we have used the parameter set known as D1S that was adjusted more than twenty years ago [7, 11] in order to reproduce the most relevant nuclear matter parameters, the binding energies of several magic nuclei and the fission barrier of ^{240}Pu. In the present context, one of the advantages of the Gogny interaction is its good performance in describing the fission phenomena [7, 12].

The quasi-particle operators have been expanded in an axially symmetric harmonic oscillator (HO) basis and special attention has been paid to the convergence of the results with the number of basis states included. The size of the basis depend upon two parameters, N_0 and q, which are related to the allowed range of the HO quantum numbers trough the relation $\frac{1}{q}n_z + (2n_\perp + |m|) \leq N_o$. Along the perpendicular direction we take N_0 shells, i.e. $2n_\perp + |m| = 0,\ldots,N_0$) and along the z direction we include up to qN_0 shells depending on the value of $2n_\perp + |m|$. On the other hand, the oscillator length parameters $b_\perp$ and b_z characterizing the HO basis have been determined in order to minimize the Routhian for each value of the octupole moment considered.

Once the HFB solution is obtained for each value of the octupole moment considered, we subtract to the HFB energy the Rotational Energy correction (REC) coming from the restoration of the rotational symmetry. In order to estimate the REC we have used the strong deformation limit to the projected energy [9] in which $E_{REC} = \langle \Delta \vec{J}^2 \rangle / (2 \mathscr{J}_Y)$, where $\langle \Delta \vec{J}^2 \rangle$ is the fluctuation on angular momentum of the HFB wave function and $\mathscr{J}_Y$ is the Yoccoz moment of inertia [9]. This moment of inertia has been computed using the "cranking" approximation in which the full linear response matrix appearing in its expression is replaced by the zero order approximation (that is, two quasiparticle energies). The effect of the "cranking approximation" in the Yoccoz moment of inertia was analyzed with the Gogny interaction for heavy nuclei in [13] by comparing it with the one extracted from an angular momentum projected calculation and the conclusion was that the approximate one is typically a factor 0.7 smaller than the exact one. Therefore, in our estimation of the REC we have included this factor of 0.7 in the energy to account for the approximate evaluation of the Yoccoz moment of inertia.

We have also subtracted to the total energy, the zero point energy (ZPE) correction $\varepsilon_0(q_3)$ associated with the octupole motion [14, 15]. This ZPE correction is given by

$$\varepsilon_0(q_3) = \frac{1}{2} G(q_3) B^{-1}_{ATDHFB}(q_3)$$

where

$$G(q_3) = \frac{M_{-2}(q_3)}{2M^2_{-1}(q_3)}$$

and

$$B_{ATDHFB}(q_3) = \frac{M_{-3}(q_3)}{M^2_{-1}(q_3)}.$$

In the above expressions we use the quantities

$$M_{-n}(q_3) = \sum_{\mu\nu} \frac{\left|(Q^{20}_{30})_{\mu\nu}\right|^2}{(E_\mu + E_\nu)^n}$$

which are defined in terms of $(Q^{20}_{30})_{\mu\nu}$ (the 20 component of the octupole operator $\hat{Q}_{30}$ in the quasiparticle representation [9]) and the quasiparticle energies E_μ obtained as the eigenvalues of the HFB hamiltonian matrix.

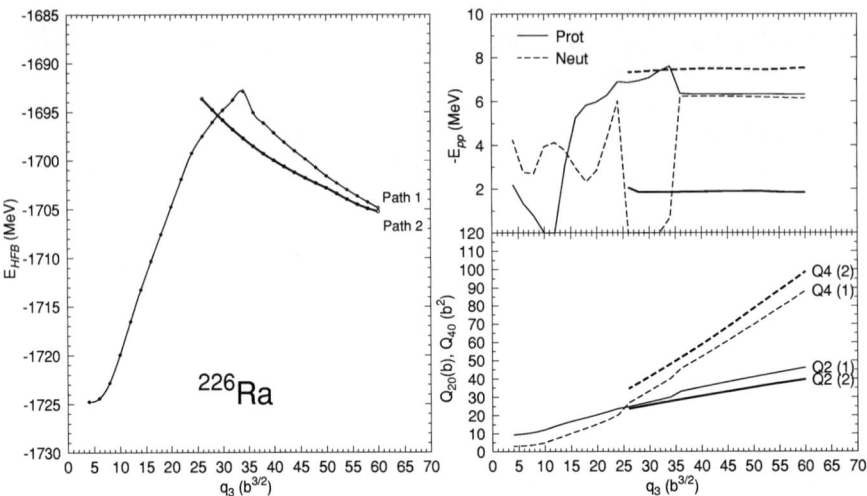

FIGURE 1. On the left hand side, the HFB energy as a function of the octupole moment q_3. On the right hand side, lower panel, the multipole moments Q_2 and Q_4 are represented as a function of q_3. In the upper panel, the particle-particle (pairing) correlation energies are also plotted as a function of q_3. Two paths are obtained: one corresponds to the emission of ^{20}O (Path 1, thin lines) and of ^{14}C (Path 2, thick lines).

RESULTS OF THE CALCULATIONS.

In this section we will present the results obtained with the methodology of the previous section for the nucleus ^{226}Ra. As a first step, HFB calculations constraining on the octupole moment $q_3 = \left\langle \Phi \left| \hat{Q}_{30} \right| \Phi \right\rangle$ (in this case and as we are breaking reflection symmetry the additional constraint on the center of mass to be at the origin has been imposed) are carried out for increasing values of the octupole moment. The HFB quasi-particle operators are expanded in a HO basis with $N_0 = 17$ and $q = 1.5$ which is big enough as to ensure a reasonable convergence of the energies and other relevant observables. The results for the energy as a function of the octupole moment are depicted in Fig. 1 along with the self-consistent values of the quadrupole and hexadecapole moments and the proton and neutron pairing energies. In all the panels of Fig. 1 we always observe two curves that correspond to two different coexisting minima. For the curve labeled as Path 1 we observe an increase of the energy as the octupole moment increases up to $q_3 = 34b^{3/2}$ and then a steady decrease that corresponds to the Coulomb repulsion of a configuration with two fragments (^{20}O + ^{206}Hg). For the curve labeled Path 2 we always have a two fragment solution (^{14}C+ ^{212}Pb) and the behavior of the energy is the one corresponding to the Coulomb repulsion of the two fragments. As can be observed on the right hand side of the figure (lower panel), the corresponding values of Q_2 and Q_4 are similar (but different) in the two paths. The proton and neutron pairing energies have two different regimes depending on the number of fragments of the system. For low octupole moments (one fragment solution) the pairing energies show a quite irregular behavior that is associated to the many level crossings that take place as

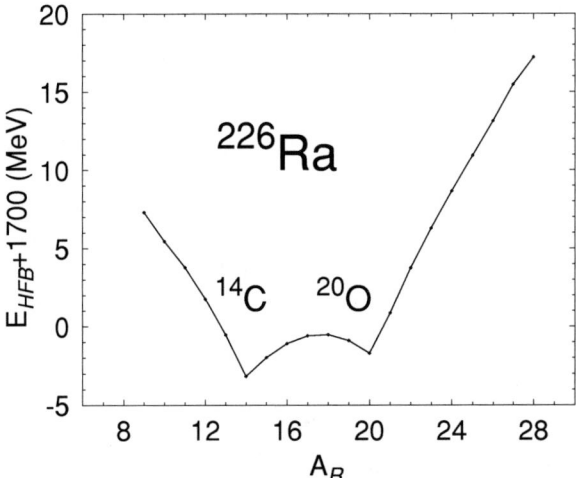

FIGURE 2. HFB energy as a function of the number of particles in the lighter daughter nuclei A_R in the cluster emission process of ^{226}Ra. Two minima corresponding to the emission of ^{20}O and ^{14}C are clearly observed. The octupole moment has been held constant at a value $q_3 = 50 b^{3/2}$.

the octupole moment is increased. However, as long as the system is in the two fragments case the pairing energies remain constant as a function of the octupole moment. This is the consequence of the fact that when the system has split in two fragments the most favorable way to generate octupole moment is just to separate the fragments keeping unaltered the intrinsic structure of the two of them.

From these results we conclude two important things: First, that a one centered HO basis with a sufficiently large number of shells is able to produce two fragments solutions. Secondly, the octupole degree of freedom can be used as the driving coordinate for the description of the "cluster emission" phenomenon and this will allow us a consistent calculation of the collective inertias and zero point energies.

In order to ensure that no other two fragments solutions are present in the system we have carried out calculations by constraining on the quantity $A_R = \langle \Phi | \hat{S}(z_0, \infty) | \Phi \rangle$ which is the mean value of the "slice operator" between the point z_0 and ∞. By choosing the parameter z_0 adequately in the mid point separating the two fragments the quantity A_R measures the number of particles on the right hand side fragment (the lighter one as we are constraining on positive values of the octupole moment). The results, obtained at an octupole deformation of $q_3 = 50 b^{3/2}$ are depicted in Fig. 2. From that picture we conclude that the two solutions found corresponding to the emission of ^{14}C and ^{20}O are the only two possible for the phenomenon of cluster emission in the nucleus ^{226}Ra.

In a second step we have subtracted to the HFB energy the rotational energy correction as well as the zero point energy correction associated to the octupole motion in order to obtain the total energy of the system. The corresponding curves (lower panel) as well as the subtracted quantities (upper panel) are plotted in Fig 3. As deduced from the upper panel curves, the effect of the zero point energy correction $\varepsilon_0(q_3)$ on the total energy is just an overall displacement. On the other hand, the effect of the rotational energy

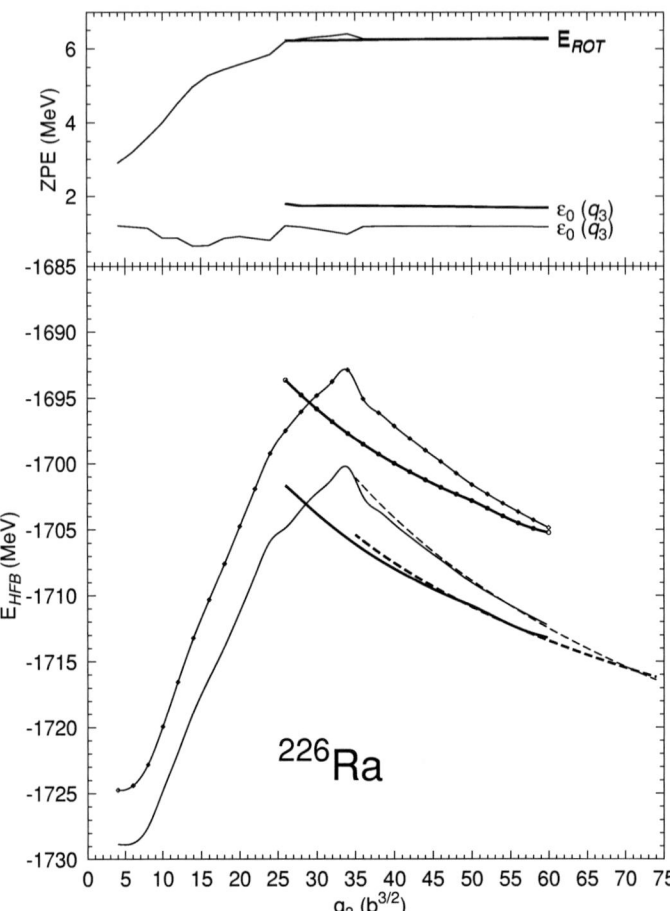

FIGURE 3. The total energy of the system (HFB minus the zero point energy corrections) for the two paths found (lower panel). In the upper panel the zero point energy correction of the octupole motion ($\varepsilon_0(q_3)$) as well as the rotational energy correction (REC) are depicted as a function of q_3 for the two paths.

correction amounts to a substantial change of the topology of the total energy curves as this quantity increases substantially in going from the lowest values of the octupole moment up to the higher ones.

As mentioned before, when the solution corresponds to two fragments, the octupole moment is generated simply by separating the two fragments. Therefore, there is a one to one correspondence between the octupole moment and the separation distance of the fragments (R) that can be easily shown to be

$$q_3 = f_3 R^3$$

in this case, where the two fragments are spherical. The constant f_3 is given in terms of the total mass number A, and the mass numbers of fragments 1 and 2 A_1, and A_2, respectively, by the expression $f_3 = \frac{A_1 A_2}{A} \frac{(A_1-A_2)}{A}$. Using this relation we can write the Coulomb repulsion energy between the fragments as a function of the octupole moment

$$V_{Coul} = e^2 \frac{Z_1 Z_2}{(q_3/f_3)^{1/3}}$$

and therefore we can model the two fragment energies as

$$V_{Coul}(q_3) - V_{Coul}(q_3^{(0)}) + V(q_3^{(0)}).$$

The result of this model is also plotted in Fig. 3 as dashed lines. As observed from this plot the two fragment's energy actually correspond to the Coulomb repulsion. From this result, we can also obtain an estimation of the Q value of the disintegration by taking the $q_3 \to \infty$ limit in the above model of the energy, obtaining

$$Q = V(q_3^{(0)}) - V_{Coul}(q_3^{(0)})$$

For the ^{14}C case we obtain a Q value of 25.18 MeV whereas for the ^{20}O case the value 41.46 MeV is obtained. These two values are in rather good agreement with the experimental results of 28.21 and 40.82 MeV, respectively.

Finally, with the total energy curve of Fig. 3 acting as collective potential and the collective masses corresponding to the octupole motion defined in the previous section we can compute the spontaneous half lives of the two emission processes using the standard WKB formula. The result obtained for the ^{14}C emission half life is $t_{1/2} = 8.8\,10^{22}$s that compares pretty well with the experimental value of $1.58\,10^{21}$ s. For the ^{20}O emission we obtain $t_{1/2} = 2.7\,10^{21}$ s which is lower than the ^{14}C half life. However, this emission mode has not been observed in the disintegration of ^{226}Ra and therefore there must be a mechanism hindering this disintegration channel. It has to be mentioned, however that the disintegration ^{228}Th$\to^{20}$O+^{208}Pb is observed experimentally and the corresponding half life is of the order of magnitude of the one obtained here. Work to unveil this paradox is under way.

CONCLUSIONS

In this paper we have explored the cluster radioactivity phenomenon in the realm of the HFB approximation with the realistic Gogny interaction. We have studied the "cluster emission" of the nucleus ^{226}Ra and our results show that a description based on the HFB theory and using the octupole moment as driving coordinate can explain from a microscopic point of view all the features of this type of exotic radioactivity. Owing to the parameter free character of the calculations we conclude that the procedure and the interaction have predictive power in this subject.

ACKNOWLEDGEMENTS

This work is partly sponsored by the Ministry of Education and Science (Spain) under Project FIS2004-06697.

REFERENCES

1. A. Sandulescu, D. Poenaru and W. Greiner, Sov. J. Part. Nucl. **11** (1980) 528.
2. H.J. Rose and G.A. Jones, Nature **307** (1984) 245.
3. D.N. Poenaru, Y. Nagame, R.A. Gherghescu and W. Greiner, Phys. Rev. **C65** (2002) 054308.
4. D.N. Poenaru, W. Greiner, K. Depta, M. Ivascu, D. Mazilu and A. Sandulescu, At. Data. Nucl. Data Tables **34** (1986) 423.
5. D.N. Poenaru and W. Greiner, in *Nuclear Decay Modes*, Institute of Physics, Bristol, 1996.
6. H. Flocard, P.H. Heenen, S.J. Krieger and M.S. Weiss, Prog. Theor. Phys. 72 (1984) 1000.
7. J. F. Berger, M. Girod and D. Gogny, Nucl. Phys. **A428** (1984) 23c.
8. J. F. Berger and D. Gogny, Nucl. Phys. A333 (1980) 302
9. P. Ring and P. Shuck, *The Nuclear Many Body Problem* (1980), Springer–Verlag Edt. Berlin.
10. J. Dechargé and D. Gogny, Phys. Rev. **C21** (1980) 1568.
11. J.F. Berger, M. Girod and D. Gogny, Comp. Phys. Comm. **63** (1991) 365.
12. M. Warda, J.L. Egido, L.M. Robledo and K. Pomorski, Phys. Rev. C66 (2002) 014310
13. J.L. Egido and L.M. Robledo, Phys. Rev. Lett. **85** (2000) 1198.
14. M.J. Giannoni and P. Quentin, Phys. Rev. C21 (1980) 2076
15. L.M. Robledo, J.L. Egido, B. Nerlo-Pomorska and K. Pomorski, Phys. Lett. **B201** (1988) 409

New results on the ternary fission of ^{243}Cm

J. Heyse*,†, C. Wagemans*, S. Vermote*, O. Serot**, P. Geltenbort‡,
T. Soldner‡ and J. Van Gils†

Ghent University, Dept. of Subatomic and Radiation Physics, 9000 Gent, BE
†*EC-JRC-IRMM, Retieseweg 111, 2440 Geel, BE*
****CEA-Cadarache, DEN/DER/SPRC/LEPh, 13108 Saint Paul lez Durance, FR*
‡*ILL, 38042 Grenoble, FR*

Abstract. Ternary fission is an important source of He and tritium gas in nuclear reactors and used fuel elements. Therefore a systematic study of the ternary fission yields for ^{4}He and tritons (t) is being performed. In recent years the influence of the excitation energy of the fissioning nucleus on the triton emission probability (t/B) has been investigated for different Cm and Cf isotopes. In this paper we report on new results on the neutron induced fission of ^{243}Cm.

Keywords: NUCLEAR REACTIONS ^{243}Cm(n,f), E = cold; measured α and t spectra; deduced ternary fission yields.
PACS: 24.75.+i

INTRODUCTION

During ternary fission light charged particles ranging from protons to Ar are emitted. However, the emission yields for α particles and tritons are by far the most important. As such, the ternary fission process is an important source of helium and tritium gas in nuclear reactors and used fuel elements. Therefore accurate ternary fission yields for ^{4}He and tritons are demanded by nuclear industry. Furthermore, from a physics point of view, the ternary particles provide interesting information about the fission process itself.

Over the last decade our group has been involved in a systematic study of the characteristics of ternary α's (often referred to as Long Range Alpha's or *LRA*) and tritons (t). Correlations have been found between the yields of these ternary particles and relevant parameters of the fissioning nucleus such as the neutron multiplicity $<v>$, the fissility parameter Z^2/A, the Coulomb parameter $Z^2/A^{1/3}$, ...

The influence of the excitation energy of the compound nucleus on the LRA and t yields has been studied by comparing the spontaneous fission of ^{248}Cm with the thermal neutron induced fission of ^{247}Cm [1], producing the same compound nucleus with an excitation energy essentially equal to the neutron separation energy. Similar experiments were performed to study ^{246}Cm(SF) and ^{245}Cm(n_{th},f), and ^{252}Cf(SF) and ^{251}Cf(n_{th},f) (see e.g. [2]). These experiments revealed an increase in the triton yield with increasing excitation energy, as can be expected. For the LRA, however, a decrease in the yield with increasing excitation energy was observed.

To enlarge the available database and to get a better insight in the systematics, similar experiments are being performed to study ^{244}Cm(SF) and ^{243}Cm(n_{th},f). In this paper we report on the results obtained for the thermal neutron induced fission of ^{243}Cm.

CP798, *Nuclear Fission and Fission-Product Spectroscopy,*
edited by H. Goutte, H. Faust, G. Fioni, and D. Goutte
© 2005 American Institute of Physics 0-7354-0288-4/05/$22.50

TABLE 1. Sample composition and contributions of the different isotopes to the neutron induced and spontaneous fission yields taking into account the detection geometry and the neutron flux during the experiment.

Isotope	Abundance (%)	Spontaneous fission (fissions/s)	Induced fission (fissions/s)
^{243}Cm	97.110 %	0.0000	137.7
^{244}Cm	1.473 %	0.0015	0.0
^{245}Cm	0.065 %	0.0000	0.4
^{239}Pu	1.352 %	0.0000	2.5
Total	100.000%	0.0015	140.6

EXPERIMENTAL CONDITIONS

The measurements have been performed at the PF1B cold neutron guide of the Institut Laue Langevin in Grenoble (France) with a 2 μg ^{243}Cm sample. The neutron flux at the sample position was about 5×10^9 neutrons/(s.cm^2). Table 1 shows the composition of the sample and the expected contributions of spontaneous and neutron induced fission for the different isotopes. Although a separation of the Pu isotopes produced by the α decay of the various Cm isotopes was done 7 months prior to the experiment, a minor correction still had to be made for the ^{239}Pu produced since then, because of its large thermal fission cross section. From the table it is clear that the contributions of spontaneous fission and neutron induced reactions on the other isotopes are negligible.

The sample was placed in the centre of a vacuum chamber at a 45° angle with the incident neutron beam. A polyimide foil was used to cover the sample in order to prevent the contamination of the chamber by recoil nuclei. Two ΔE-E telescopes, each consisting of a thin ΔE and a thick E silicon surface barrier detector, were placed on both sides of the sample at 4 cm distance and perpendicular to the beam. For the ternary fission measurements, the telescope facing the sample was covered with a 30 μm aluminium foil in order to stop the fission fragments and the α's produced by the decay of the Cm isotopes. On the other side, the decay α's and the fission fragments were stopped by the 30 μm aluminium sample backing.

MEASUREMENTS AND RESULTS

The telescope facing the target, consisting of a 25 μm ΔE and a 500 μm E detector, was used to measure the LRA yield relative to the binary fission yield B. For the LRA measurement, the alpha's were identified using the relation $T/a = (E + \Delta E)^{1.73} - E^{1.73}$ with T the thickness of the ΔE detector and a a particle and material specific constant [3]. A similar relation was used to correct the particle energy loss in the aluminium foil. For the binary fission measurement, this foil was removed and the ΔE detector was replaced by an empty dummy detector with the same dimensions, keeping the same detection geometry as for the ternary fission measurement.

The other telescope, consisting of a 50 μm ΔE and a 1500 μm E detector, was used

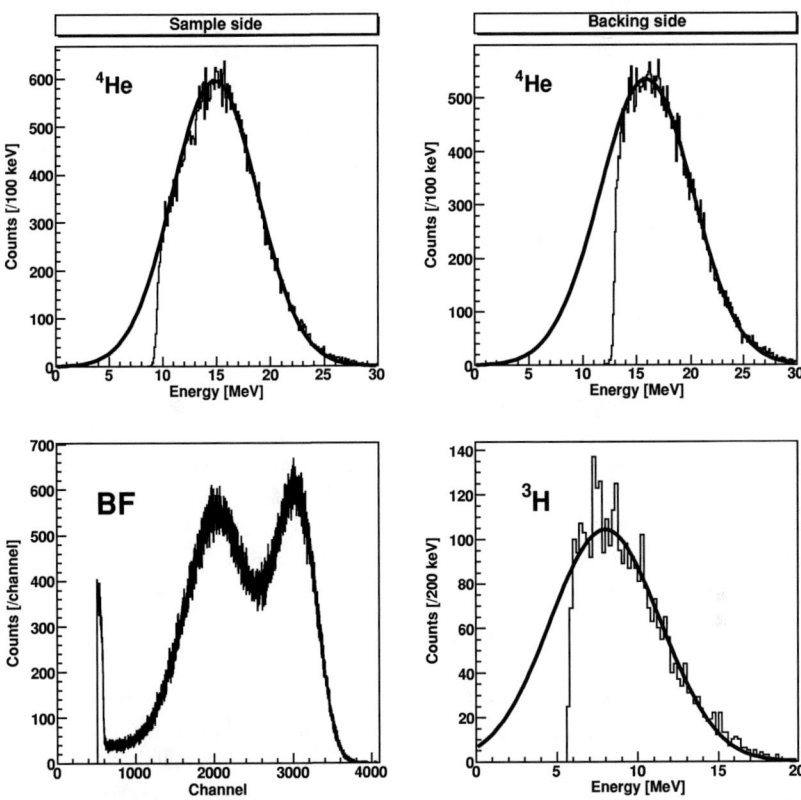

FIGURE 1. Spectra for the LRA/B (lefthand side) and t/B (righthand side) measurements. The solid curves represent a Gaussian fit to the data (see text for more details).

to measure the t and LRA yields simultaneously. A thicker ΔE detector allows a better separation between ternary tritons and α particles, but has the disadvantage of raising the energy threshold at which particles reach the E detector.

Figure 1 shows the spectra for the LRA and B measurements on the lefthand side and for the LRA and t measurement on the righthand side. For both LRA spectra, a Gaussian fit to the spectra was made, starting at an α energy of 14 MeV. As described in [4], for the α's, (6 ± 1) % has to be added to the area under the Gaussian curve in order to take into account the contribution of the non-Gaussian low-energy tail. For the tritons the yield is simply given by the area of a Gaussian fit. However, due to the quite high energy threshold of the spectrum and the limited statistics, the mean energy of the triton distribution has been fixed at 8 MeV in order to fit the distribution with a Gaussian function. In the binary fission spectrum the fission fragments could be rather nicely separated from the alpha pile-up.

TABLE 2. Ternary alpha and triton emission probabilities for the ^{243}Cm(n$_{th}$,f) reaction (preliminary values). These values are calculated taking into account a (6±1) % non-Gaussian low-energy tailing for the alpha emission probability. The corresponding values for ^{245,247}Cm(n$_{th}$,f) are given for comparison [5].

	(LRA/B)$_{tot}$ (x10^3)	(t/LRA)$_{tot}$ (%)
^{243}Cm(n$_{th}$,f)	2.89 ± 0.30	7.48 ± 1.47
^{245}Cm(n$_{th}$,f)	2.28 ± 0.16	8.11 ± 0.72
^{247}Cm(n$_{th}$,f)	1.96 ± 0.11	9.39 ± 0.77

Table 2 shows preliminary values for the ternary alpha and triton emission probabilities obtained for ^{243}Cm(n$_{th}$,f). These values are compared with the corresponding data for ^{245}Cm(n$_{th}$,f) and ^{247}Cm(n$_{th}$,f). As expected based on the systematics, an increase in the $(LRA/B)_{tot}$ value and a decrease in the $(t/LRA)_{tot}$ value is observed in comparison with the heavier Cm isotopes. However, a large uncertainty is quoted for the triton yield, mainly due to the high energy threshold for the triton distribution.

CONCLUSION

For the first time, experimental LRA/B and t/LRA values for the ^{243}Cm(n$_{th}$,f) reaction are reported. Additional measurements will be performed to improve the accuracy and to lower the energy threshold for the tritons. As a next step, these results will be compared with the results of a ^{244}Cm(SF) measurement to determine the dependence of t/B and LRA/B on the excitation energy of the fissioning nucleus ^{244}Cm.

REFERENCES

1. O. Serot, C. Wagemans, J. Wagemans and P. Geltenbort, "Influence of the excitation energy on the ternary triton emission probability", in *Proc. 3rd Int. Conf. on Fission and Properties of Neutron-Rich Nuclei - Sanibel Island, USA*, edited by G.H. Hamilton, A.V. Ramayya and H.K. Carter, World Scientific, Singapore, 2003, pp. 543.
2. O. Serot, C. Wagemans, J. Heyse, J. Wagemans and P. Geltenbort, "New results on the ternary fission of Cm and Cf isotopes", in *Proc. Seminar on Fission - Pont d'Oye V*, edited by C. Wagemans, J. Wagemans and P. D'hondt, World Scientific, Singapore, 2004, pp. 151.
3. F.S. Goulding, D.A. Landis, J. Cerny and R.H. Pehl, *Nucl. Instr. Meth.* **31**, 1 (1964).
4. C. Wagemans, J. Heyse, P. Janssens, O. Serot and P. Geltenbort, *Nucl. Phys.* **A 742**, 291 (2004).
5. O. Serot, C. Wagemans and J. Heyse, "New results on helium and tritium gas production from ternary fission", in *Proc. Nuclear Data For Science and Technology ND2004 - Santa Fe, USA*, edited by R.C. Haight, M.B. Chadwick, T. Kawano and P. Talou, 2005, in press.

Studies On Particle-Accompanied Fission Of ^{252}Cf(sf) And ^{235}U(n_{th},f)

Yu N Kopatch[1,2], V Tishchenko[3], M Speransky[3], M Mutterer[4,10], F Gönnenwein[5], P Jesinger[5], A M Gagarski[6], J von Kalben[4], I Kojouharov[1], E Lubkiewics[7], Z Mezentseva[3], V Nezvishevsky[8], G A Petrov[6], H Schaffner[1], H Scharma[9], W H Trzaska[10], H-J Wollersheim[1]

1) GSI, 64291 Darmsstadt, Germany *2) Frank Lab JINR, 141980, Russia*
3) Flerov Lab JINR, 141980 Dubna, Russia *4) IKP TU Darmstadt, 64289 Darmstadt, Germany*
5) U Tübingen, 72076 Tübingen, Germany *6) PNPI, 188300 Gatchina, Russia*
7) Jagiellonian U, 30059 Cracow, Poland *8) ILL, 38042 Grenoble, France*
9) FZ Rossendorf, 01314 Dresden, Germany *10) Jyväskylä U, 40014 Jyväskylä, Finland*

Abstract. In recent multi-parameter studies of spontaneous and thermal neutron induced fission, ^{252}Cf(sf) and ^{235}U(n_{th},f) respectively, the energies and emission angles of fission fragments and light charged particles were measured. Fragments were detected by an energy and angle sensitive twin ionization chamber while the light charged particles were identified by a series of ΔE-E_{rest} telescopes. Up to Be the light particle isotopes could be disentangled. In addition, in the ^{252}Cf(sf) experiment, gammas emitted by the fragments were analyzed by a pair of large-volume segmented clover Ge detectors. Here the main interest is to study the γ-decay and the anisotropy of gammas emitted by fragments and light particles. On the other hand, the high count rates achieved in the U-experiment performed at the high flux reactor of the ILL, Grenoble, should allow to explore fragment-particle correlations in very rare events like quaternary fission. At the present stage of data evaluation, yields and energy distributions of light particles are available. For the present contribution in particular the yields of Be-isotopes for the two reactions studied are compared and discussed. For ^{252}Cf(sf) these isotopic yields were hitherto not known.

Keywords: Ternary Fission, Quaternary Fission, Yields of Light Charged Particles in Fission
PACS: 24.75.+i, 25.85. -w

INTRODUCTION

Studies of ternary fission have a long history (1,2). Especially spontaneous and thermal neutron induced fission of actinides has been extensively investigated searching for the decay of a nucleus into two fission fragments (FF) and, additionally, one or more light charged particles (LCP). Nevertheless many questions are still open. The experiments to be presented here focus on two different issues in ternary fission.

In a first experiment on ^{252}Cf(sf), besides both FFs and a further LCP, the gammas emitted in ternary fission have been analyzed. Of interest are here the angular anisotropies of gammas relative to the fission axis for both, binary and ternary fission. The comparison of anisotropies in the two processes should give some insight whether the distributions of fragment angular momentum are disturbed by the emission of LCPs. But also the LCPs themselves may be produced in an excited state and emit gammas. The comparison of yields for LCPs being born either in the ground or excited state may give a clue to the excitation mechanism at scission. Further, most of the gammas, if not all, will be emitted in flight while the LCPs are heading towards the detectors. The study of Doppler shifts then allows to assess the lifetimes of possibly long-lived scission configurations. Finally, for spontaneous fission the yields of individual LCP isotopes of an element are only known for elements up to $Z = 3$ (Li). For comparison with theory it is desirable to extend this list to heavier elements.

A second experiment was performed for the reaction $^{235}U(n_{th},f)$ at the high flux reactor of the Institut Laue-Langevin in Grenoble/France. The aim was here to accumulate a large body of ternary data and to inspect details of the correlations between FF and LCP properties. For example, it has been suggested that the mass distributions of FFs become narrower the heavier the LCPs are. In experiment the difficulty here is that one has to cope with the increasingly smaller yields for heavier LCPs. Another question which has not found a satisfactory answer is how, for a given mass fragmentation, the LCP yield depends on the kinetic energy of the FFs. Finally, for the very rare quaternary fission process, it would be interesting to inspect the FF mass and energy distribution.

LAYOUT OF EXPERIMENTS

A schematic drawing of the detector setup for the study of ternary fission in the spontaneous decay of $^{252}Cf(sf)$ is given in the left part of figure 1. A Cf source on a thin backing is placed at the center of the cathode of a twin back-to-back ionization chamber. As a specific feature of this chamber called CODIS 2 the cathode is segmented into 8 sectors which permits to not only measure the energies of both FFs but also their polar and azimuthal angles of emission relative to the chamber axis. In addition, two rings with 12 ΔE-E_{res} telescopes each are installed in the chamber. They

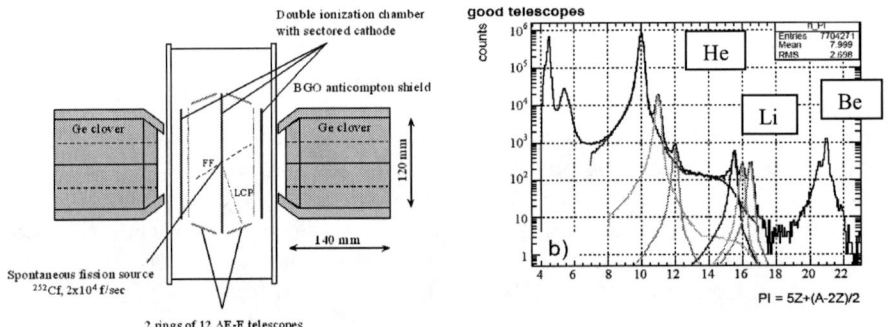

FIGURE 1. Left panel: detector arrangement. Right panel: LCP identification

are designed to identify the LCPs emitted in coincidence with the FFs. Data for LCPs ranging from helium up to carbon could be taken, and for helium up to beryllium even the individual isotopes were resolved. This is shown in the right panel of Figure 1. The yields are plotted as a function of the parameter PI = 5Z + (A - 2Z) /2. For He and Li the deconvolution into isotopes is indicated. Further, for detecting the gammas emitted by FFs and LCPs two segmented large-volume Super-Clover Ge-detectors were aligned on the chamber axis in close geometry. Each detector has 4 large Ge crystals, 14 cm in length and 6 cm in diameter. More details are to be found in (3).

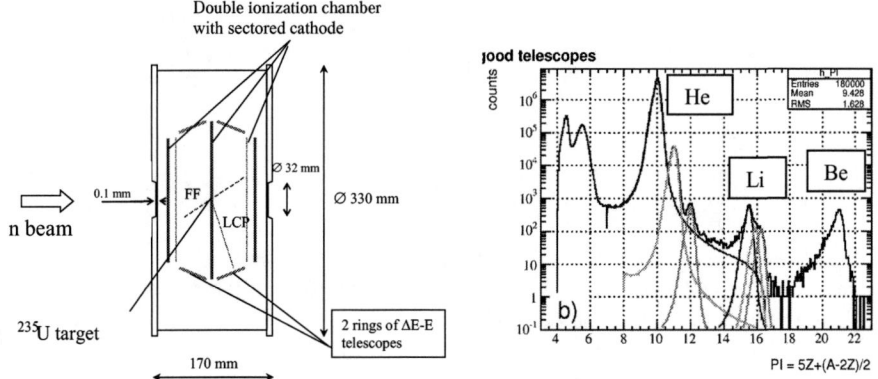

Figure 2: Left panel: detector arrangement. Right panel: LCP identification

The detector assembly for the experiment on ternary fission of ^{236}U* following capture of cold neutrons is shown on the left of Figure 2. A thin (50 µg/cm²) and highly enriched ^{235}U target was placed in a cold neutron flux of $3\cdot10^9$ neutrons/(cm²·s). The same double ionization chamber CODIS 2 as in the experiment with ^{252}Cf was used but modifications had to be made to adapt it to the high neutron flux. In actual practice a count rate capability of $2.5\cdot10^5$ / s was achieved for binary fission without compromising the overall performance. The count rate for ternary fission was 70 / s and for quaternary fission $7\cdot10^{-4}$ / s. More details are given in (4).

FIRST RESULTS

In the first step of the evaluation it has to be made sure that the energy distributions and yields of LCPs obtained are reliable and consistent. The evaluation is delicate because, as brought to evidence in Figures 1 and 2, it is not self-evident how to disentangle the yields of the individual LCP isotopes. Several prescriptions are conceivable how this should be done. Some fine tuning of the procedures is still under study and, therefore, the yields presented below should be considered as preliminary. Though, we are confident that no major changes will be incurred in the final results.

Once the isotopes have been identified their kinetic energy distributions can be determined. A sample of energy spectra for the LCP isotopes ^{4}He, ^{7}Li and ^{10}Be in ternary fission of ^{252}Cf(sf) is on display in Figure 3. The histograms are experimental data while smooth curves are Gaussian fits. A difficulty common to many detector experiments in ternary fission is evident from the figure. Since the solid state detectors

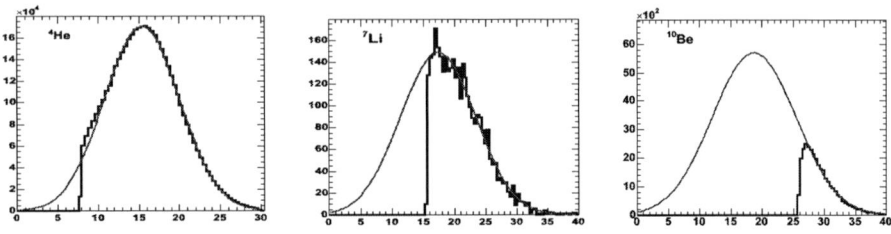

Figure 3: Sample of LCP energy spectra for ^{4}He, ^{7}Li and ^{10}Be from ^{252}Cf(sf)

employed in experiments have to be protected from the intense flux of FFs, thin absorber foils are usually installed in front of the detectors. In the present case the gas of the ionization chamber served this purpose. In consequence, due to the cutoff introduced, energy distributions of LCPs cannot be taken down to the lowest energies. As to be observed in Figure 3, the heavier the LCPs are, the higher is the fraction of particles which are lost for detection. The question then arises how to find the full energy distribution. From experimental data where the full energy spectra could be measured, it is established that the LCP distributions are well represented by Gaussians. This is without any doubt true for those LCPs where there is no sidefeeding from neighboring isotopes. A prominent example for sidefeeding is ^{4}He where it is established that about 20% of the yield observed comes from primary ^{5}He which emits a neutron and is converted into ^{4}He before the particles reach the detectors (5). The ^{5}He contribution becomes visible as an indication of a shoulder in the distribution at low energies (6). Yet, the error introduced by assuming an overall Gaussian also in this case does not significantly impair the quality of the fit. In the present work, therefore, all energy distributions were assumed to be satisfactorily described by Gaussians. As may be inferred from Figure 3, this ansatz should be considered to be sufficient to find the full distribution whenever the measured data exhibit a maximum, as it is the case for ^{4}He and ^{7}Li. However, for the Be isotope ^{10}Be (and similarly for other cases) the ansatz may still lead to ambiguous results and a further assumption has to be made. Again based on experimental evidence, it is known that the energy distributions of LCPs extend down to energy zero. More precisely, in the energy window encompassing energy zero, ternary LCP events are consistently found. In the evaluation this feature is taken into account by artificially imposing for the Gaussian to carry (1-2) % of the maximum yield at energy zero. Evidently this recipe introduces a systematic error which has to be accounted for.

Another difficulty for ΔE-E$_{rest}$ detectors as used here is to cope with the full range of energy distributions from the lightest ternary particles (hydrogen isotopes) up to the heaviest ones. In the present experiment the solid state detectors were not thick enough to stop the most energetic hydrogen ions. Therefore, neither energies nor yields are readily determined. The hydrogen isotopes require, hence, further analysis.

The hitherto unknown isotopic energy distributions for Be and heavier elements from ^{252}Cf(sf) are of special interest. For Be a row of isotopes ranging from ^{9}Be up to ^{12}Be could be analyzed. For ^{10}Be the distribution was already shown in Figure 3, while for ^{9}Be and ^{11,12}Be the energy spectra are on display in Figure 4. Both, the average energies and the widths of the distributions are seen to be very similar.

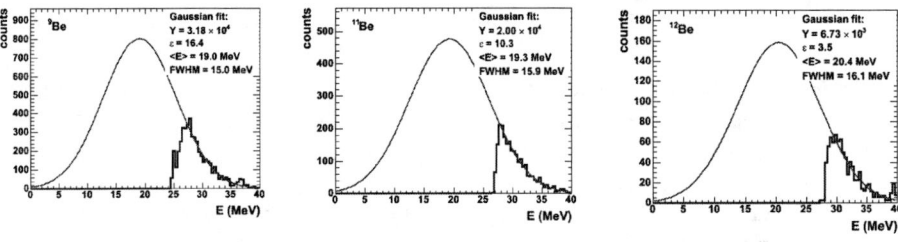

Figure 4: Energy distributions of Be isotopes from ^{252}Cf(sf)

Once the distributions have been determined, the yields of the ternary LCPs can be evaluated. We will in the following not give a complete list of all results obtained but rather focus on the Be yields which in the following are of interest for discussion. In Table 1 the Be yields for ^{236}U* following cold neutron capture from different experiments have been put together with those for ^{252}Cf(sf).

TABLE 1: Ternary Be yields from 235(Un,f) and 252Cf(sf) per 10^4 α-particles

	8Be	9Be	10Be	11Be	12Be	
236U*	0.8(3)	2.0(4)	30(3)	- -	- -	Present Experiment
236U*	- -	2.9(3)	32(3)	2.0(3)	1.5(6)	A Vorobyov 1972 (7)
236U*	- -	3.1(1)	27.6(1)	2.4(4)	1.2(1)	W Baum 1992 (8)
252Cf	12(6)	17(3)	123(10)	10(4)	4(2)	Present Experiment

All yields are normalized to 10^4 α-particles from the respective reaction. The ^{8}Be yields quoted have been obtained by the authors in previous experiments which were devoted to the study of quaternary fission (9). There it was found that, besides quaternary fission proper with the simultaneous emission of two heavy fragments and two light charged particles, also sequential decays with ^{8}Be as a primary ternary particle are to be observed. The ^{8}Be nucleus decays into two α-particles before hitting detectors and, hence, simulates a quaternary process. The yields of simultaneous and sequential quaternary events could be disentangled based on the angular correlations between α-particles being different for the two cases. Missing entries in the table indicate that the corresponding yields were not measured or, as in the case for ^{11}Be and ^{12}Be from the present experiment on ^{235}U(n,f), are not yet evaluated. The agreement between the results reported by different groups for the $^{9\text{-}12}$Be yields from fission of ^{236}U* has to be considered acceptable in view of the difficulties of the experiments. As to the Be yields, in published former work only the cumulative yields for ^{9}Be up to ^{12}Be could be determined. Cumulative yields reported in the references (10), (11) and (12) range from 126(30) to 164(9) and 185(20), respectively, in reasonable agreement with 154(12) from the present experiment, all yields being relative to 10^4 α-particles.

DISCUSSION

The yields of light charged particles from ternary fission are discussed in the literature starting from very different point of views concerning the mechanism of LCP emission. In microscopic models preformation probabilities in particular for α-particles have been considered with the α`s being set free in a sudden approximation at rupture of the neck (10, 11). Thermodynamic models describe the emission of LCPs as an evaporation from the neck region which is heated up locally when in fission the viscous nuclear fluid flows from the saddle to the scission point (12). LCP yields depend in this model mainly on the exponential factor $\exp\{[B_{A,Z} - Z\varepsilon_p - (A-Z)\varepsilon_n] / \theta\}$ with $B_{A,Z}$, ε_p, ε_n and θ the binding energy of the LCP, the chemical potential of the proton and neutron, and the temperature, respectively. By contrast, in a model proposed by Pik-Pichak it is claimed that LCPs are emitted adiabatically at scission (13). The probability for the ratio of LCP to α-particle yields is calculated by perturbation theory. The ratio is found to depend exponentially on the differences of Q-values and Coulomb energies of the scission configuration. In a further dynamic model the appearance of a third particle has been traced to a double rupture of the neck. Following rupture, the primary light particles exchange nucleons with the fragments to find their final composition which is governed by a Boltzmann equilibrium (14).

Several models are in the main phenomenological. One of the first models was proposed by Halpern (15). There it is argued that ternary fission is a sequential process in the sense that, first, the nucleus undergoes a binary division and, second, a light particle forming part of one of the fragments is placed midway between the two main fragments. The energy costs E_{cost} for this transition from binary to ternary fission are calculated from the reaction Q-values and the changes in Coulomb energy for the scission configurations involved. The probability for LCP emission is finally set proportional to $\exp(-E_{cost}/\Gamma)$ where Γ is a temperature-like parameter. Remarkably, the final formula for assessing LCP yields is very similar to the one derived by Pik-Pichak (13).

Let us finally quote an empirical model which has the virtue to fit surprisingly well experimental data (8). Apart from a normalization constant the yields are simply fitted to the function $\exp[(Q_{ter} - V_{sci})/\Gamma]$ with Γ a universal parameter. In the formula Q_{ter} is the Q-value for the ternary fission reaction and V_{sci} is the Coulomb energy of the scission configuration. This heuristic ansatz obviously borrows ingredients from the Pik-Pichak (13) and/or the Halpern model (15).

The challenge for all these models is to reproduce fairly well both, the overall trends of yields from the very lightest to the very heaviest LCPs observed, as well as the pattern of yields for a given element, including odd-even effects. The latter aspect is reviewed for the Be yields from the present study. In Figure 5 the experimental findings are compared to the predictions from different models.

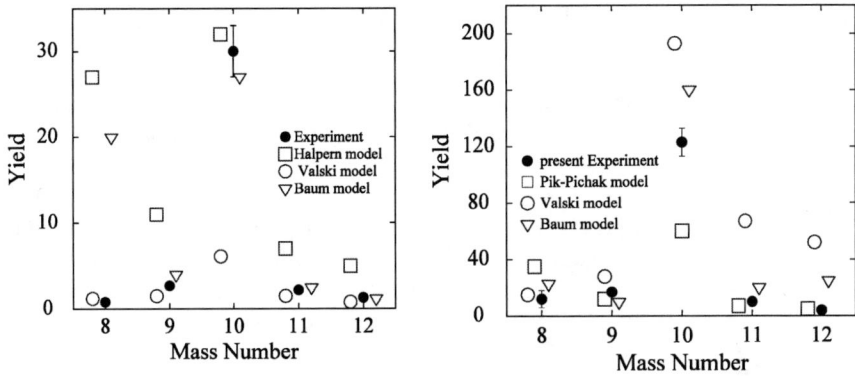

Figure 5: Comparison of Be yields between experiment and theoretical models for ^{235}U(n$_{th}$,f) in the left diagram and for ^{252}Cf(sf) in the right diagram. The experimental U yields shown are average values from Table 1.

First it should be pointed out that the scales for the Be yields plotted relative to 10^4 ternary α-particles from the two reactions are quite different. Taking the partly large errors into account, the Be yields from fission of the heavier nucleus ^{252}Cf are about a factor of 5 larger than for the lighter nucleus ^{236}U*. This rise in heavy LCP yields observed for increasingly heavier nuclei undergoing fission is well known and once more established in the present study for Be (16). The two reactions have, however, in common that the yields for ^{10}Be are by far dominating and, the change in scale taken apart, the patterns of the measured yields for the different Be isotopes are quite similar.

Yet, what is catching the eye in Figure 5 is that none of the models is satisfactory. In addition, the discrepancies among the model predictions are that vast to make even the figures to become confusing. Starting with the phenomenological Halpern model (15) for ^{236}U* in the left part of Figure 5, only the ^{10}Be yield is correctly reproduced. This should not be too surprising because very probably the agreement with experiment has simply to be ascribed to the overall scaling factor which is fitted to the bulk of LCP yields. At a closer look, the model fails seriously to predict the trends in the probability to emit the different Be isotopes. In particular the ^{8}Be yield is not well accounted for. Similar remarks pertain in the right part of Figure 5 to the Be yields for ^{252}Cf(sf) from the Pik-Pichak model (13). Nevertheless, the overall fits of experimental yields from the lightest to the heaviest LCPs ranging from He up to Si (not shown here) are quite satifactory in both models.

The thermodynamical model by Valski (12) traces the uranium data quite well except for ^{10}Be which is surprisingly down by a factor of 5 compared to experiment. For californium it is the other way round. The Be yields become from ^{10}Be to ^{12}Be increasingly too large. The author is well aware of this fact, arguing that for heavier and heavier LCPs the model is expected to run into difficulties simply because the supply of nucleons in the neck is fading out. The notion of a heat bath becomes, hence, pointless. It has to be stressed, though, that the model is not only useful for light charged particles but that also the yields of neutrons evaporated as ternary particles at scission may be calculated.

Finally the phenomenological model introduced by Baum (8) is seen to give the best overall fit for both the ^{235}U(n$_{th}$,f) and the ^{252}Cf(sf) reaction. The only exception is the yield calculated for ^{8}Be in fission of ^{236}U* which is far off, in contrast to the ^{8}Be yield in ^{252}Cf(sf) which is reasonably well accounted for. This disparity is amazing. It is conjectured that like in other models also here the proper choice of the ternary scission configurations, which are required to calculate Coulomb energies, is of crucial importance. Different choices for the configurations may lead to very different results. In view of the shortcomings established theories of ternary fission are facing, it should be worthwhile to also elaborate on the details of heuristic models.

In summary, already the first steps in the evaluation of the two complex experiments having been performed have brought interesting new results. With the yields and energy distributions of light charged particles being now available the more involved correlations of LCPs with gammas and/or fission fragments may be tackled.

ACKNOWLEDGMENTS

The present studies were supported by the BMBF in Bonn, Germany, under contract numbers 06DA913 and 06TU699, the RFBR in Moscow, Russia and by INTAS in Brussels, Belgium, under contract numbers 99-0299 and 03-51-6417. The technical help by K. Schmidt in preparing the present report is gratefully acknowledged.

REFERENCES

1. M. Mutterer, J.P. Theobald, in *Nuclear Decay Modes*, Bristol, England, IOP, ed. D. N. Poenaru, Chapt. 12
2. C. Wagemans in *The Nuclear Fission Process*, Boca Raton, USA, CRC Press, ed. C. Wagemans, *Chap. 12*.
3. Yu. N. Kopatch et al., Proc. of *Symposium on Nuclear Clusters*, EP Systema, Debrecen, Hungary, 2003, 273
4. M. Mutterer et al., *Nucl. Phys. A738, 122 (2004)*
5. Yu. N. Kopatch et al., *Phys. Rev. C 65*, 044614 (2002)
6. C. Wagemans et al., *Nucl. Phys. A 742, 291 (2004)*
7. A. A. Vorobyov et al., *Phys. Lett. 40B, 102 (1972)*
8. W. Baum, PHD thesis, Technical University Darmstadt, Germany, 1992
9. P. Jesinger et al., *Eur. Phys. J. A24, 379 (2005)*
10. O. Tanimura and T. Fliessbach, *Z .Phys. A328, 475 (1987)*
11. O. Serot, N. Carjan and C. Wagemans, *Eur. Phys. J. A8, 187 (2000)*
12. G. V. Valsky, *Phys. At. Nucl. 67, 1264 (2004)*
13. G. A. Pik-Pichak, Phys. At. Nucl. 57., 906 (1994)
14. V. A. Rubchenya and S. G. Yavshits, *Z. Phys. A329, 217 (1988)*
15. I. Halpern, *Ann. Rev. N ucl. Sci. 21, 245 (1971)*
16. F. Gönnenwein, *Nucl. Phys. A734, 213 (2004)*

Sudden Emission of Nucleons at Scission

N. Carjan*,†, P. Talou*, D. Strottman*, O. Serot** and H. Goutte‡

*Theoretical Division, Los Alamos National Laboratory, Los Alamos, NM 87545, USA
†CENBG, Le Haut Vigneau, 33175 Gradignan Cedex, France
**CEA - Cadarache, DEN/DER/SPRC/LEPh, Bât.230, 13108 St-Paul-lez-Durance, France
‡CEA-Bruyères-le-Châtel, DPTA/SPN, BP12, 91680 Bruyères-le-Châtel, France

Abstract. At a certain finite neck radius during the descent of a fissioning nucleus from the saddle to the scission point, the attractive nuclear forces can no more withstand the repulsive Coulomb forces producing the neck rupture and the sudden absorption of the neck stubs by the fragments. At that moment the nucleons, although still characterized by their pre-scission wave functions, find themselves in the newly created potential of their interaction with the separated fragments. Their wave functions become wave packets with components in the continuum. The probability to populate such states gives evidently the emission probability of nucleons at scission. In this way we have studied scission neutrons and protons for the fissioning nucleus ^{236}U, using two-dimensional realistic nuclear shapes.

Keywords: Fission dynamics, scission protons and neutrons
PACS: 25.85.Ec

INTRODUCTION

Emission of scission nucleons has been always an intriguing subject. Based on energetic considerations they should be the most probable among the light particles that accompany the fission process. In particular they should outnumber the alpha particles. Experimentally however the emission rate for scission protons represents only few percent of that for scission alphas [1] while the existing results on scission neutrons are very contradictory [1], [2]. There is no doubt that understanding this paradox is a central piece of the nuclear fission puzzle.

Attempting to describe this process, one cannot ignore the fact that, if the neck ruptures at finite radius, a violent transition between two quite different nuclear configurations occurs. The corresponding large difference in the potential energy of deformation is made available to the internal degrees of freedom, that is, mainly to the nucleons.

This sudden release was proposed as an emission mechanism by Halpern [3] without being quantitatively developed. The only attempt so far to carry out calculations using the sudden approximation was done later using a simplistic one-dimensional model [4] for the emission of scission alpha particles. A gradual potential change and its effect on the energy transferred to the nucleons was studied by Fuller [5] and by Boneh and Fraenkel [6].

In the present paper we have generalized the formalism from ref. [4] to two dimensions in order to include realistic nuclear shapes and have applied it to the emission of single nucleons during $(n_{th}, f)^{235}U$ reaction.

FORMULA FOR THE SUDDEN EMISSION PROBABILITY

The single-particle wave functions for an axially symmetric fissioning nucleus have the general form:

$$\Psi(\rho,z,\phi) = u(\rho,z)e^{i(\Omega-\frac{1}{2})\phi}|\uparrow> + d(\rho,z)e^{i(\Omega+\frac{1}{2})\phi}|\downarrow> \tag{1}$$

where $u(\rho,z)$ and $d(\rho,z)$ contain the spatial dependence of the two components: spin up and down respectively. Ω is the projection of the total angular momentum along the symmetry axis and is a good quantum number. Since we are dealing with symmetric fission the parity π is also a constant of motion.

If the neck rupture is followed by a sudden absorbtion of the neck protuberances the nucleonic wave functions will find themselves in the new potential but unchanged. It makes then sense to develop a given eigenstate Ψ^i of the "just before scission" potential into the complete set Ψ_m^f of eigenstates of the "immediately after scission" potential:

$$|\Psi^i> = \sum_{all\ states} a_m^{if}|\Psi_m^f> \tag{2}$$

where

$$a_m^{if} = <\Psi_m^f|\Psi^i> = 2\pi \int\int (u^i u_m^f + d^i d_m^f)\rho d\rho dz. \tag{3}$$

One can notice that only $|\Psi_m^f>$ states with the same (Ω,π) values as $|\Psi^i>$ will have non-zero contributions.

The emission probability of the nucleon which had occupied that initial state, Eq.(2), is

$$P_{em}^i = 1 - \sum_{bound\ states} |a_m^{if}|^2. \tag{4}$$

For neutrons we included in the sum all states with negative energy and for protons all states with energy smaller than 10 MeV (the average Coulomb barrier height).

The emitted part of the wave function

$$|\Psi_{em}> = |\Psi_i> - \sum_{bound\ states} a_m^{if}|\Psi_m^f> \tag{5}$$

gives the distribution of the emission points relative to the fragments, $<\Psi_{em}|\Psi_{em}>$.

NUMERICAL RESULTS

For exemplification we have chosen the fissioning nucleus ^{236}U. We have calculated the single-particle wave functions [7] at two deformations, $\varepsilon_i = 0.985$ and $\varepsilon_f = 1.001$, between which the sudden transition is supposed to occur. An idea of the shapes involved is given (Fig.1) by the neutron equipotential lines corresponding to $V_0/2$, V_0 being the depth of the Woods-Saxon interaction. The neck radius at rupture is seen to be 1.6 fm.

For each type of nucleons, we have selected 3 states around the Fermi level of the initial configuration and calculated their expansion coefficients, Eq.(3), for all bound

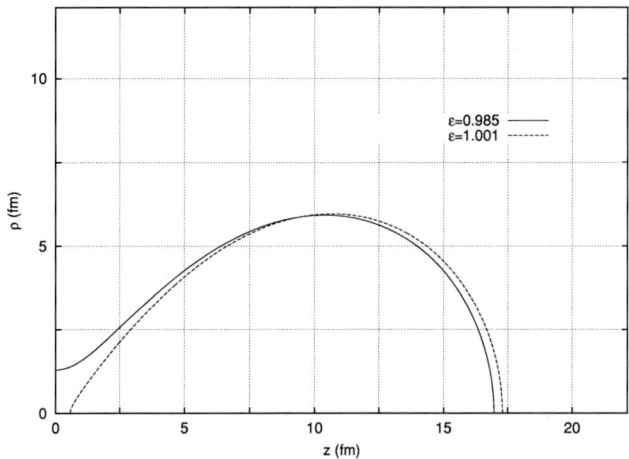

FIGURE 1. Comparison between the equipotential lines V=-22.1 MeV of the neutron potential for just-before-scission and immediately-after-scission configurations.

TABLE 1. Emission probabilities for different initial states.

Nucleon type	$\Omega\pi$	$P_{em}(\%)$	Experimental value [1]
n	5/2 +	1.058	
	1/2 -	7.374	2.7
	7/2 -	1.947	
p	1/2 +	1.436	
	5/2 +	0.037	0.002
	5/2 -	0.037	

states of the final configuration. The results are plotted in Figs.2 and 3 for neutrons and protons respectively. In all cases the initial strength is distributed over a relatively small number of the final states reflecting the fact that the fragments are already preformed before scission. In other words there is no major change in the single-particle states from ε_i to ε_f. Consequently, there is always one dominant component that takes most of the strength: about 60% for the 1/2 states and more than 90% for the others. This particular final state is the equivalent of the initial state, i.e., obtained by adiabatic transition.

The a_m^{if} coefficients were then used in Eq.(4) to estimate the probability of a nucleon, that just before scission had occupied one of the initial states, to be emitted during scission (see Table 1). The experimental values were included only for orientation. In order to compare with data, the calculations should take into account all possible initial states (not only 3) and their occupation probabilities.

Figs.4 and 5 show the distribution of the emission points for scission neutrons corresponding to two different initial states: 1/2− and 7/2− respectively. We can see that

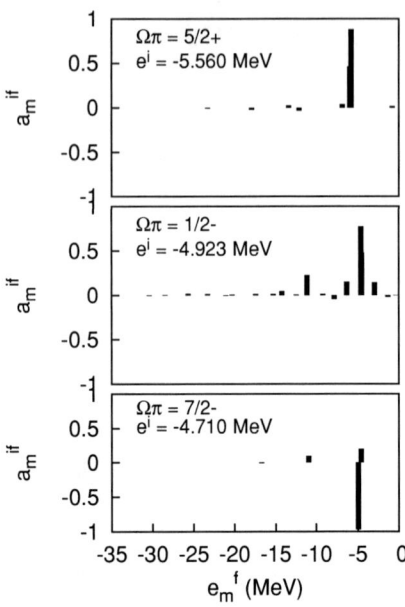

FIGURE 2. Histograms of the expansion coefficients of 3 initial neutron states into the final states.

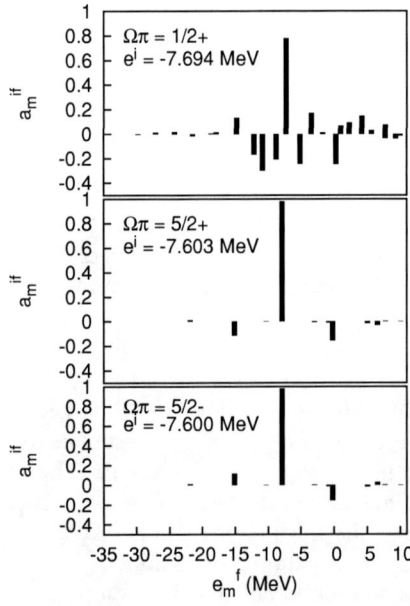

FIGURE 3. The same as in Fig.1 but for protons.

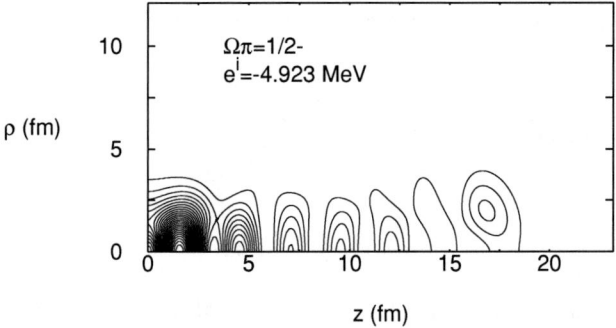

FIGURE 4. Contour plot of the emitted part of the 1/2− neutron wave function.

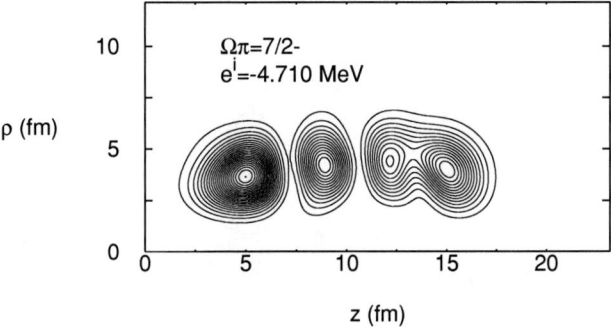

FIGURE 5. The same as in Fig.4 but for the 7/2− state.

only for the 1/2− state the emission takes mainly place between the fragments. The wave functions with high Ω values, having only high angular momenta components are positioned away from the z-axis and cannot be present in the neck.

CONCLUSIONS

We have presented a first-attempt calculation for the emission of scission nucleons using the sudden approximation. The order of magnitude for the neutron rate was correctly reproduced (10^{-1}) while the proton rate was overestimated (10^{-4} instead 10^{-5}). There is however a lot of place for improvements in these preliminary calculations. It is also known that the proton polar emission, that was not included in the experimental value, is quite important.

There are two characteristics of the sudden emission that emerge from our calculations:

1. The rate of emission is determined by the width of a_m^{if} distribution and hence by the difference between just-before-scission and immediately-after-scission configurations.

2. The distribution of the emission points is strongly dependent on the Ω value of the ocuppied states, leading to nearly isotropic emission at scission.

REFERENCES

1. C. Wagemans, *The Nuclear Fission Process*, CRC Press, Boca Raton, USA, 1991.
2. H. R. Bowman et al., *Phys. Rev.*, **126** (1962) 2120.
3. I. Halpern, *First Symposium on Physics and Chemistry of Fission*, Salzburg, 1965 (IAEA, Vienna, 1965) vol.II, 369.
4. O. Serot, N. Carjan, C. Wagemans, *Eur. Phys. J. A*, **8** (2000) 187.
5. R. W. Fuller, *Phys. Rev.*, **126** (1962) 684.
6. Y. Boneh, Z. Fraenkel, *Phys. Rev. C*, **10** (1974) 893.
7. V. V. Pashkevich, *Nucl. Phys. A*, **169** (1971) 275.

SPECTROSCOPY OF NEUTRON RICH NUCLEI

Nuclear structure studies of neutron-rich Cu and Zn isotopes produced by means of proton-induced fission of ^{238}U

J.-C. Thomas*,†, H. De Witte*, M. Gorska*,**, M. Huyse*, K. Kruglov*, Y. Kudryavtsev*, D. Pauwels*, N.V.S.V. Prasad*, K. Van de Vel*,‡, P. Van Duppen*, J. Van Roosbroeck*, S. Franchoo§,¶, J. Cederkall§, H.O.U. Fynbo§,‖, U. Georg§, O. Jonsson§, U. Köster§, L. Weissman††, W.F. Mueller††, V.N. Fedoseyev‡‡, V.I. Mishin‡‡, D. Fedorov§§, A. De Maesschalck¶¶, N.A. Smirnova¶¶, the IS365 collaboration§ and the ISOLDE collaboration§

*Instituut voor Kern- en Stralingsfysica, University of Leuven, Celestijnenlaan 200D, B-3001 Leuven, Belgium
†GANIL, B.P. 55027, F-14076 Caen Cedex 5, France
**GSI, Planckstrasse 1, D-64291 Darmstadt, Germany
‡VITO, IMS, Mol, Belgium
§ISOLDE, CERN, 1211 Genève 23, Switzerland
¶IPN Orsay, 15 rue G. Clémenceau, F-91406 Orsay Cedex, France
‖Department of Physics and Astronomy, University of Århus, Dk-8000 Århus C, Denmark
††National Superconducting Cyclotron Laboratory, Michigan State University, 48824-1312 MI, USA
‡‡Institute of Spectroscopy, Russian Academy of Sciences, 142092 Troitsk, Russia
§§St. Petersburg Nuclear Physics Institute, 188350 Gatchina, Russia
¶¶Vakgroep Subatomaire en Stralingsfysica, Universiteit Gent, Proeftuinstraat 86, B-9000 Gent, Belgium

Abstract. The neutron-rich nuclei ^{72}Ni and ^{72}Cu have been produced in the proton-induced fission of ^{238}U at the LISOL and ISOLDE facilities of Louvain-La-Neuve and CERN. Partial β-decay schemes are presented, giving some information about the nuclear structure of their daughter nuclei ^{72}Cu and ^{72}Zn. The lifetime of ^{72}Cu was determined to be $T_{1/2} = 6.63(3)$ s, in line with previous measurements. No β-decaying isomeric states were identified in ^{72}Cu, in contrast to ^{70}Cu. The spin and the parity of several states in ^{72}Cu are tentatively assigned and a comparison is made with different shell-model predictions.

Keywords: ^{72}Ni, ^{72}Cu, ^{72}Zn, proton-induced fission, resonant laser ionization, β decay, nuclear structure, residual interaction, shell-model calculations
PACS: 21.10.-k, 23.40.-s, 25.85.Ec, 27.50.+e, 32.80.Fb

INTRODUCTION

The β-decay study of neutron-rich nickel and copper isotopes offers a unique opportunity to probe the evolution of the nuclear structure in the Z=28 and N=40 to N=50 region. As recently exemplified in the β-decay study of ^{74,76,78}Cu [1], the subshell closer at N=40 clearly modifies the level spectrum of the neutron-rich Zn isotopes, questionning the interplay between collective nucleon-nucleon forces and shell effects around N=40. On the other hand, earlier works have shown the dominant role of the residual π-ν interaction in the nuclear structure of the neutron-rich Cu isotopes, as illustrated by the monopole migration phenomenon of low-lying states in ^{69,71,73}Cu [2] and by the observation of three β-decaying isomeric states in ^{70}Cu [3, 4]. Different theoretical approches including large-scale shell-model calculations [4] as well as schematic forces [1, 5] may be tested in order to improve our understanding of the residual π-ν interaction. It emphasizes the relevance of nuclear structure studies of neutron-rich nuclei in the vicinity of the doubly-magic nucleus ^{78}Ni.

In the following, we present briefly the results obtained in the β-decay study of ^{72}Ni and ^{72}Cu, produced by means of proton-induced fission of ^{238}U at the facilities of LISOL (CRC Louvain-la-Neuve, Belgium) and CERN-ISOLDE (Geneva, Switzerland). More details about this work will be given in [6].

RESULTS

Experimental setups and analysis procedure

^{72}Cu was produced at the CERN-ISOLDE facility using a 1 GeV proton beam of 2 μA in average, impinging on a thick UC$_x$/graphite target. After a selective laser ionization and a mass separation were performed, up to 10^7 ^{72}Cu ions per μC were implanted on a movable tape surrounded by two ΔE β detectors associated with two HPGe detectors. Singles and β-gated on-resonance spectra were taken, allowing the identification of γ transitions in the β decay of ^{72}Cu to ^{72}Zn. A second setup consisting in a 4π β-detector surrounding a movable tape was devoted to the lifetime measurement of ^{72}Cu. The obtained value T$_{1/2}$ = 6.63(3) s is in agreement with the value of 6.6(1) s reported in [7]. In a separate run, some beam time was dedicated to the in-source laser spectroscopy of ^{72}Cu. In contrast to ^{70}Cu [3], no β-decaying isomeric states were identified.

^{72}Ni was produced at LISOL using a 20 MeV proton beam of several μA impinging on two thin ^{238}U targets located inside an Ar gas cell. The nickel fission fragments were selectively ionized, mass separated and implanted on a movable tape surrounded by three HPGe detectors associated with ΔE β detectors. About 10 ^{72}Ni ions per μC were collected. β-gated γ spectra were accumulated with and without a selective ionization of nickel ions. Such a procedure allowed the identification of resonant γ lines in the β decay of ^{72}Ni to ^{72}Cu.

Partial level schemes were deduced for ^{72}Cu and ^{72}Zn. They were derived from an analysis of the γ-γ coincidences and on the basis of energy matching considerations. The high production rate of ^{72}Cu obtained at ISOLDE allowed to check for consistency the time behavior of the β-delayed γ-lines in ^{72}Zn. More details about the experimental conditions can be found in [4, 8] for the β-decay works on neutron-rich copper isotopes at ISOLDE and in [2, 9] for the β-decay studies of neutron-rich nickel isotopes at LISOL.

β-decay of ^{72}Cu

Figure 1 presents the partial level scheme of ^{72}Zn, deduced from the analysis of the β-gated γ rays observed in the decay of the selectively laser ionized ^{72}Cu ions.

Absolute γ-ray intensities were extracted using GEANT [10] simulated photo-peak efficiencies, corrected for the true summing of cross-over transitions. Absolute β branching ratios and log(ft) values are given assuming no direct feeding of ^{72}Zn ground state and must be considered as upper and lower limits, respectively.

The spin and parity assignment of several excited states in ^{72}Cu is partly derived from a comparison with earlier works of Runte et al. [7], Hudson and Glover [11] and Wilson et al. [12]. More details will be available in [6].

The (2$^+$) spin and parity assignment of ^{72}Cu ground state is based on a study by Mach et al. [13]. As dicussed in the next section, it presumably corresponds to a $\pi 2p_{3/2}^{+1} \nu(2p_{1/2}^{-1} 1g_{9/2}^{+3})$ configuration. Its main decay branch (B.R.=16 (4)%, log(ft)=6.5) feeds the 2$_1^+$ excited state at 653 keV in ^{72}Zn, which can be associated with the transformation of the uncoupled $\nu 2p_{1/2}$ neutron into an additional $\pi 2p_{3/2}$ proton, leading to a $\pi 2p_{3/2}^{+2} \nu(2p_{1/2}^{-2} 1g_{9/2}^{+4})$ configuration. Assuming that the 2$_1^+$ state at 885 keV in ^{70}Zn has a similar two neutron-particle two neutron-hole configuration [4], the lowering of about 230 keV of the 2$_1^+$ state in ^{72}Zn can be interpreted as the weakening of the neutron pairing energy as the $\nu 1g_{9/2}$ orbital is filled by an additionnal pair of neutrons.

β-decay of ^{72}Ni

Figure 2 presents the partial level scheme of ^{72}Cu. The β-delayed γ rays were identified by comparing β-gated γ spectra obtained with and without a selectively laser ionization of ^{72}Ni ions. The decay scheme was built on the basis of γ-γ coincidences and energy matching considerations. For comparison, the partial level scheme obtained in an earlier work [13] is shown in the right part of the figure.

Absolute γ-ray intensities per 100 β decays are reported. They were corrected for the true summing of cross-over transitions, for the contribution of γ lines of similar energies related to the decay of ^{72}Cu and for the contribution of non-resonant γ lines associated with the implantation of contaminant ions (^{144}Ba,^{144}La). Absolute β branching ratios and log(ft) values are given assuming no direct feeding of ^{72}Cu ground state. Some of them were slightly corrected to account for the contribution of the electronic conversion process to the electromagnetic decay of low-energy excited

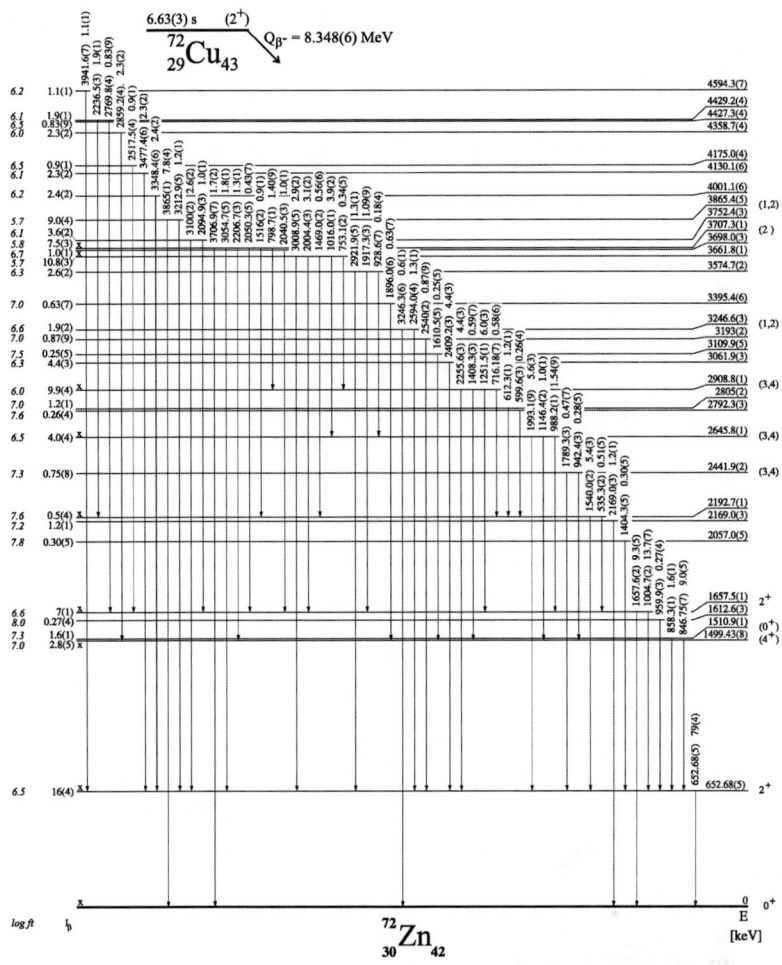

FIGURE 1. Partial level scheme of ^{72}Zn. Levels labeled with an x symbol were observed in previous experiments [14]. Absolute γ-ray intensities and β branching ratios are given per 100 β decays, assuming no direct feeding of ^{72}Zn ground state.

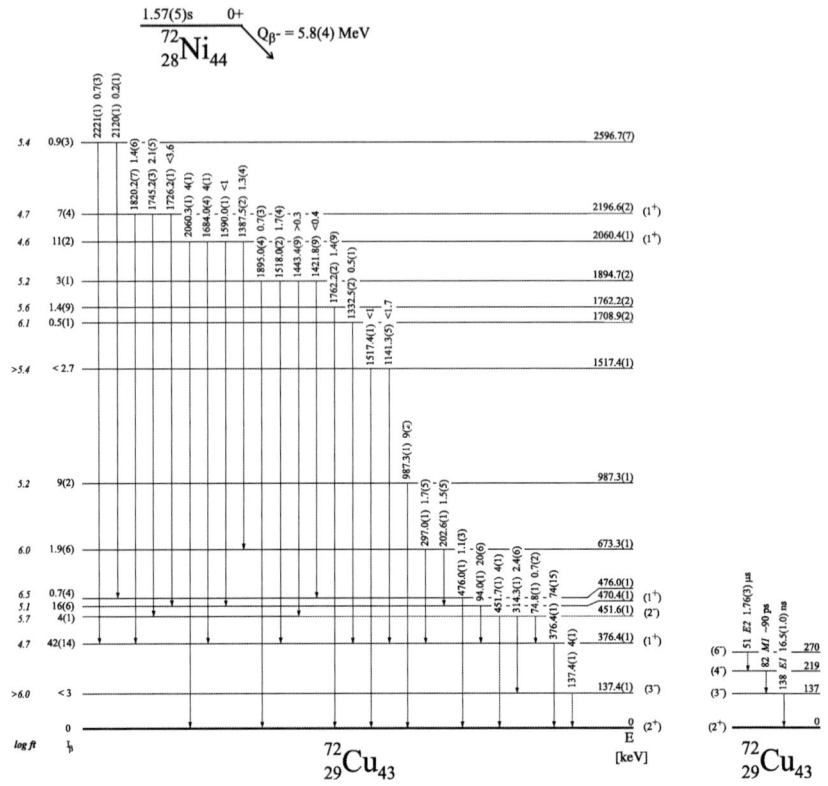

FIGURE 2. Partial level scheme of ^{72}Cu. Spin and parity assignments were partly derived from a previous study [13]. Results from the latter work are presented in the right part of the figure. Absolute γ-ray intensities and β branching ratios are given per 100 β decays, assuming no direct feeding of ^{72}Cu ground state.

states. This correction was performed on the basis of the spin and parity assignments discussed below, considering the most probable multipolarity for each transition.

The spin and parity assignments of (2^+) and (3^-) for the ground state of ^{72}Cu and the excited state at 137 keV were adopted from [13]. A spin of (1^+) was attributed to the excited states at 376, 2060 and 2197 keV due to the rather low log(ft) values associated to the feeding of these states. The excited state at 452 keV is the only one that decays to the low-energy (3^-) state. The same situation occurs in ^{70}Cu [4] in which a (3^-) state at 101 keV is only fed in the γ decay of a (2^-) state lying at 369 keV. We therefore assigned a spin of (2^-) to the excited state at 452 keV in ^{72}Cu.

Two shell-model calculations were performed in order to understand the level scheme of ^{72}Cu. In the first approach, a large-scale calculation was performed using the ANTOINE code [15] in the $(2p_{3/2}1f_{5/2}2p_{1/2}1g_{9/2})$ shell space outside the doubly magic ^{56}Ni core. We used the realistic effective interaction derived by Hjorth-Jensen, Kuo and Osnes [16], which monopole part was modified according to Nowacki [17]. In a schematic approach detailed in [1], the coupling between the 29th proton and the valence neutrons occupying respectively the $\pi(2p_{3/2}1f_{5/2}2p_{1/2}1g_{9/2})$ and $\nu 1g_{9/2}$ orbitals was computed by means of a δ force and a quadrupole-quadrupole interaction. Figure 3 compares the predictions of the different shell-model approaches to the low energy spectrum of ^{72}Cu observed experimentally.

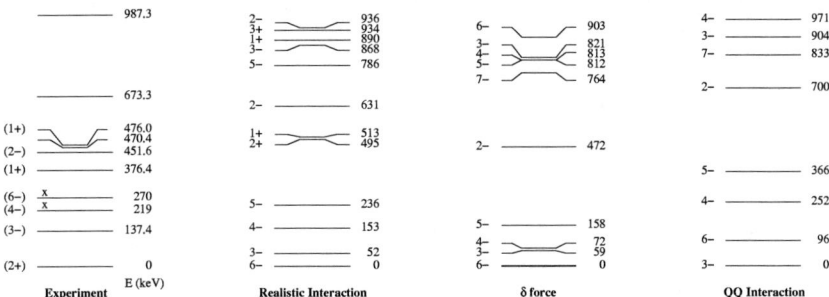

FIGURE 3. The level scheme of ^{72}Cu derived from the present work and from [13] (states labeled with an x symbol) is compared to the theoretical spectra obtained with (i) a large-scale shell-model calculation performed within the $(2p_{3/2}1f_{5/2}2p_{1/2}1g_{9/2})$ shell space and using a realistic interaction, and with (ii) a schematic approach using two types of residual proton-neutron interactions (a δ force and a quadrupole-quadrupole interaction).

All calculations predict the $(3\text{-}6)^-$ multiplet of states resulting from the $\pi 2p_{3/2}^{+1}\nu 1g_{9/2}^{+3}$ configuration to lie at low energy and to be of rather pure configuration (see [6] for details). Experimentally, three of the negative parity states have presumably been observed with a reasonable excitation energy, thus confirming the dominant contribution of the $\pi 2p_{3/2}\nu 1g_{9/2}$ coupling to the low-energy level spectrum of neutron-rich even copper isotopes [4].

The first 2^- state associated with a dominant $\pi 1f_{5/2}\nu 1g_{9/2}$ configuration [4, 6] is expected at a higher excitation energy than the one observed experimentally at 452 keV. It may indicates that the monopole corrected single-particle energy adopted for the $\pi 1f_{5/2}$ proton orbital is too high or that the configuration mixing of this 2^- state is not well reproduced theoretically.

The $(1\text{-}2)^+$ doublet associated to a dominant $\pi 2p_{3/2}^{+1}\nu(2p_{1/2}^{-1}1g_{9/2}^{+4})$ configuration is expected to lie at around 500 keV, with a very small energy splitting. It is not predicted within the schematic approach because of the truncation of the neutron shell space. Experimentally, the 2^+ state is presumably the ground state of ^{72}Cu while the 1^+ state lies at 376 keV, giving rise to a rather large energy splitting of the doublet. This feature was not observed in the case of ^{70}Cu [4] and it remains unexplained.

CONCLUSION

The nuclear structure of ^{72}Cu and ^{72}Zn was studied at the LISOL and ISOLDE facilities in the β decay of the selectively ionized nuclei ^{72}Ni and ^{72}Cu, produced by means of proton-induced fission of ^{238}U. The in-source laser spectroscopy study of ^{72}Cu, which lifetime $T_{1/2} = 6.63(3)$ s was measured, revealed no β-decaying isomeric states, in contrast to ^{70}Cu [3].

Shell-model calculations using realistic as well as schematic interactions allowed to identify the leading configurations of several states in ^{72}Cu, stressing the relevance of further investigations of the nuclear structure of neutron-rich copper isotopes in order to improve our understanding of the residual π-ν interaction in the N=40 to N=50 region.

ACKNOWLEDGMENTS

We gratefully thank J. Gentens and P. Van den Bergh for running the LISOL separator and the ISOLDE technical group for assistance during the experiment performed at CERN. This work was supported by the Inter-University Attraction Poles (IUAP) Research Program nr. P5/07 and the European Union (contract ERBFMGEECT980120). K.V.d.V. is Research Assistant of the FWO-Vlaanderen.

REFERENCES

1. Van Roosbroeck, J., et al., *Phys. Rev. C*, **71**, 054307 (2005).
2. Franchoo, S., et al., *Phys. Rev. C*, **64**, 054308 (2001).
3. Van Roosbroeck, J., et al., *Phys. Rev. Lett.*, **92**, 112501 (2004).
4. Van Roosbroeck, J., et al., *Phys. Rev. C*, **69**, 034313 (2004).
5. Smirnova, N. A., et al., *Phys. Rev. C*, **69**, 044306 (2004).
6. Thomas, J.-C., et al., to be submitted to PRC.
7. Runte, E., et al., *Nucl. Phys. A*, **399**, 163 (1983).
8. Van Roosbroeck, J., Ph.D. thesis, *Systematic Nuclear-Structure Study of Even-Mass Zn and Cu Isotopes between N=40 and 50*, Katholieke Universiteit Leuven, Leuven, Belgium, 2002.
9. Franchoo, S., Ph.D. thesis, *Evolution of Nuclear Structure towards ^{78}Ni Investigated by the β Decay of Laser-Ionized $^{68-74}$Ni*, Katholieke Universiteit Leuven, Leuven, Belgium, 1999.
10. GEANT - Detector Description and Simulation Tool, http://wwwasd.web.cern.ch/wwwasd/geant/.
11. Hudson, F. R., and Glover, R. N., *Nucl. Phys. A*, **189**, 264 (1972).
12. Wilson, A. N., et al., *Eur. Phys. J. A*, **9**, 183 (2000).
13. Mach, H., "Structure of the exotic nuclei near ^{70}Ni probed by advanced time-delayed coincidence methods: the interpretation of the 1.76 μs isomer in ^{72}Cu," in *International Symposium on Nuclear Structure Physics: Celebrating the Career of Peter von Brentano, University of Götingen, Germany, 5-8 March 2001*, World Scientific, Singapore, 2001, p. 379.
14. Chou, W.-T., and King, M. M., *Nuclear Data Sheets*, **73**, 215 (1994).
15. Caurier, E., and Nowacki, F., *Acta Physica Polonica*, **30**, 705 (1999).
16. Hjorth-Jensen, M., Kuo, T. T. S., and Osnes, E., *Phys. Rep.*, **261**, 125 (1995).
17. Nowacki, F., Ph.D. thesis, *Description microscopique des processus faibles dans les noyaux sphériques*, Université Louis Pasteur, Strasbourg, France, 1996.

Recent Results and Future Prospects for Nuclear Structure Studies at the ILL

G.S. Simpson*, J. Genevey†, J.A. Pinston†, I. Tsekhanovich* and W. Urban**

*Institut Laue-Langevin, 6 rue Jules Horowitz, F-38042 Grenoble Cedex 9, France
†Laboratoire de Physique Subatomique et de Cosmologie, IN2P3-CNRS/Universite Joseph Fourier, F-38026 Grenoble Cedex, France
**Institute of Experimental Physics, Warsaw University, Ul. Hoza 69, PL-OO-681 Warsaw, Poland

Abstract.
The Lohengrin fission-product spectrometer at the ILL offers some unique opportunities to study the structure of very rare neutron-rich nuclei near ^{132}Sn, in the mass A~100 region and in the A~160 region using microsecond isomeric gamma rays. Techniques to measure isomeric gamma rays, conversion electrons, excited-state lifetimes in the pico-to-nanosecond domain and g factors will be presented. New results on an isomeric state in ^{98}Zr will also be discussed.

A 2.5 μs isomeric 17^- state at 6.5 MeV has recently been measured with the Lohengrin spectrometer in ^{98}Zr. This spherical isomeric state is believed to have a $(\pi(p_{1/2}^{-2}g_{9/2}^2)\nu(g_{7/2}^1 h_{11/2}^1))17^-$ configuration and decays into two collective bands. One of the collective bands has a vibrational character, which becomes rotational at higher spins.

An idea to build a neutron guide for nuclear structure experiments on very neutron-rich nuclei at the ILL is currently being discussed. A brief outline of a recent neutron-guide experiment will be presented.

PACS: 21.10.Tg, 23.20.Lv, 25.85.EC, 27.60.+j

INTRODUCTION

The Lohengrin fission-product spectrometer remains at the forefront of nuclear structure research in spite of almost thirty years of operation. Recent improvements in instrumentation have enabled much progress to be made in the study of the structure of neutron-rich around ^{132}Sn and in the mass 100 region. The main strength of the instrument is the ability to study excited states of very weakly produced nuclei, which are far from stability. The energies of the excited states are measured using gamma-ray spectroscopy, and conversion-electron spectroscopy is used to determine the multipolarity of the transitions. Recently new techniques have been introduced at the instrument including fast timing, which allows the lifetimes of excited states in the pico-to-nanosecond regime to be measured, and g-factor experiments, which examine the magnetic properties of excited nuclear states.

Lohengrin uses a thin fissile target, several milligrammes in mass, placed close to the high-flux reactor core of the Institut Laue-Langevin. The fission rate at the target is $\sim 10^{12}$ fissions per second. In thermal-neutron-induced fission the kinetic energy given to the fission fragments is ~ 100 MeV for the light fragment and ~ 70 MeV for the heavy fragment, and as the target is thin they leave the target with an energy loss of just a few

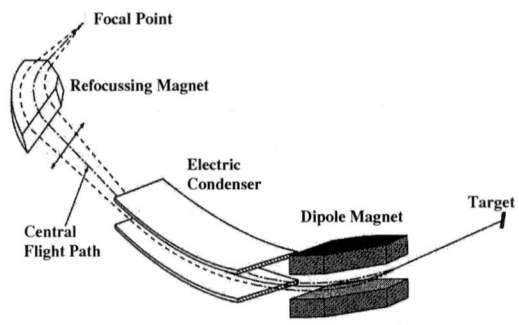

FIGURE 1. Schematic representation of the Lohengrin spectrometer.

MeV. The spectrometer (shown in figure 1) uses a combination of crossed magnetic and electric fields to analyse the mass, ionic charge and velocity of any fission fragments entering the spectrometer. The focal point of the spectrometer is $\sim 6 \times 1.5$ cm. As the flight time through the spectrometer is about 2 μs, it is possible to observe the decay of excited states populated in fission, using microsecond isomers.

GAMMA-RAY SPECTROSCOPY OF MICROSECOND ISOMERS

Gamma-ray spectroscopy of μs isomers is performed at the Lohengrin focal point in conjunction with an ionisation chamber. A split-anode ionisation chamber, filled with several tens of millibars of isobutane, is employed to partially identify, and give a precise arrival time, of individual fission fragments. As Lohengrin separates the fission fragments according to their mass, ionic charge and velocity then measurements of the energy loss of the fission fragments in the chamber allows their mass to be determined, and in the case of low-mass fission fragments their atomic number, as for example in [1]. Towards the end of the chamber an array of PIN diodes is mounted. The gas pressure in the chamber is tuned so that the ions are stopped in the PIN diodes, with an energy loss of a few MeV. This allows the position of the implanted fission fragment to be identified, improving the mass resolution and also defines the position of the stopped ions allowing the position of the gamma ray detectors to be optimised. The ionisation chamber is designed to be compact, so that two "Clover" germanium detectors, or other gamma-ray detectors can be placed as close as possible to the PIN diodes, maximising the solid angle coverage. For the setup shown in figure 2 the gamma-ray detection efficiencies are $\sim$10 % and $\sim$4 % for gamma-rays of 100 keV and 1 MeV respectively.

As decays from isomeric states are typically of low multiplicity (3-4) then it is possible to use only a few detectors in close geometry, without having problems with "sum" peaks, arising from two, or more, gamma rays from the same decay cascade hitting the same germanium crystal.

FIGURE 2. Typical gamma-ray spectroscopy set up with a Clover germanium detector (lower) and Miniball cluster (upper) in place.

FIGURE 3. An expanded view of 2 showing the proximity of the Clover detector to the ionisation chamber.

The arrival of a fission fragment in the ionisation chamber opens a window in the data acquisition system. Any gamma rays detected within a few tens of microseconds after the arrival of the fragment are recorded. Using a narrow time window, and mass selection, allows isomeric decays to be observed in very weakly produced fission fragments. Indeed the first observation of excited states in several very neutron-rich nuclei has recently been performed at Lohengrin [1, 2].

CONVERSION-ELECTRON SPECTROSCOPY

In addition to the observation of gamma rays it is also possible to measure conversion electrons, from decays of isomeric states. Again as in the gamma-ray spectroscopy setup an ionisation chamber is used to identify the masses of the fission fragments. The fragments are stopped in a 12 μm-thick mylar foil at the end of the chamber. The pressure of the gas in the chamber is adjusted so that the fission fragments stop in the last 1 or 2 μm of the mylar foil. A Si(Li) conversion-electron detector, 6 x 2 cm^2 in area, is placed close behind the mylar foil, giving an efficiency of $\sim$25 %, as described in [2]. Germanium detectors are then placed perpendicular to the beam, in approximately the same geometry as the gamma-ray spectroscopy setup. Generally conversion-electron spectra have to be analysed in combination with coincident gamma rays, so that they are clean enough to allow transitions to be measured unambiguously.

FAST TIMING

The fast-timing technique [3] was recently introduced at Lohengrin. This technique uses $\gamma - \gamma$ coincidences between BaF$_2$ detectors to measure directly the lifetimes of excited states below microsecond isomers. Measurements of the lifetimes of excited nuclear states can give information on the purity of single-particle excitations or the deformation in collective excitations. The arrival of a gamma ray in one detector starts a time-to-amplitude convertor (TAC) and the detection of a second gamma ray, within a range of $\sim$50 nanoseconds, stops the TAC. These coincidences are time-correlated with the arrival of a mass-separated fission fragment in the ionisation chamber to improve selectivity. BaF$_2$ detectors have poor energy resolution ($\sim$10 %), but this is not so important as isomeric decays can be correlated with the arrival of a particular fragment at the Lohengrin focal point. Lifetimes as short as $\sim$10 ps have been measured with this technique. First experiments have been performed in the ^{132}Sn region and the data are awaiting publication.

G-FACTOR MEASUREMENTS

Gyromagnetic factors can be sensitive probes of the nuclear wavefunction. It is possible to measure the g factors of excited states in nuclei at Lohengrin using two different techniques, integrated perturbed gamma-ray angular correlations (IPAC) and time-dependent perturbed gamma-ray angular distributions (TDPAD). The TDPAD technique is used in the microsecond lifetime domain where the lifetime of the state of interest is significantly longer than the time resolution of the germanium detectors (tens of nanoseconds) used to measure its decay. For shorter lifetimes, in the nanosecond domain the IPAC technique is used. Both techniques have recently been introduced and data are under analysis.

A NEW HIGH-SPIN MICROSECOND ISOMER IN ^{98}ZR

A new 2.5 μs isomer at an energy of 6.6 MeV and proposed spin of 17$^-$ has recently been measured at the Lohengrin mass spectrometer. The isomer was detected using the techniques described previously. Analysis of the intensities of coincident gamma rays then allowed the level scheme below the isomer to be constructed.

The isomeric state, which is thought to be single-particle in nature, has a proposed $(\pi(p_{1/2}^{-2} g_{9/2}^2) \nu(g_{7/2}^1 h_{11/2}^1))17^-$ configuration. The isomeric state decays into two collective bands. The majority of the positive-parity part of the level scheme below the isomer agrees with that published by Urban [4] and Hamilton [5].

Recently the spin of the yrast states in ^{98}Zr was extended to 20 $\hbar$ by Wu et al. [6], using the reaction $^{238}U(\alpha, f)$. These new data allow the E-GOS (E-Gamma Over Spin) of the yrast band to be plotted against energy, as seen in figure 5 and hence the collective nature of this band can be determined. For a vibrator, the value of this ratio gradually decreases to zero with increasing spin, while for an axially symmetric rotor it approaches a constant, $(4\hbar^2/2J)$. Below spin 14 $\hbar$ the E-GOS fit drops steadily, allowing the band to be described as vibrational in nature. Above 14 $\hbar$ the slope of E-GOS is less steep, and the band becomes rotational in character.

The negative-parity band has not been previously observed, except for some of the low-lying transitions [4] [5]. This band is built on top of a 3064 keV 5$^-$ state, which is octupole-quadrupole in nature. The spin of this state was recently measured, using gamma-ray angular correlations, to be 5$^-$ by Urban using the data from [4]. The slope of the E-GOS plot is consitent with this band being an anharmonic vibrator.

The configuration of the 6.6 MeV, 2.5 μs isomeric state is very likely $\pi(p_{1/2}^{-2} g_{9/2}^2) \nu(g_{7/2}^1 h_{11/2}^1)$, as this configuration is present at approximately this excitation energy. The isomer decays by a 63 keV E2 transition. The nature of this transition was determined by looking at the gamma-ray intensities decaying in to and out of the 6540 keV level. These intensities allow an internal-conversion coefficient, $\alpha \; (= \frac{I_e}{I_\gamma})$ of 5.2 to be measured for this level. This is consistent with a fairly pure E2 transition ($\alpha_{E2} = 6.0$). Similarly, using the experimentally measured lifetime of 2.5 μs gives a value of $\frac{B(E2)_{exp}}{B(E2)_W} = 1.2$, again suggesting a rather pure E2 transition.

The 952 keV transition decaying out of the 6540 keV level is not the same transition as the 950 (952) keV transition decaying oput of the 16$^+$ level observed by Wu et al. [6] and Hamilton [5] respectively. This is because the 6540 keV transition decays by a 952 keV gamma ray and a 820 keV gamma ray, with almost equal intensity. The 820 keV gamma ray was not observed by Hamilton or Wu, and this level is likely to be a particle(s)-hole(s) excitation, as shown by the purity of the E2 transition. Combining all these facts allows a spin of 15$^-$ to be proposed for the 6540 keV level. Hence a spin of 17$^-$ is proposed for the isomer. This isomer is a quite pure single particle excitation and is near-yrast, which is unexpected.

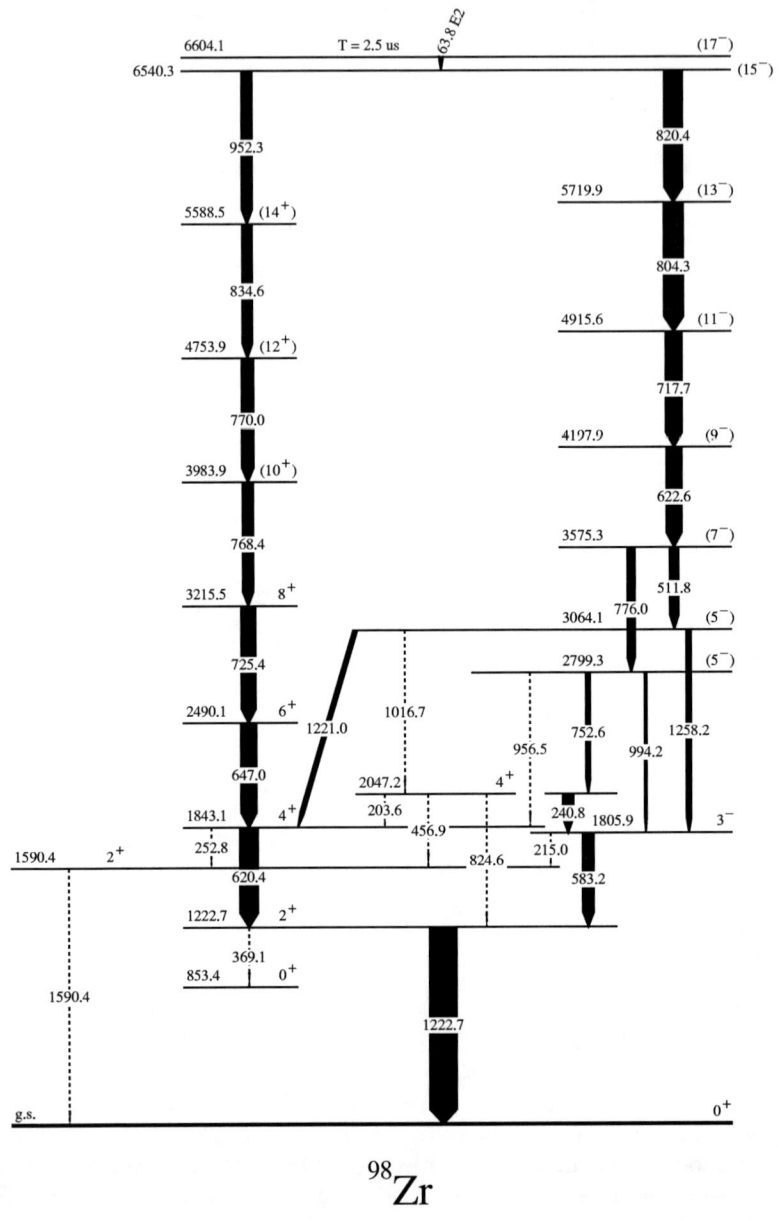

FIGURE 4. Level scheme of the 2.5 μs, 6.6 MeV, spin 17⁻ isomer in ^{98}Zr.

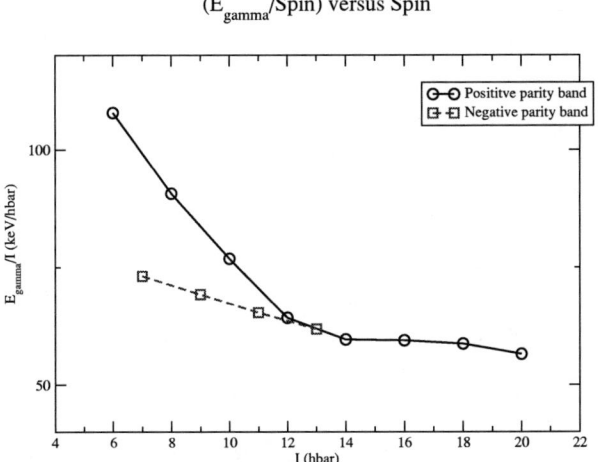

FIGURE 5. E-GOS versus spin, for the two collective bands observed below the isomer in ^{98}Zr.

NUCELAR STRUCTURE EXPERIMENTS USING NEUTRON GUIDES AND FISSION

In March 2005 an experiment was performed to search for isomeric states in the lifetime range 100 ns - 2 μs. This experiment was performed at the PF1B neutron guide, which has a thermal-equivalent flux of 1.3×10^{10} n/cm^2/s. Neutron guides have a very low component of fast neutrons, enabling germanium detectors to be placed in close proximity. A thin target of $\sim$10 mg of ^{235}U was placed in the neutron beam and one arm of the FiFi spectrometer [7] was placed close to the target, and perpendicular to the neutron beam, see figure 6. Opposite the FiFi spectrometer was placed an evacuated tube $\sim$0.7 m in length, containing a stopper foil. Around this stopper foil was placed an array of seven 60 % germanium detectors, a Si polarimeter and a Euroball germanium Cluster. Any fission fragments entering the FiFi spectrometer were identified by mass, using a combination of time-of-flight and energy deposited in a Bragg ionisation chamber. This allowed limits to be set on the mass of complementary fission fragments stopping in the foil of the evacuated tube, which was at the center of the gamma-ray detector array. Preliminary analysis of this data has shown that known microsecond isomers can be identified (for example in ^{134}Te and ^{136}Xe) and it is hoped further analysis will identify many new isomeric states in neutron-rich nuclei.

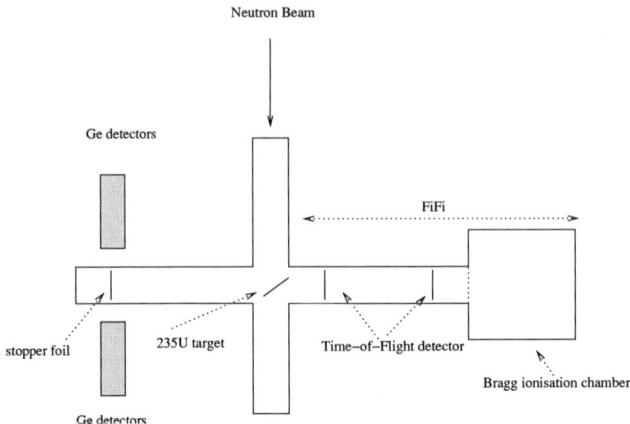

FIGURE 6. Schematic drawing of the experiment at the PF1B neutron guide using the FiFi spectrometer.

CONCLUSION

Even after 30 years of operation the Lohengrin mass spectrometer is still producing interesting physics, thanks to the recent upgrade of the gamma-ray detection efficiency. Other techniques to examine various properties of the structure of nuclei have also been introduced recently, including fast timing and g-factor measurements. A high-spin isomer has recently been observed in ^{98}Zr, at 6.6 MeV and spin (17^-), which is single-particle in nature. This isomer decays into two collective (vibrational) bands. In addition to ^{98}Zr new results from the studies of isomeric states in the mass 100 and ^{132}Sn region, using the the Lohengrin spectrometer, have been presented elsewhere at this conference in articles by J. Genevey and J.A. Pinston.

REFERENCES

1. J. Genevey, R. Guglielmini, R. Orlandi, J. Pinston, A. Scherillo, G. Simpson, I. Tsekhanovich, and N. Warr, *submitted to Phys. Rev. C* (2005).
2. J. Genevey, F. Ibrahim, J. Pinston, H. Faust, T. Friedrichs, M. Gross, and S. Oberstedt, *Phys. Rev. C*, **59**, 82–89 (1999).
3. H. Mach, R. Gill, and M. Moszynski, *Nucl. Instr. Meth. A*, **280**, 49 (1989).
4. W. Urban, J. Durell, A. Smith, W. Phillips, M. Jones, B. Varely, T. Rzaca-Urba, I. Ahmad, L. Morss, M. Bentaleb, and N. Schulz, *Nucl. Phys. A*, **689**, 605–630 (2001).
5. J. Hamilton, A. Ramayya, Z. S.J., G. Ter-Akopian, Y. T. Oganessian, J. Cole, J. Rasmussen, and M. Stoyer, *Prog. Part. Nucl. Phys.*, **35**, 635–704 (2001).
6. C. Wu, H. Hua, D. Cline, R. Teng, R. Clark, P. Fallon, A. Goergen, O. Macchiavelli, and K. Vetter, *Phys. Rev. C*, **70**, 064312 (2004).
7. C. J. Pearson, B. J. Varley, W. R. Phillips, and J. L. Durell, *Rev. Sci. Instrum.*, **66**, 3367–3373 (1995).

Neutron-Rich In and Cd Isotopes Close to the Doubly-Magic ^{132}Sn

A. Scherillo*,§, J. Genevey¶, J.A. Pinston¶, A. Covello#, H. Faust*, A. Gargano#, R. Orlandi*,¥, G.S. Simpson*, I. Tsekhanovich*

*Institut Laue-Langevin, B.P. 156, F-38042 Grenoble Cedex 9, France
§Institut für Kernphysik, Universität zu Köln Zülpicherstr.77, D-50937 Köln Germany.
¶ Laboratoire de Physique Subatomique et de Cosmologie, IN2P3-CNRS/Université Joseph Fourier, F-38026 Grenoble Cedex, France
#Dipartimento di Scienze Fisiche, Università degli studi di Napoli Federico II and INFN, Complesso Universitario di Monte Sant'Angelo, via Cintia, I-80126 Napoli, Italy
¥Department of Physics and Astronomy, University of Manchester, Brunswick Street, M13 9PL, United Kingdom

Abstract. Microsecond isomers in the In and Cd isotopes, in the mass range A = 123 to 130, were investigated at the ILL reactor, Grenoble, using the LOHENGRIN mass spectrometer, through thermal-neutron induced fission reactions of Pu targets. The level schemes of the odd-mass $^{123-129}$In are reported. A shell-model study of the heaviest In and Cd nuclei was performed using a realistic interaction derived from the CD-Bonn nucleon-nucleon potential.

PACS: 21.10.Tg, 23.20.Lv, 25.85.Ec, 27.60.+j

INTRODUCTION

Experimental progress is currently being made in the region around doubly-magic ^{132}Sn. However, nuclear structure information is more complete for nuclei above the Z=50 shell-closure [1] than for the nuclei below the shell-closure (such as the Cd and In isotopes), which are much more difficult to produce. In the present work we searched for and studied the decay of µs isomers in the neutron-rich mass A = 123 to 130 nuclei with the LOHENGRIN spectrometer at the ILL reactor in Grenoble. The aim was to complete the previous data on the heavy Cd and In isotopes.

The low-spin levels up to 13/2 in $^{123-127}$In were previously investigated from the β-decay of Cd isotopes [2,3], and, very recently, high-spin ms isomers in $^{125-129}$In were discovered [4,5]. Krautzsch et al [6], have also obtained some spectroscopic information on the $^{126-128}$Cd isotopes. Preliminary reports were presented by Hellström et al [7,8] on the search for µs isomers in the heavy Cd and In isotopes at the FRS spectrometer at GSI, but no level schemes were proposed.

EXPERIMENT AND DISCUSSION

Our new results on odd-In isotopes have been published recently [9,10]. In ^{127}In we have observed a 9(2) µs isomer which decays by a cascade consisting of a strongly-converted E2 transition of 47 keV, and two γ-rays of 221 and 233 keV in coincidence one with the other (fig. 1,2). The two γ-rays were the same as those first observed by Hellström et al [7,8], and the reported value of the half-life 13(2) µs is in rough agreement with that found in this work. The coincidence spectra presented on fig. 1

was obtained with the two new clover detectors and the new chamber. It shows the improvement of the γ-efficiency of our setup.

The Si(Li) spectrum obtained in coincidence with any of these two lines shows the characteristic indium X-rays, and the K and L conversion electrons of the isomeric transition. The multipolarity of the transition is E2, and B(E2)=0.3 W.u. was deduced.

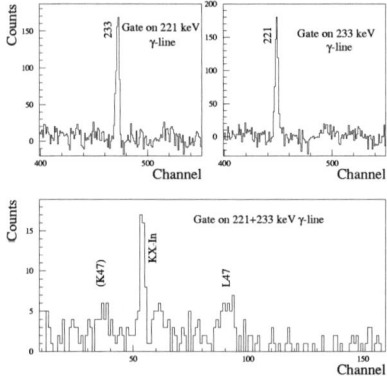

FIGURE 1. Coincidence spectra gated on the γ-ray of 233 or 221keV and Si(Li) spectrum obtained in coincidence with the γ-rays of 221 and 233keV.

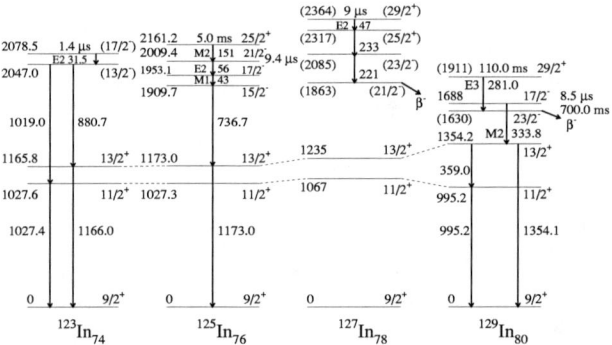

FIGURE 2. Level schemes of ^{123,125,127,129}In. The level energies in parenthesis are deduced from β-spectra [5] have large uncertainties.

The level schemes of $^{123-129}$In shown in fig.2 are the result of the synthesis of different works: the ms isomer experiments performed at the OSIRIS mass separator [4,5], and the μs isomer experiments performed with the FRS at GSI [7,8] and the LOHENGRIN spectrometer[1,9,10]. All the reported levels are in the vicinity of the yrast line, therefore the low-spin levels fed in previous works by β-decay experiments are not shown in fig. 2.

The heavy In and Cd nuclei, with neutron and proton holes inside the ^{132}Sn core, are characterized by the presence of two high-spin states, $\pi g_{9/2}^{-1}$ and $\nu h_{11/2}^{-1}$ at low excitation energy. The very strong p-n interaction in the ($\pi g_{9/2}^{-1} \nu h_{11/2}^{-1}$) configuration

146

is expected to produce very perturbed yrast line and give rise to long-lived high spin isomers.

We have performed calculations to test the ability of the shell model to describe the heavy Cd and In isotopes, with proton and neutron holes outside the ^{132}Sn core. In this work, a realistic effective interaction derived from the CD-Bonn nucleon-nucleon potential [11] is used. Similar calculations were performed in Ref. [12] for nuclei with proton particles and neutron holes around ^{132}Sn, and in Ref. [9] for ^{129}In. In both cases good agreement with the experimental data was found.

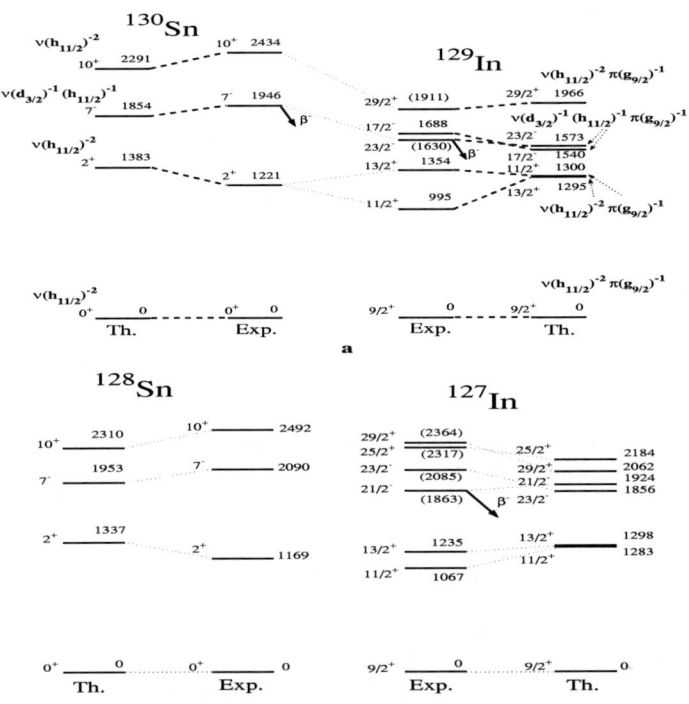

FIGURE 3. Experimental and calculated energies for ^{129}In and ^{130}Sn (a), and for ^{127}In and ^{128}Sn (b).

In Fig. 3a experimental levels of ^{129}In and ^{130}Sn are shown together with the calculated ones. For ^{129}In all the experimental levels, except the 1/2⁻ at 369 keV, are reported, while only some selected yrast level of ^{130}Sn are shown. The dominant configurations of all these levels are also indicated. The excitation energies of ^{130}Sn are rather well reproduced by the shell-model calculations. However, it is interesting to note that the first 2⁺ state is overestimated by the calculation by 162 keV. This is a common feature for this state in this region and it is probably an effect of the truncation of space. The experimental levels of ^{129}In are expected to result from the coupling of a $\pi g_{9/2}^{-1}$ hole with the reported two neutron hole states of ^{130}Sn.

The comparison of the experimental levels in Fig. 3b shows that the 29/2⁺ and 23/2⁻ in ^{127}In are closer to the 10⁺ and 7⁻ in ^{128}Sn respectively, than in Fig. 3a. This effect could be explained by a decrease of p-n interaction from ^{129}In to ^{127}In. Another feature, possibly related to the effects of the p-n interaction is the inversion of the 29/2⁺ and 25/2⁺ and 23/2⁻ and 21/2⁻ levels, respectively, in the calculated spectrum of ^{127}In.

In the vicinity of the two closed shells of ^{132}Sn the µs isomers are very abundant and disappear rapidly far from them [1]. However, below Z=50 they disappear suddenly for the Cd isotopes, no isomers having been identified up to now in the even-mass ones. The calculations for ^{126,128}Cd predict short half-lives (~10 ns) for the 8^+ states, which could explain why they have not been observed in the present work. In ^{130}Cd no excited levels are experimentally known up to now. The shell model calculations predict an isomer 8^+ of 0.6 µs, very close to the LOHENGRIN detection limit. The low yield estimate for this element makes a measurement very difficult.

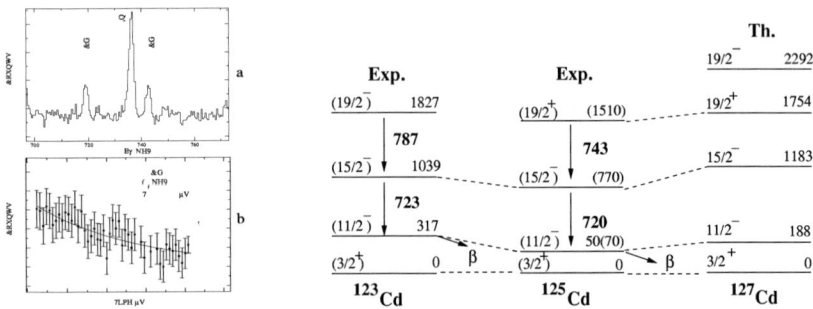

FIGURE 4. γ-spectrum obtained in delayed coincidence with mass A=125 (a). Time spectrum of ^{125}Cd(b). Experimental levels of ^{125}Cd in comparison with ^{123}Cd and a theoretical calculation of ^{127}Cd.

In the odd-nuclei, only the isomer of ^{125}Cd, which decays by two γ-lines of the same intensity, as seen in Fig. 4, is known. Its half-life is 19(2) µs, in rough agreement with the value reported previously [7,8]. A tentative level scheme is proposed by analogy with its neighbour ^{123}Cd and a shell model calculation on ^{127}Cd.

CONCLUSION

The present µs isomeric γ rays and conversion electron measurements of fission products have allowed us to enrich the level schemes of the In isotopes in the mass range A = 123 to 129. The realistic shell-model calculations provide a satisfactory interpretation for ^{129}In, but some discrepancies occur for ^{127}In.

REFERENCES

1. Pinston, J.A and Genevey, J., *J. Phys.* **G 30**, R57 (2004)
2. Hoff, P.,et al., *Nucl. Phys.* **A459**,35 1986).
3. Huck, H., et al., *Phys. Rev.* **C39**, 997 (1989).
4. Fogelberg, B., et al., *Proc. 2nd Intern. Workshop on Fission and Fission-Product Spectroscopy (Seyssins, France)* (AIP Conf. Proc 447) Ed. by G. Fioni, H. Faust, S. Oberstedt, F. Hambsch, p. 191
5. Gausemel, H., et al., *Phys. Rev.* **C69**, 054307 (2004).
6. Krauzsch, T., et al., *Eur. Phys. J. A* **9**, 201(2000).
7. Hellström, M., et al., *Proc. 3rd Conf. on Fission and Properties of Neutron-Rich Nuclei (Sanibel Island,FL)* 2002, Ed. by J. Hamilton , A. Ramayya, H. Carter (Singapore: World Scientific), p.22.
8. Hellström, M., et al., *Proc. of Intern. Workshop XXXI on Gross Properties of Nuclei and Nuclear Excitations (Hirscheg, Austria)* Ed. by H. Feldmeier, GSI 2003, p.72.
9. Genevey, J., et al., *Phys. Rev.* **C67**, 054312 (2003).
10. Scherillo, A., et al., *Phys. Rev.* **C70**, 054318 (2004), and PhD Thesis, Cologne, Germany 2005.
11. Machleidt, R., et al., *Phys. Rev.* **C63**, 024001 (2001)
12. Corragio, L., et al., *Phys. Rev.* **C66**, 064311 (2002).

Shape Coexistence In Odd And Odd-Odd Nuclei In The A~100 Region

J. A. Pinston[1], J. Genevey[1], G. Simpson[2], and W. Urban[3]

[1]*Laboratoire de Physique Subatomique et de Cosmologie, F-38042 Grenoble Cedex, France*
[2]*Institut Laue-Langevin, F-38042 Grenoble Cedex, France*
[3]*Institute of Experimental Physics, Warsaw University Ul. Hoza 69, PL-OO-681 Warazawa, Poland*

Abstract. In the even-even nuclei around $A=100$ a transition from spherical to deformed shapes occurs from $N=58$ to $N=60$. The isotones with $N=59$ are of special interest, because they are just at the border between the two regions. Very recently, we have studied odd-neutrons and odd-odd nuclei with $N=59$, by means of prompt γ-ray spectroscopy of the spontaneous fission of ^{248}Cm, using the EUROGAM 2 multi-detector, and by measurements of μs isomers produced by fission of ^{239,241}Pu with thermal neutrons at ILL (Grenoble). In the latter case, the detection is based on time correlation measurements between fission fragments detected by the LOHENGRIN mass spectrometer and γ-rays or conversion electrons from the isomer decay. It was found that three shapes coexist in the odd ^{97}Sr and ^{99}Zr and two shapes coexist in the odd-odd ^{96}Rb. A simple explanation of the shape-coexistence mechanism is proposed. It is based upon the Nilsson diagram and stresses the fundamental importance of the unique parity states.

Keywords: Exotic nuclei, Shape coexistence.
PACS: 23.20.Lv, 25.85.Ca, 25.85.Ec, 21.60.Ev

INTRODUCTION

The region of neutron-rich nuclei near $A=100$ is distinctive for the sudden change in the ground-state (*g.s.*) properties of nuclei. In particular, for the even ^{98}Sr and ^{99}Zr isotopes a sudden onset of strong deformation at $N=60$ is observed, whereas the lighter isotopes up to $N=58$ are rather spherical. Consequently, the isotones with $N=59$ are of special interest because they are just at the border between the two regions. A good knowledge of the spectroscopic properties of these nuclei should allow a better understanding of the origin of the deformation and the nature of the shape-coexistence phenomena in this mass region. Unfortunately, the $N=59$ isotones of interest are far from the stability line and are rather difficult to study. Consequently, for several decades, only β-decay experiments were performed and only low-spin states were measured. The main progress in this mass region was recently obtained by Urban *et al.* [1] who was able to observe for the first time well developed rotational bands in ^{97}Sr and ^{99}Zr. In this work, the $N=59$ isotones were produced in the spontaneous fission of ^{248}Cm and the prompt γ-rays were measured using the EUROGAM 2 Ge array. The level scheme of ^{99}Zr is shown in figure 1. These data confirmed that the *g.s.* and first two excited states are the neutron $s_{1/2}$, $d_{3/2}$ and $g_{7/2}$ shell model levels, but in addition, two well developed rotational bands, based upon the intrinsic configurations $v[411\ 3/2^+]$ and $v[541\ 3/2^-]$, were also found at about 600 keV excitation. The observation of these regular bands has firmly established the shape coexistence in these two $N=59$ isotones. More interesting, the quadrupole moments of these two bands were also measured and a mean deformation of $\beta_2=0.32(2)$ was deduced. It is interesting to note that this value is well below the deformation measured for the g.s. band in ^{98}Sr ($\beta_2=0.41(2)$) [1], which is expected to be the maximum deformation in this region.

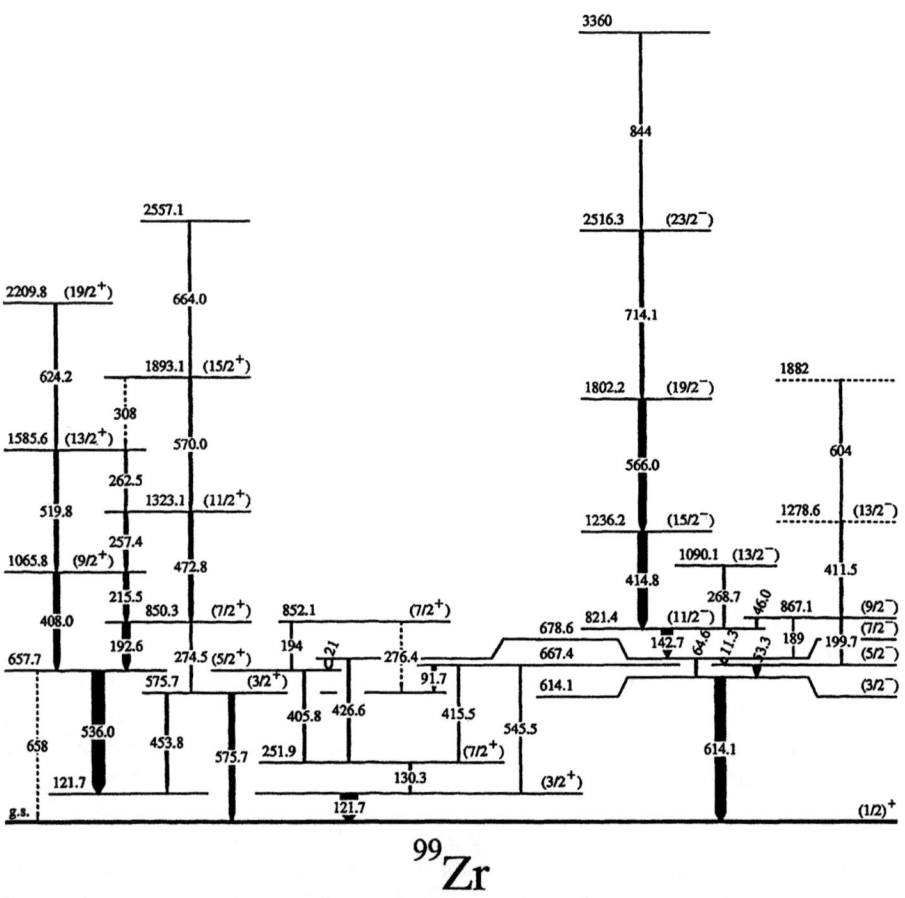

FIGURE 1. Level scheme of ^{99}Zr from EUROGAM 2 experiment. The *g.s.* and two first excited states are spherical, while two rotational bands based on $v[411\ 3/2^+]$ and $v[541\ 3/2^-]$ configurations, respectively, are present at about 500 keV excitation (from [1]).

This first success triggered a more complete study of the nuclei of this mass region. For this purpose, a combination of two different techniques was used. In the first one, the nuclei of ^{99}Zr, ^{97}Sr and ^{101}Zr, were produced by spontaneous fission of ^{248}Cm and the prompt γ rays were detected with the EUROGAM 2 array, while in the second one, microsecond isomers in ^{95}Kr, ^{97}Sr and ^{96}Rb, were produced by fission of ^{239,241}Pu and studied with the LOHENGRIN spectrometer at ILL. In the last experiments, the detection is based on time correlation between fission fragments selected by the spectrometer and the γ rays and conversion electrons from the isomers. More details on this experimental setup can be found in Ref. [2, 3].

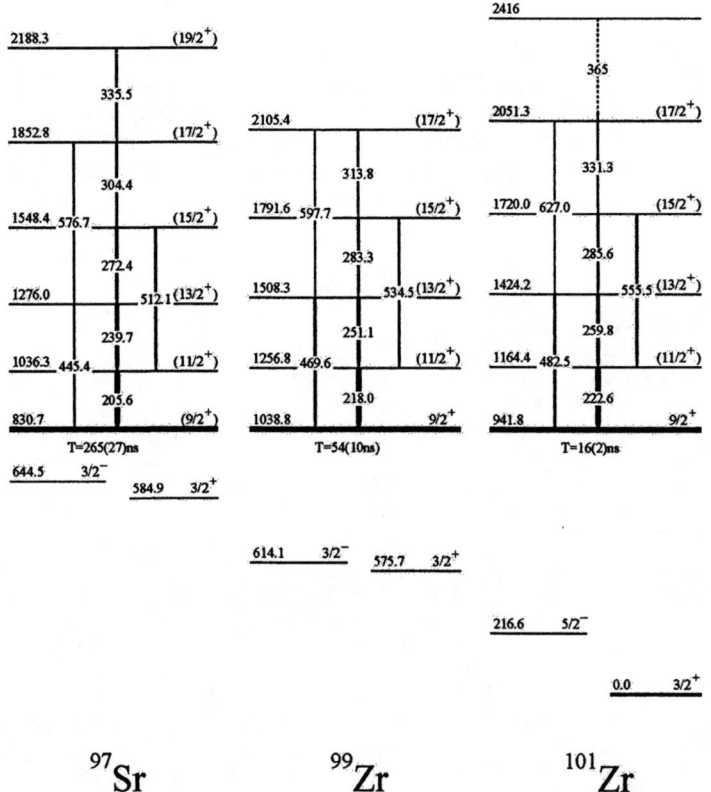

FIGURE 2. The level schemes of the strongly deformed [404 9/2$^+$] bands in ^{97}Sr, ^{99}Zr and ^{101}Zr. The levels below, mark the position of the other intrinsic configurations of moderate deformations (from [5]).

THE [404 9/2$^+$] BAND IN ^{99}ZR, ^{97}SR AND ^{101}ZR

A third rotational band was latter observed in ^{99}Zr from the data of EUROGAM 2 experiments [4]. The band head at 1038.8 keV excitation is a K isomer with a half life $T_{1/2}$=54 ns, which decays to several lower energy states. The spin and parity assignment of the band g.s., I^π=9/2$^+$, was derived from the intensities of the various partial branching ratios and angular-correlation measurements. For the first time, the [404 9/2$^+$] band was observed in this mass region. Soon after, this band was also observed in ^{97}Sr [5,6,7] and ^{101}Zr[5]. In these three nuclei, the band head is a K isomer. However, for ^{97}Sr, where the half life was re-measured very recently, the new value found (526(13) ns) was substantially longer than the one reported by Hwang et al. (265(27) ns) [6].

In figure 2, the level schemes of the three [404 9/2$^+$] bands are shown and the intrinsic excitations of the 3/2$^+$, 3/2$^-$ and 5/2$^-$ levels, of moderate deformation, are also indicated [5]. The intraband-transition energies show very strong similarities in these three nuclei and the energy spacings in these bands have a strongly coupled character. The quadrupole moments Q_0 deduced from the experimental γ branching ratio ΔI=1 to ΔI=2 for intraband transitions lead to a mean deformation of β_2=0.41(3), for these three nuclei.

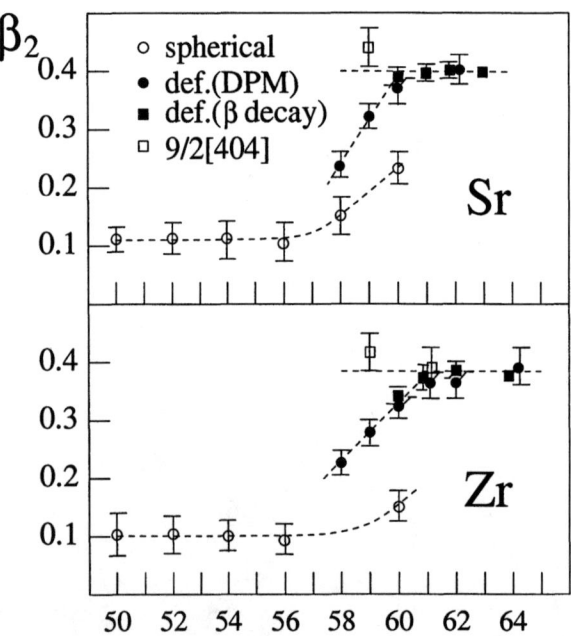

FIGURE 3. Systematic of deformations for various configurations in Sr and Zr isotopes. For N=59, isotones three shape coexist simultaneously. Above N=59, the deformation saturates at a value β_2=0.41. Note that this value is comparable to the deformation found for the [404 9/2$^+$] band in ^{97}Sr, ^{99}Zr and ^{101}Zr (from [5]).

SHAPE COEXISTENCE IN $N=59$ ISOTONES

In figure 3, the β_2 values for the three known [404 9/2$^+$] bands in ^{99}Zr, ^{97}Sr and ^{101}Zr are compared with the values found for the other bands in strontium and zirconium isotopes in $A=100$ mass region [5]. This picture shows that three shapes coexist for $N=59$ isotones. In these nuclei, the ground state and first two excited states are the neutrons $s_{1/2}$, $d_{3/2}$ and $g_{7/2}$ shell-model levels. At about 600 keV, two rotational bands, based upon the ν[411 3/2$^+$] and ν[541 3/2$^-$] configurations are present and have a mean deformation $\beta_2=0.32(2)$. In contrast, the [404 9/2$^+$] bands at about 1 MeV are strongly deformed and their deformations are comparable with the maximum value found in this mass region $\beta_2=0.41(2)$. This large value is observed for several even and odd Sr and Zr nuclei above $N=59$ and is remarkably constant.

The two-neutron unique-parity states, $\nu g_{9/2}$ of the $N=4$ shell and $\nu h_{11/2}$ of the $N=5$ shell, play a dominant role in the mechanism of shape coexistence observed in this region. Figure 4 shows a portion of the Nilsson diagram, around $N=58$ region [5]. It is interesting to note that the occupation of the [404 9/2$^+$] orbital, with its strong upwards slope, will favor a spherical equilibrium shape, whereas the state originating from the $g_{9/2}$ level slopes downward and thus favors deformation. For the moderate deformation bands at $N=59$, the odd unpaired neutron has the configurations ν[541 3/2$^-$] or ν[411 3/2$^+$]. Consequently, in both bands the core has the same structure with one pair in the deformation-driving ν[550 ½$^-$] orbital and the other pair in the up-sloping ν[404 9/2$^+$] orbital, working against deformation. This configuration produces the moderate deformation measured in ^{99}Zr and ^{97}Sr nuclei.

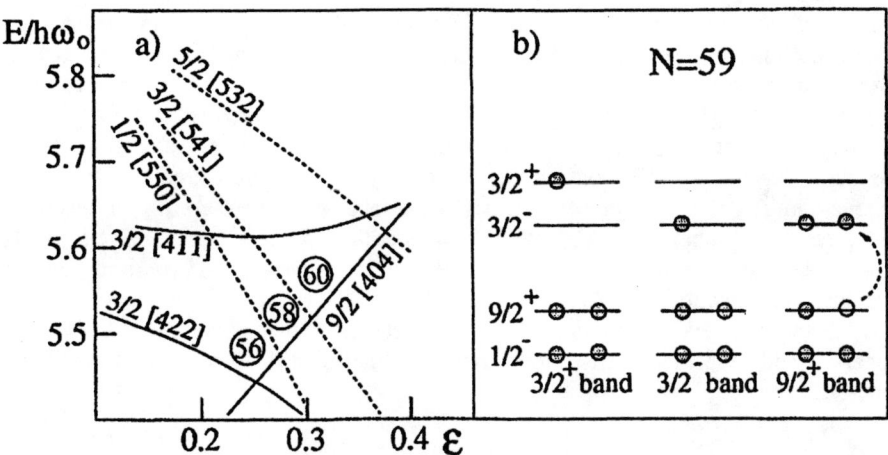

FIGURE 4. a) Portion of the Nilsson neutrons levels, b) Schematic representation of the deformed configurations in Sr and Zr isotopes (from[5]).

Promotion of one neutron of the ν[404 9/2$^+$]^2 pair into a ν[541 3/2$^-$] orbital changes dramatically the situation. In this case, the core has four neutrons occupying the down sloping $\nu[550\ 1/2^-]$ and $\nu[541\ 3/2^-]$ orbitals while the $\nu[404\ 9/2^+]$ is completely depleted. Now, the neutron spectator orbiting around the core is in the $\nu[404\ 9/2^+]$ orbital. In this mass region, this configuration produces the deformation limit, already at $N=59$. A comparable mechanism was proposed a long time ago by Kleinheinz et al., [8,9] to explain shape coexistence in the lanthanide region. In this case, the unique parity neutron states $\nu h_{11/2}$ and $\nu i_{13/2}$ play a role analogous to the one of $\nu g_{9/2}$ and $\nu h_{11/2}$ in the $A=100$ region. However, it is interesting to note that the effect is simpler in the latter region because just a fewer levels are active and the deformation change is also more dramatic.

SHAPE COEXISTENCE IN ODD-ODD NUCLEI

To increase nuclear-structure information in the $N=59$ isotones, the odd-odd nuclei were also investigated. For this reason, ^{98}Y was recently revisited by Brant et al. [10] and we have also reinvestigated ^{96}Rb. The latter nucleus was previously measured with the LOHENGRIN spectrometer [11]. However, the efficiency of the γ detection was too weak to build a reliable level scheme. More recently, this efficiency was strongly improved and the γ and conversion electrons de-exciting the isomer were studied. The new level scheme is shown in figure 5. This very neutron-rich nucleus has a structure comparable to the previously known ^{98}Y. Both nuclei show rather spherical levels at low energy, while deformed states appear at about 500 keV excitation. The $(\pi[431\ 3/2^+]\ \nu[541\ 3/2^-])3^-$ and $(\pi[422\ 5/2^+]\ \nu[541\ 3/2^-])4^-$ intrinsic configurations were found for ^{96}Rb and ^{98}Y nuclei, respectively. All these neutron and proton orbitals originate from the $\pi g_{9/2}$ and $\nu h_{11/2}$ spherical unique parity states. A lower limit, $\beta_2>0.28$, was deduced for these two nuclei, from the experimental γ branching ratio $\Delta I=1$ to $\Delta I=2$ for intraband transitions. Moreover, the comparable behavior observed for the decay of the isomer in both nuclei, suggests that they have the same $(\pi g_{9/2}\ \nu h_{11/2})10^-$ configuration. Consequently, in the same nucleus, these unique-parity states are present in spherical and deformed configurations. The presence of a spherical yrast trap in competition with spherical levels is the consequence of a strongly attractive n-p interaction, because n and p are in coplanar orbitals. Very recently, we observed an $I^\pi=17^-$ isomer in ^{98}Zr [12], of spherical origin and about 6.5 MeV excitation. This result shows that the competition between spherical and deformed states, along the yrast line, is still active at high excitation energy.

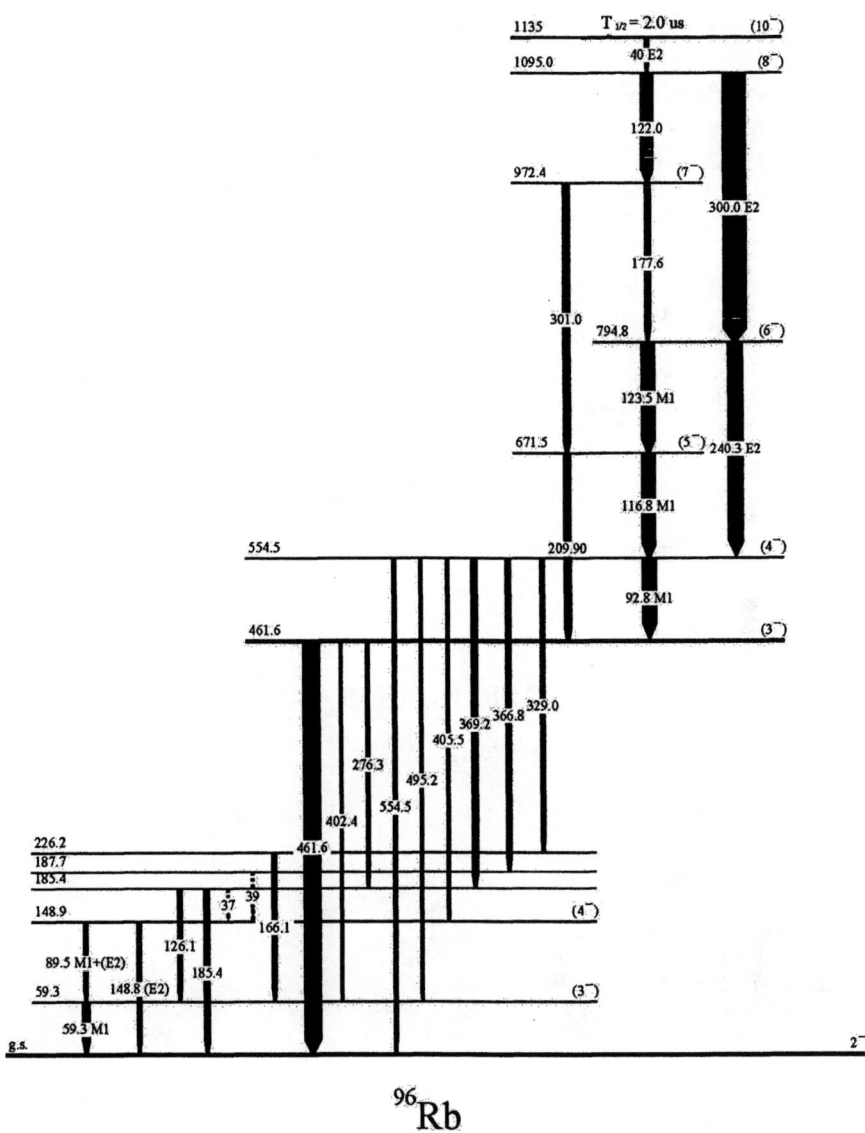

FIGURE 5. Level scheme of the 2.0 μs isomer in ^{96}Rb. The *g.s.* with spin and parity $I^\pi=2^-$, the low-lying levels and the 1135 keV isomer have rather spherical configurations, while a well developed rotational band is seen at 461.6 keV excitation.

CONCLUSIONS

In this work odd and odd-odd $N=59$ istones were investigated, using a combination of two experimental techniques. It is now well established that three shapes coexist in odd ^{97}Sr and ^{99}Zr, while two different shapes were seen in odd-odd ^{98}Y and ^{96}Rb. These new data demonstrate that the spectroscopy of odd-A and odd-odd nuclei, provides much more information on the structure of the different shapes than even-even nuclei. The theoretical interpretation of these collective excitations is based on the Nilsson diagram in the $N=58$ region and it shows that the spherical unique-parity state plays a very important role in the shape coexistence mechanism. In conclusion, a great wealth of nuclear structure information was recently gained for these odd and odd-odd $N=59$ isotones and we hope that these new results will trigger new calculations for this mass region.

ACKNOWLEDGMENTS

We thank Pr. Asghar for stimulating discussions and a careful reading of the manuscript.

REFERENCES

1. Urban W. et al., Nucl. Phys. A **689,** 605 (2001).
2. Pinston J. A. and Genevey J., J. Phys. G **30**, R57 (2004).
3. Genevey J. et al., Eur. Phys. J. A **9,** 191 (2000).
4 Urban W. et al., Eur. Phys. J. A **16,** 11 (2003).
5. Urban W. et al., Eur. Phys. J. A **22,** 241 (2004).
6. Hwang J.K. et al., Phys. Rev. C **67**, 054304 (2003).
7. Wu C. Y. et al., Phys. Rev. C **70**, nn064312 (2004).
8. Kleinheinz P. et al., Phys. Rev. Lett. 32, 68 (1974).
9. Kleinheinz P. et al., Nucl. Phys. A 283, 189 (1977).
10. S. Brant et al., Phys. Rev. C **69**, 034327 (2004).
11. Genevey J. et al., Phys. Rev. C **59**, 82 (1999).
12. Simpson G. et al., talk presented at this conference.

Disposition of Legacy Materials at LBNL and Reuse of Valuable Items in Target Preparation

Ilham AlMahamid[1], Dawn A. Shaughnessy[2], Ralf Sudowe[1]

1: Lawrence Berkeley National Laboratory, 1Cyclotron Road, Berkeley, California 94720, USA
2: Lawrence Livermore National Laboratory, 7000 East Avenue, Livermore, California 94550, USA

Abstract. One of the most demanding missions of the United States Department of Energy (DOE) is to manage the radioactive waste. Lawrence Berkeley National Laboratory (LBNL) gets its share of radioactive materials and waste but at a much smaller scale than other national labs. Disposition of legacy materials at LBNL includes characterization and purification, reuse of valuable materials, and submitting waste items in accordance with federal and state regulations. During the clean up of legacy materials at LBNL, several items were identified for use in collaboration with the CEA/ DSM under the Mini-Inca program to perform transmutation studies. Cm and Np targets were prepared using an electroplating technique. Additional targets of Pu, Am, and Cf are planned for the next set of experiments.
An overview of the clean up operation of two facilities will be given. An outline of items assigned to the collaboration and a brief description of the target preparation and electrodeposition set-up will be provided.

Keywords: Legacy Materials, LBNL, Transmutation, Electroplating.

INTRODUCTION

Over the past five years, LBNL conducted a significant effort to deal with radioactive materials that were generated by research projects during the 1950's and the 1960's (Banghart and Rothermich, 2005). These radioactive materials were categorized as "Legacy Materials". An inventory of the Legacy Materials at the Department of Energy (DOE) facilities is described in reference 2. In this paper we will outline the clean up effort for two facilities: the Heavy Element Research Laboratory (HERL) and the Environment, Health and Safety Storage Facility. We disposed of over 1000 items within these two facilities. This operation included characterization and purification of samples, reuse of valuable materials, and submission of waste items to the Hazardous Waste Handling Facility (HWHF) in accordance with federal and state regulations. We found reuse options for more than 250 items. The majority of them were shipped within the United States. A very important aspect of working with legacy materials was to coordinate efforts of various groups at LBNL to include research groups, HWHF, Analytical Services, Health Physics, Radiation Protection, and the shipping department. In some cases, we hired outside contractors to deal with some of the hazardous materials found such as reactive gases.

DESCRIPTION OF LEGACY MATERIALS

Due to the varieties of research projects conducted at LBNL, the legacy materials found were manifold in types and radioactivity levels. We found old, custom-made gas cylinders containing hazardous compounds, contaminated glove boxes, contaminated resins, corroded vials, alpha contaminated metal plates, inorganic and organic complexes of actinide elements, liquid waste, dry active waste packages containing liquid and solid materials, etc. The activity level of the materials found varied from a few pCi to several Ci. 225 items were identified as transuranic waste (TRU). Transuranic waste is defined as radioactive waste containing more than 100 nCi of alpha-emitting transuranic isotopes (Z>92) per gram of waste materials with an isotope half life above 20 years. Radioactive materials were handled in hoods, glove boxes or lead-shielded boxes depending on the radioactivity level.

To dispose of materials we needed to perform radiological and chemical characterizations. While radiological characterization was relatively easy and could be performed at LBNL, chemical characterization posed a big challenge. Many items could not be accepted at the analytical laboratories due to their levels of radioactivity, which exceeded the analytical lab's permit limits. To overcome this chemical characterization issue, a great deal of effort was placed on collecting process knowledge from researchers who handled or were familiar with the legacy materials. Many work sessions were scheduled with retired researchers and technicians who contributed significantly to the process knowledge identification. Figures 1 to 3 show some examples of gas cylinders found in the HERL. Figures 4 to 6 show other waste examples of legacy items.

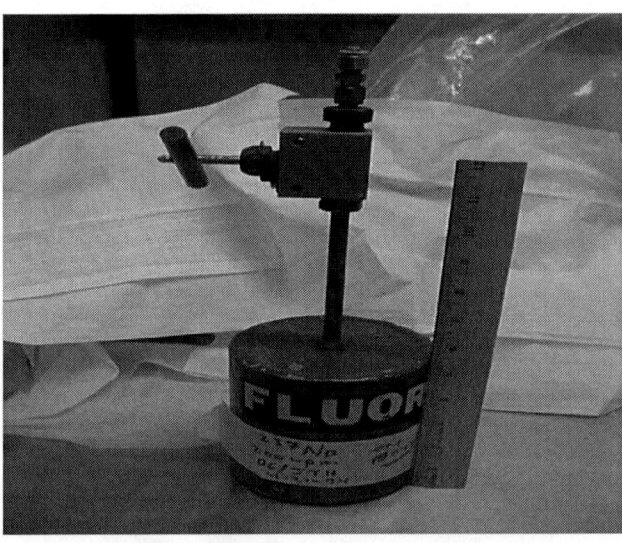

FIGURE 1. Fluorine Cylinder Contaminated with Np-237.

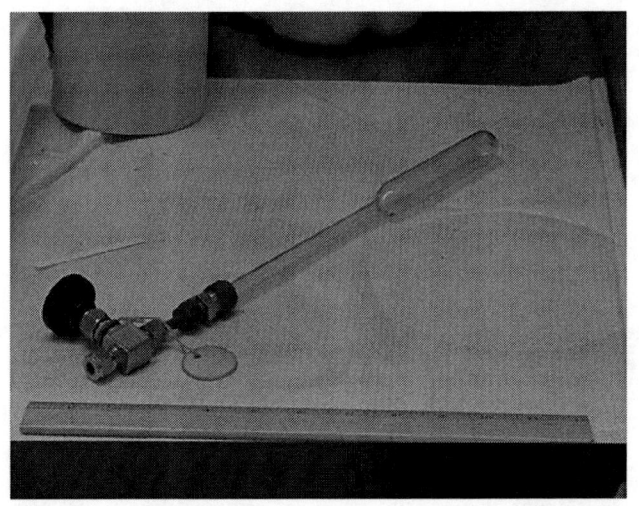

FIGURE 2. Osmium Fluoride, LBNL Custom-Made Cylinder.

FIGURE 3. Tellurium Fluoride, LBNL Custom Made Cylinder.

FIGURE 4. Spilled Liquid from a Corroded Container.

FIGURE 5. Degraded Organic Compounds of Pa-231 in Sealed Tubes.

FIGURE 6. Contaminated Glove Box Containing Hazardous Reactive Compounds.

REUSE OF VALUABLE MATERIALS

About 250 items of valuable materials were transferred to researchers for reuse options. Examples of valuable materials that were reused are shown in figures 7 and 8. The majority of valuable items were sent to other labs and facilities within the USA.

Some of the materials were used in scientific collaboration with the French Atomic Energy Authority (CEA). The initial purpose of this collaboration was to find reuse options for some of the valuable materials found during the Legacy Materials clean-up operation. The collaboration was initiated by Dr. Gabriele Fioni from the CEA in 2000 and consists of preparing actinide targets at LBNL to be used for transmutation studies at the Institut Laue-Langevin (ILL) in Grenoble, France. Later on, Drs. Frederic Marie and Alain Letourneau from the ILL joined the collaboration. Transmutation is a technique in which long-lived isotopes will be converted to short-lived isotopes by neutron capture reactions and neutron-induced fission. Several targets were prepared and are under study such as Cm-243, Np-237 and Th-232. The next set of targets will include Cf, Am and Pu isotopes.

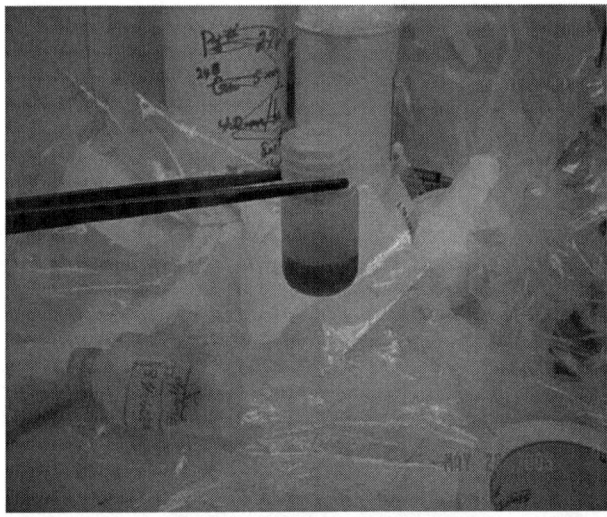

FIGURE 7. 300 mg of Np-237 in Acid Solution.

FIGURE 8. 500 mg of Highly Enriched U-235.

Target Preparation

The actinide targets were prepared by electrodeposition from organic solutions using a technique adapted from Aumann and Müllen (3). The electroplating cell was constructed following a design developed by Dr. Kenneth Gregorich within the Nuclear Science Division, LBNL in 1992.

The experimental set-up consists of a glove box, a high voltage plating supply, an electroplating cell, a heating block, a tube furnace, a stirrer, and a hot plate. Figure 9 shows the glove box in which targets were prepared.

The baking material was made of Ti with a diameter of 12 mm and a thickness of 6 µm. The diameter of the deposit area was 5 mm. Actinide stock solutions were evaporated to dryness in HNO_3. The residue was dissolved in isopropanol. The electrodeposition was carried out at 300-600 V for about 30 min. The backing material was heated at 600 °C to convert the deposit to the oxide form.

FIGURE 9. Electroplating Set-up in the Glove Box.

CONCLUSION

Two facilities at LBNL were cleaned up from Legacy Materials. We disposed of over 1000 items. 250 items were recovered for reuse and were distributed to researchers mainly within the United State. Some of the materials were purified and used as targets for transmutation studies in collaboration with the French Atomic Energy Authority (CEA).

ACKNOWLEDGMENTS

The disposition of legacy materials was performed with the help of Jim Floyd and Bette Muhammad from the Technical Services Group; John Ahlquist, James Haley, Elio Gusti, Suzanne Hargis, and Gerald English from the Material Characterization and Decontamination Group; Maram Kassis, Nancy Rothermich, and Chan Ho Yi from the Hazardous Waste Handling Facility; Dawn Banghart and John Van Wart from the Radiation Protection Group; John Seabury from Industrial Hygiene Group; David Shuh, Jerome Bucher, and Wayne Lukens from the Chemical Sciences Division

REFERENCES

1. D. Banghart and N. Rothermich, "Accountable Material Reduction at the Lawrence Berkeley National Laboratory", 38th Annual Health Physics Society Midyear Symposium, February 2005, New Orleans, LA, USA
2. U.S. Department of Energy, "Linking Legacies: Connecting the Cold War Nuclear Weapons Production Processes to their Environmental Consequences", DOE/EM-0319, January 1997.
3. D.C. Aumann and G. Müllen, *Nuc. Ins. Meth.* **115**, 75-81 (1974).

RESONANCES, BARRIERS, AND FISSION TIMES

Fission Modeling, Data Evaluation, and Integral Data Testing at Los Alamos

Mark B. Chadwick, Phillip G. Young, Robert E. MacFarlane, Patrick Talou, and Toshihiko Kawano

PADNWP Program Office, and T-16, Nuclear Physics Group, Los Alamos National Laboratory

Abstract. We describe recent nuclear model calculations and evaluations of neutron reactions on actinides in the forthcoming ENDF/B-VII US nuclear data library, with particular attention to the uranium isotopes $^{232-241}$U in the keV - 30 MeV energy range. This work makes use of extensive sets of measurements for fission, elastic, inelastic, (n,xn) and capture, as well as fission probability data. Nuclear reaction model calculations were performed for the whole suite of uranium isotopes to allow us to take advantage of the systematical properties from isotope-to-isotope, which is especially useful for nuclides where few measurements exist. In addition to improving the neutron cross sections and energy-angle distributions, new prompt fission neutron spectra and prompt/delayed neutron multiplicity evaluations are included for several isotopes. We also give some examples of integral data testing using MCNP simulations of critical assemblies, for both k-eff calculations and fission, n2n, and capture reaction rate calculations.

INTRODUCTION

The US will be issuing a new evaluated nuclear data library, ENDF/B-VII, at the end of 2005. This new database includes many improvements: improved actinide cross sections; photonuclear evaluations; new high-energy evaluations for many materials; and improved fission product data. In this work we discuss the actinide improvements, using the uranium isotopes as an illustrative example.

We have carried out a program of systematic analysis of the nuclear reaction data on uranium isotopes from A = 232 to 241. New or revised evaluations of neutron data files for all the uranium isotopes have been completed covering the neutron energy range 10^{-5} eV to 30 MeV. The evaluations are based on experimental data and theoretical model calculations, combined with existing resonance-parameter analyses. The isotopes with extensive experimental databases result in evaluations closely linked to measurements (^{233,235,238}U). Several of the isotopes ($^{232, 234, 236}$U) have weaker experimental databases, and a few evaluations are based almost entirely on theoretical calculations and/or systematics (^{237,239,240,241}U). Reports describing the ^{233}U, ^{232,234}U, and the ^{238}U evaluations have been published.[i,ii,iii]

Major features of this analysis include: systematic accumulation of all relevant experimental data; re-normalization of the neutron data to modern standards; assessment of the applicability of several recent optical model potentials for actinide calculations; interpretation of experimental results in terms of nuclear theory to allow interpolation and extrapolation of the data into unmeasured regions; and assemblage of the experimental and theoretical results into formal evaluated nuclear data files that can be processed for use in applied nuclear programs. In the course of the analysis, the standard ^{235}U(n,f) cross section was updated for new experimental data and applied to all the U-isotope evaluations, as well as to ^{237}Np and ^{239}Pu evaluations. All the work described here will be incorporated into Version VII of ENDF/B.

THEORY AND MODELS

There were several objectives for our theoretical analysis. To provide unified descriptions of reactions for all the uranium isotopes, we needed consistent analyses over the energy range 10 keV to 30 MeV of the measurement-rich systems in order to infer model parameters for the unmeasured systems. We relied on the theoretical calculations for all the isotopes for energy-angle correlated emission spectra, for elastic and inelastic neutron angular distributions, for $(n,n'continuum)$ cross sections, and in most cases for (n,xn) cross sections. Our analysis consisted of developing suitable coupled-channels optical model potentials for each U isotope, and performing Hauser-Feshbach/statistical, preequilibrium, and direct reaction calculations.

We investigated several existing coupled-channels optical model potentials, which are based largely on ^{238}U experimental data, in the uranium-isotope analysis. The potentials include (1) a new global actinide potential developed by Vladuca et al.[iv] that spans the incident neutron energy range from 1 keV to 20 MeV; (2) a new potential by Maslov et al.[v] covering the same energy range, developed by fitting s-wave strength functions and experimental neutron data; (3) an earlier potential (and variations) for E_n = 1 keV to 30 MeV by Young and Arthur,[vi] which was utilized for ENDF/B-VI evaluations; (4) a new ^{238}U + n potential developed by Ignatyuk et al.[vii] covering the energy range 1 keV – 150 MeV; (5) a new potential derived by Sukhovitskij et al.[viii] by fitting ^{238}U + n and ^{238}U + p scattering angular distributions and neutron total cross sections up to 150 MeV; (6) an extensive modification[ix] of the earlier ^{238}U + n potential from Los Alamos[vi] mentioned in item (3) above, covering the energy range out to 200 MeV; and, (7) a new potential by Maslov et al.[x] that covers the incident nucleon energy range from 1 keV to 200 MeV.

Optical model calculations were performed with the 1996 version of the ECIS coupled-channels optical model code by Raynal[xi] to produce spin-dependent transmission coefficients for the Hauser-Feshbach calculations (below), as well as total and scattering cross sections to low-lying states. For most of the potentials, we coupled the lowest 3-7 ground-band rotational states into the calculations and included compound nucleus competition from uncoupled states plus a continuum of (n,n') states. The discrete levels were taken mainly from the International Atomic Energy Agency (IAEA) RIPL-2 database.[xii]

Hauser-Feshbach statistical calculations were performed both with ECIS96 and with the GNASH statistical/preequilibrium theory code[xiii] using the same optical potentials and level density parameters. The GNASH code includes a double-humped fission barrier model using uncoupled oscillators for the barrier representation, as described by Arthur.[xiv] Each compound nucleus formed in the calculations is permitted to decay through the fission channel and by neutron and gamma-ray emission. Gamma-ray transmission coefficients are obtained from strength functions calculated with the generalized Lorentzian model of Kopecky and Uhl.[xv] Transmission coefficients for fission are calculated from the fission model summarized here and detailed in ref. xiii. Competition from fission was included in the ECIS96 compound nucleus calculations by appropriate scaling using the evaluated fission cross sections.

Continuous level density functions based on Fermi gas and constant temperature models[xvi] were used, with level density parameters chosen to reproduce s-wave level spacings, $<D_0>$, measured at the neutron separation energy.[xii] Multiplicative factors are applied to the level density functions to account for enhancements in the fission transition state densities at the fission barriers due to increased asymmetry conditions, and the continuum level densities are combined with discrete fission transition states at each barrier. The discrete fission transition state spectra are calculated from bandhead information taken from calculations and compilations by Britt.[xvii]

A semiclassical exciton model is used to simulate preequilibrium particle emission. The matrix element normalization constant that describes the competition between precompound particle emission and internal transitions to higher exciton states in preequilibrium emission was typically fixed in the range 120-150 MeV^3. Similarly, the nuclear single-particle state densities were typically set to A/13 MeV^{-1} in the asymptotic limit where shell effects are washed out. Gamma-ray strength functions were normalized to experimental information[xii] on $2\pi\Gamma_\gamma/<D_0>$, sometimes with a slight renormalization to optimize agreement of calculated (n,γ) cross sections with experimental data.

Initial values of fission barrier parameters were taken from the work of Britt,[xvii] which were then optimized by comparing calculated fission cross sections from GNASH with experimental data. In most cases fits to experimental fission cross sections were used directly in the evaluations. In our calculations of the $(n,n'continuum)$, $(n,2n)$, $(n,3n)$, and $(n,4n)$ reactions, we utilized Kalbach[xviii] angular distribution systematics to obtain correlated energy-angle distributions for the continuum reactions.

Delayed-neutron multiplicities were updated in the ^{235}U and ^{238}U evaluations, and new fission neutron spectra were incorporated using calculations with the Los Alamos model.[xix]

EXPERIMENTAL DATA

There is a great deal of experimental data available for neutron-induced reactions on U isotopes, particularly for fission cross sections. We obtained all experimental data from either the EXFOR/CSISRS database at the National Nuclear Data Center at Brookhaven National Laboratory or the Nuclear Energy Agency in Paris.

Much of the fission cross-section data is relative to ^{235}U. The cross section ratio measurements were converted to absolute values using the new ENDF/B-VII standard cross section, which has been produced by Alan Carlson's group working under the auspices of CSEWG, an IAEA Coordinated Research Program, and the NEA/WPEC. This cross section is about ~1% higher in he fast region than the ENDF/B-VI standard, and differs significantly to the old standard above 14 MeV, where the new evaluation was improved through the availability of new higher energy measurements, including Los Alamos measurements by Lisowski et al.

Most of the prompt neutron multiplicities from fission (nubar) are also in the form of ratios, frequently relative to the very accurately known ^{252}Cf nubar value. All prompt nubar measurements were adjusted to conform to ENDF/B-VI standards, where possible.

EVALUATION FEATURES

Total Cross Sections

We performed coupled-channels optical model calculations for all the uranium isotopes. In cases where experimental data were available (^{233,235,238}U), the evaluated total cross sections are based on a combination of calculations and measurements. For ^{235}U and ^{238}U, results from a covariance analysis of the measurements were included in the analysis. For the remaining U isotopes, coupled channels optical calculations were used for the total cross sections.

Detailed assessments of the optical potentials[i,iii] show that most of the potentials give reasonable results at most energies, and the calculated total cross sections for the various uranium isotopes behave similarly with neutron energy. For our evaluations, the Los Alamos optical model potentials[i,vi] (or modifications thereof) were selected for our base coupled-channels calculations. In Fig. 1, the 1992 potential,[vi] used for the ^{238}U evaluation, is shown with the other potentials to agree reasonably with total cross section experimental data.

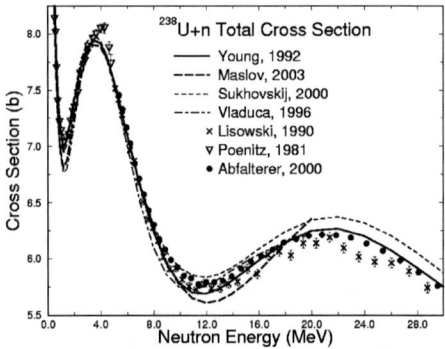

FIGURE 1. Comparison of Measured Total Cross Sections for ^{238}U with Optical Model Calculations.

Fission Cross Sections and Nubar

Most of the uranium isotopes (^{232,233,234,235,236,238}U) have useful experimental (n,f) cross section data for the evaluations. Two of the isotopes (^{237,239}U) have fission probability data from (t,pf) surrogate reactions[xx] that can be used in the evaluations. GNASH reaction theory calculations were optimized to the experimental data, so that reliable calculations could be performed of the (n,n') and (n,xn) reaction channels. Fission barrier parameterizations were interpolated/extrapolated for GNASH calculations of the completely unmeasured target nuclei [240,241]U.

The evaluated ^{233}U(n,f) cross section is compared to measurements in Fig. 2 (l.h.s.), and indeed the evaluation was based on our statistical analysis of measured data, and the new fission cross section contributes to a significant improvement in the critical assembly data testing for ^{233}U. In Fig. 2 (r.h.s.), the GNASH calculation for the ^{239}U(n,f) cross section is compared to experimental data inferred from surrogate measurements.[xx] An overview of the evaluated (n,f) cross sections for the U isotopes is given in Fig. 3. The systematic trend of decreasing fission cross section with increasing mass is striking. This can be understood in a straightforward way: the fission probability is largely influenced by the extent to which the compound nucleus excited after neutron absorption is above, or below, the fission barrier. The fission barriers for the uranium isotopes do vary from isotope to isotope, but only by a small amount. The neutron binding energy, on the other hand, systematically decreases with increasing target A. This leads to the decreasing fission cross section for increasing A for the uranium isotopes, as seen in Fig 3.

Prompt fission neutron multiplicities were determined from experimental data on the [233,235,238]U isotopes. For the remaining isotopes, previous evaluations and systematics were utilized for nubar.

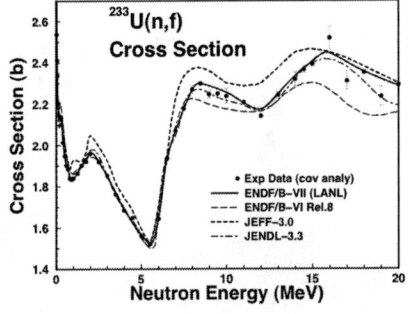 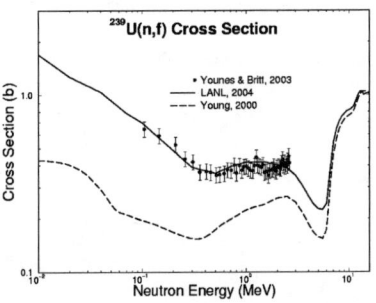

FIGURE 2. Evaluated and Measured ^{233}U(n,f) Cross Sections (l.h.s.); Calculated Surrogate ^{239}U(n,f) Cross Sections (r.h.s.)

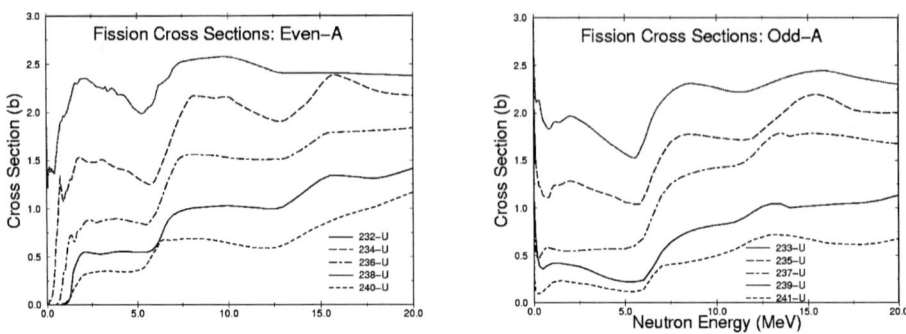

FIGURE 3. Evaluated U-Isotope Fission Cross Sections.

Elastic & Inelastic Neutron Scattering

The evaluated elastic cross sections mainly result from coupled-channels optical calculations with the 1992 LANL potential,[vi] although formally they were obtained by differencing the total and nonelastic cross sections. In most of the evaluations, we also utilize elastic angular distributions from the coupled-channels calculations In the case of ^{238}U, however, we discovered empirically that using angular distributions from the Maslov evaluation[v] below 10 MeV resulted in systematic improvement in calculations of several reactor benchmark experiments, so these were adopted.

Neutron inelastic cross sections and angular distributions for discrete and continuum reactions were generally taken from the theoretical analyses, although several of the calculated ^{238}U(n,n') cross sections were adjusted slightly to better match measurements. A comparison of measured and evaluated 2.5-MeV neutron elastic and inelastic $(E_x=45$ keV$)$ angular distributions for ^{238}U is given in Fig. 4.

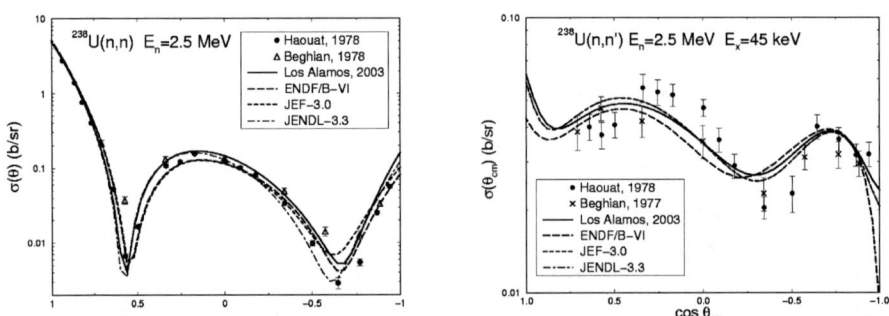

FIGURE 4. Evaluated and Measured ^{238}U(n,n) and ^{238}U(n,n') Angular Distributions for E_n=2.5 MeV.

In both the ^{235}U and ^{238}U evaluations, we assume a set of $J^\pi = 3^-$ and 2^+ states at excitation energies between E_x = 1-4 MeV that are used to approximate unmeasured collective states. Similar to an earlier analysis of neutron emission spectra from ^{184}W by

Marcinkowski et al.,[xxi] it is not possible to account for the ^{238}U neutron emission spectrum measurements of Baba et al.[xxii] at E_n = 14 MeV by assuming only compound nucleus and preequilibrium reactions. We obtained cross sections and angular distributions for these states from direct reaction calculations using ECIS96, with deformation parameters chosen to match Baba's 14-MeV data. These assumptions lead to significantly improved neutron emission spectra from ^{238}U, as well as to improved simulations of time-of-flight neutron distributions from pulsed-sphere experiments[xxiii] in calculations with the MCNP Monte Carlo code.[xxiv] Evaluated and measured double-differential spectra at E_n=14.05 MeV (θ=90°), and E_n=18.0 MeV (θ=120°) are compared in Fig 5.

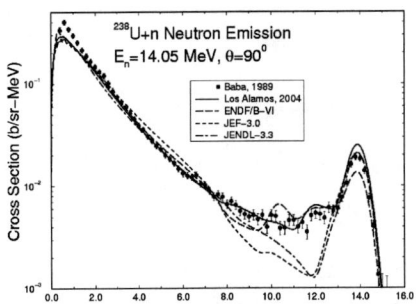
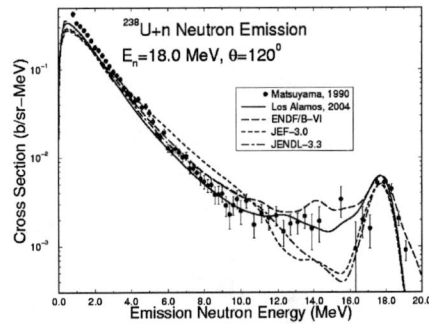

FIGURE 5. Double-Differential Neutron Emission Spectra from n+ ^{238}U Reactions Compared to Measurements.

(n,xn) Cross Sections

Theoretical calculations played an important role in the *(n,2n)* and *(n,3n)* cross sections for all the uranium isotopes. The evaluated *(n,2n)* cross sections for ^{235}U and ^{238}U are compared to measurements in Fig. 6. The LANL ^{235}U*(n,2n)* results are entirely from the GNASH calculation; the ^{238}U*(n,2n)* evaluation is from GNASH at energies below 7.5 MeV and follows covariance analysis results at higher energies. The behavior of the ^{238}U*(n,2n)* cross section near threshold was validated in MCNP simulations[xxv] of critical assembly experiments having different degrees of hardness in their neutron spectra – see Figure 7 and the accompanying text in the next section. The energy-angle neutron emission distributions from the GNASH analysis were used directly for all *(n,xn)* reactions, utilizing Kalbach[xviii] angular systematics.

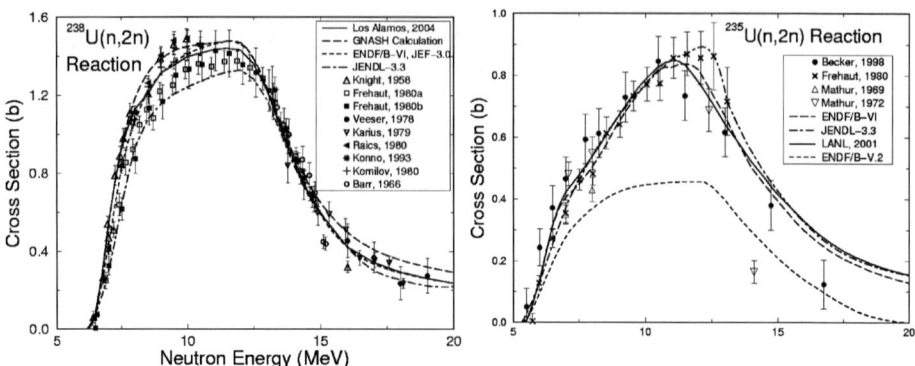

FIGURE 6. Evaluated and Measured (*n2n*) Cross Sections for ^{235}U and ^{238}U.

INTEGRAL DATA TESTING

It is crucial to validate the accuracy of a new nuclear data library before it is released to applied users. Furthermore, validation tests using simulations of critical assembly experiments is important during the evaluation stage of the library, as it can point to deficiencies in the fundamental cross sections that must be resolved. At Los Alamos, we have benefited from a close collaboration between the nuclear data evaluators & theorists, and the nuclear data testing researchers.

We provide some illustrative examples of integral data testing for validation below. Comparisons are made between calculated and measured reaction rates in critical assemblies. Here, the critical assembly was used to provide a neutron spectrum, within which reaction rates such as fission, n2n, and capture are measured. By measuring these rates at different locations, one can probe different neutron spectra – for instance, positions at the center of an assembly involve the hardest spectra, whilst positions at large radial locations (in a "traverse") involve softer spectra. Because these assemblies have simple geometries, we believe we are able to very accurately model the neutron spectrum shape with our MCNP transport code, and therefore comparisons between simulation results and measurements really are a test of the cross section accuracy.

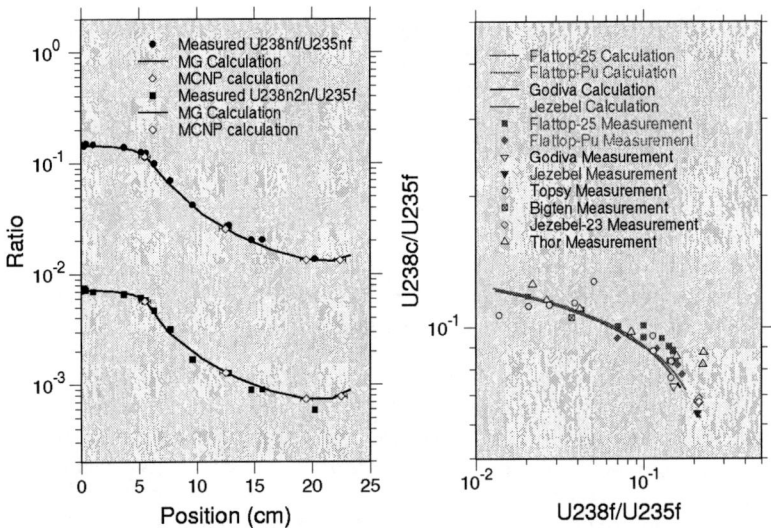

FIGURE 7 l.h.s.: Comparison of measured and calculated reaction rates for 238U fission (upper curve) and 238U(n,2n) lower curve. The positions represent different locations in a traverse of the Flattop assembly, the large radial positions corresponding to soft neutron spectra and the inner positions to hard spectra. MG indicates multigroup calculations, and the accuracy of these has been validated by the MCNP continuous energy Monte Carlo calculations shown; **r.h.s.**: Comparison of measured and calculated reaction rates for 238U capture. The measurements represent different locations in critical assemblies. The hardness of the neutron spectrum at various locations is represented by the 238U/235U fission rate ratio given on the x-axis.

Figure 7 shows comparison of experiment and calculation for a traverse of the Flattop assembly. The l.h.s. upper curve is for ^{238}U fission, and the l.h.s. lower curve for n,2n. In both cases, the rates are in ratio to ^{235}U fission, which is known precisely. These comparisons provide an integral validation of our fundamental data, and also support the shapes of our cross sections as a function of incident neutron energy. A similar comparison is provided in Figure 7 for ^{238}U radiative capture (r.h.s.). In this comparison, the x-axis values for different positions are replaced by the spectral index value, that is, the 238U/235U fission ratio, which is a good measure of the hardness of the neutron spectrum since ^{238}U is a threshold fissioner. Reasonably good agreement is seen between the calculation and the simulation. However, since the calculation lies on the lower side of the bulk of the data for the hardest spectra (by about 5%), this suggests that our capture evaluation may be too low by about 5% at the higher neutron energies.

We also note that nuclear criticality (k-eff) comparisons have been made for a very large range of critical assemblies, ranging from fast, to intermediate, to thermal. We obtain excellent agreement with the measured values (k-eff = 1), with major improvements compared to the previous ENDF/B-VI.8: (1) The fast assemblies Jezebel (^{239}Pu), Godiva (^{235}U), and Jezebel-23 (^{233}U) are now within 0.03% of experiment; (2) A longstanding reflector bias in the reflected Flattop fast assemblies has been largely eliminated, indicating better ^{238}U reflection properties; (3) Thermal assemblies involving ^{238}U now are modeled much more accurately, due to our improved ^{238}U inelastic scattering as well as a new ^{238}U resonance evaluation from Oak Ridge; (4) ^{233}U is now much more accurately modeled – previously Jezebel-23 had a significant k-eff underprediction; (5) The intermediate spectra assemblies (Bigten and the ZPRs) are now much more accurately modeled.

CLOSING REMARKS

We have completed new evaluations of neutron-induced reactions for 10 uranium isotopes over the incident neutron energy range 10^{-5} eV to 30 MeV. The new data files are part of the evaluation effort for Version VII of ENDF/B. The data have been extensively tested against critical assembly measurements for both reaction rate measurements and criticality (k-eff). Excellent performance is generally observed.

There still remain, though, some unsolved problems that will be the focus of our research in the coming years: (1) ^{235}U capture data in the 10-100s of keV region are not well understood, and significant (10%) discrepancies exist between various databases (*e.g.* ENDF and JENDL). New high-precision capture measurements using the LANSCE/DANCE detector may resolve this problem; (2) Likewise, for ^{238}U there are similar questions about the capture cross section at the higher energies (above a few 100 keV – 1 MeV) – as noted in the text above; (3) Further work is needed to more precisely determine fission cross sections such as ^{237}U, ^{239}U, ^{240}Am; (4) We believe that the neutron spectrum in critical systems may be too soft, since our calculated spectral indices for threshold fissioners (*e.g* ^{238}U(n,f)/^{235}U(n,f) are calculated ~3-5% low. This suggests deficiencies in inelastic scattering and/or the prompt fission spectrum; (5) Prompt fission spectra need to be measured as a function of energy for some major actinides, including plutonium (where the existing database is sparse).

REFERENCES

i. Young, P. G, Chadwick, M. B., MacFarlane, R. E., Talou, P., LA-UR-03-1617 (2003).
ii. Young, P. G, *et* al., LA-UR-03-3205 (2003).
iii. Young, P. G, *et* al., LA-UR, TBP (2004).
iv. Vladuca, G., *et* al., *Rom. J. Phys.* **41**, 515 (1996).
v. Maslov, V. M., *et* al., INDC(BLR)-014 (2003).
vi. Young, P. G., and Arthur, E. D, Jülich Conf., 13-17 May 1991 [Springer-Verlag, Germany (1992)]
vii. Ignatyuk, A V., Lunev, V. P., Shubin, Yu. N., Gai, E. V., *et* al., *Nucl. Sci. Eng.* **136**, 340 (2000).
viii. Sukhovitskij, E. S., Iwamoto, O., Chiba, S., and Fukahori, T., *J. Nucl. Sci. & Tech.* **37**, 120 (2000).
ix. Young, P. G., personal communication, 2001.
x. Maslov, V. M., *et* al., Nucl. *Phys.***A736**, 1 (2004).
xi. Raynal, J, CEA-N-2772 (1994).
xii. To be issued as IAEA-TECDOC (2004).
xiii. Young, P.G. *et al.* Wksp *Nucl. Reaction Data and Nucl. Reactors*, ICTP, Trieste, 15 Apr -17 May 1996 [World Sci. Publ. Co., Sing. (1998)] p. 227.
xiv. Arthur, E. D., NEANDC-158 "U" (1982) p. 145.
xv. Kopecky, J. *et al.*, *Phys. Rev. C* **42**, 1941 (1990)
xvi. Gilbert, A. *et al.*, *Can. J. Phys.* **43**, 1446 (1965).
xvii. Britt, H. C., personal communication, 1982.

xviii. Kalbach, C., *Phys. Rev. C* **37**, 2350 (1988).
xix. Madland, D. G., Nix, J. R., *Nucl. Sci. Eng.* **81**, 213 (1982).
xx. Younes, W., Britt, H.C., *Phys. Rev. C* **68**, 034610 (2003)
xxi. Marcinkowski, A., Demetriou,, P., Hodgson, P.E., *J. Phys. G [Nucl. & Part. Phys.]* **22**, 1219 (1996).
xxii. Baba, M., *et al.*, JAERI-M-89-143 (1989).
xxiii. Hansen, L. F., *et al.*, Nucl. Sci. Eng. **72**, (1979).
xxiv. Frankle, S., personal communication (2004); J. F. Briesmeister, ed., LA-13709-M (2000).
xxv. M. MacInnes, personal communication (2004

FISSION BARRIERS OF EXOTIC NUCLEI

A. Kelić and K.-H. Schmidt

GSI, Planckstr. 1, D-64291 Darmstadt, Germany

Abstract. Using available experimental data on fission barriers and ground-state masses a detailed study on the predictions of different models concerning the isospin dependence of saddle-point masses is performed.

Keywords: Fission barrier; macroscopic models; neutron-rich nuclei
PACS: 24.75.+i, 25.85. w, 27.90.+b

INTRODUCTION

Experimental information on the height of the fission barrier is only available for nuclei in a rather narrow region of the chart of the nuclides. Therefore, in any theoretical model the constraint on the parameters defining the dependence of the fission barrier on neutron excess is rather weak. This imposes a large uncertainty in estimating the fission barriers of exotic nuclei, which are relevant in some astrophysical scenarios, e.g. the r-process. Recently, important progress has been made on developing a full microscopic approach to nuclear masses (see e.g. [1]). Nevertheless, due to the complexity, this type of calculations is difficult to apply to heavy neutron-rich nuclei, where one is still to deal with semi-empirical models. Often used models are of macroscopic-microscopic type. In this paper, we consider several of such models and study the behaviour of the macroscopic part when extrapolating to very neutron-rich nuclei. This study is based on the approach of Bjørnholm and Lynn [2] and Dahlinger et al. [3], where the predictions of theoretical models are examined by means of a detailed analysis of the isotopic trends of ground-state masses and fission barriers.

In the present work we consider the following models: 1.) Droplet model (DM) [4], which is a basis of often used results of the Howard-Möller fission-barrier calculations [5], 2.) Finite-range liquid drop model (FRLDM) [6,7], 3.) Thomas-Fermi model (TF) [8,9] and 4.) Extended Thomas-Fermi model (ETF) [10]. Fig. 1 shows the predictions of these models for the macroscopic part of the fission barriers for different uranium isotopes. Important disagreement between different models, especially for very neutron-rich nuclei, is clearly seen from the figure. In order to test the self-consistency of these models, we study, as suggested in [2,3], the difference between the experimental total mass at the saddle point ($E_f^{exp} + M^{exp}$) and the macroscopic part of the total calculated mass at the saddle ($E_f^{macro} + M^{macro}$), with E_f being the height of fission barrier and M the ground-state mass:

Such defined quantity would correspond to the shell corrections at the barrier, and should show only local structure; any general trend should be included in the macroscopic model. Therefore, any general trend in δU_{sad} with respect to the neutron excess would indicate severe shortcomings of the model in extrapolating to nuclei far from stability.

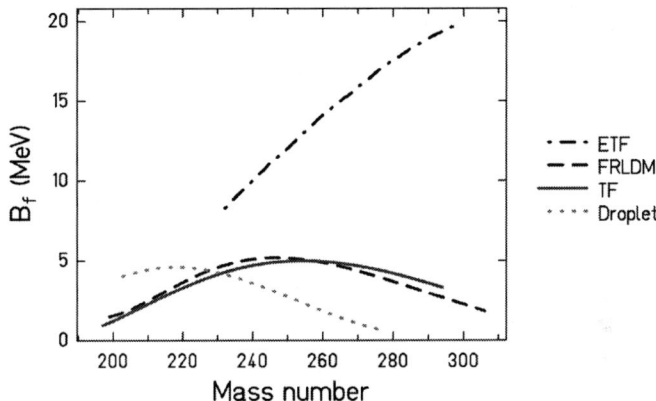

FIGURE 1. Macroscopic part of the fission barrier calculated for different uranium isotopes using: the extended Thomas-Fermi model [10] (dashed-dotted line), the finite-range liquid-drop model [6,7] (dashed line), the Thomas-Fermi model [8,9] (full line), and the droplet model [4] (point-line).

Fig. 2 shows a survey of nuclei used for the present study on a chart of the nuclides. The experimental fission barriers for these nuclei are taken from the compilation of Dahlinger et al. [3] and the experimental ground-state masses from the Audi and Wapstra 1995 compilation.

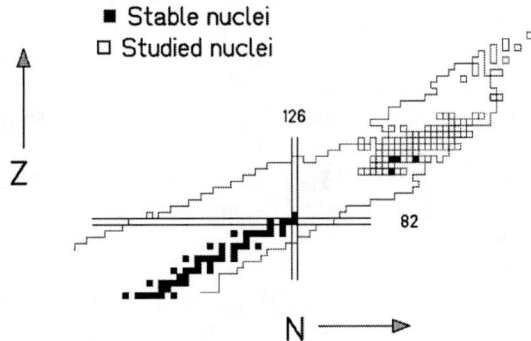

FIGURE 2. Blue squares represent the nuclei that were studied in the present work.

RESULTS AND CONCLUSIONS

For the case of uranium isotopes, the variable δU_{sad} as given by Eq. (1) is shown in Fig. 3 as a function of the neutron number. Results from the Droplet model (DM) show that δU_{sad} increases strongly with the neutron number, while the ETF model predicts a decrease. FRLDM and Thomas-Fermi model result in a quite similar behavior of δU_{sad} with almost a zero slope.

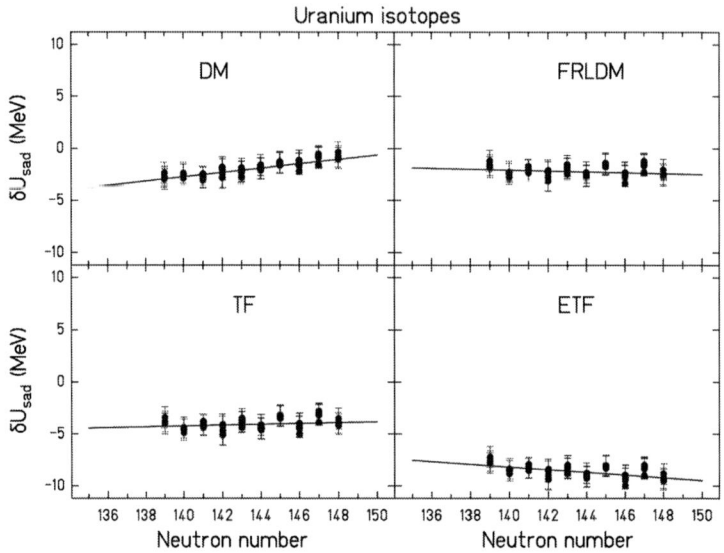

FIGURE 3. Difference between the total experimental energy at the saddle point (points) and the macroscopic part of the total calculated energy at the saddle point (lines) calculated with the droplet model, the finite-range liquid-drop model, the Thomas-Fermi model and the extended Thomas-Fermi model for different uranium isotopes.

We applied the same procedure for all nuclei indicated in Fig. 2. The extracted slopes (A_1) of δU_{sad} as function of the neutron excess are shown in Fig. 4 as a function of the nuclear charge number. Similar behavior seen in case of uranium is also seen for other nuclei. The droplet model predicts for all studied nuclei an increase in δU_{sad} as a function of the neutron excess. The value of the mean slope averaged over the studied Z is 0.15 MeV, and this indicates that a consistent description of the isospin dependence of nuclear masses within the droplet model is not possible. The same conclusion was obtained in Ref. [3], and shed a doubt on the applicability of the Howard-Möller tables of fission barriers [5] in regions far from the stability. In case of the ETF model we had available only the barriers for the uranium isotopes. Already in this case we see that over the range of 10 studied isotopes a clear correlation exists between δU_{sad} and N. For other nuclear charge numbers the values of the slopes were extrapolated from the uranium value based on the behavior of the FRLDM and the TF model. The average value of such obtained slopes is -0.12 MeV also indicating possible problems in the consistency of the ETF model in describing nuclear masses. Of course, for a definite conclusion one should perform the analysis with dedicated

ETF calculations. The FRLDM and the TF model result both in rather small slopes of δU_{sad} functions, with average values of -0.04 MeV and 0.05 MeV, respectively. Although not zero, these average values are much lower than in case of the DM or the ETF model, and make FRLDM and the TF model to be preferential for the description of the fission barriers of exotic nuclei.

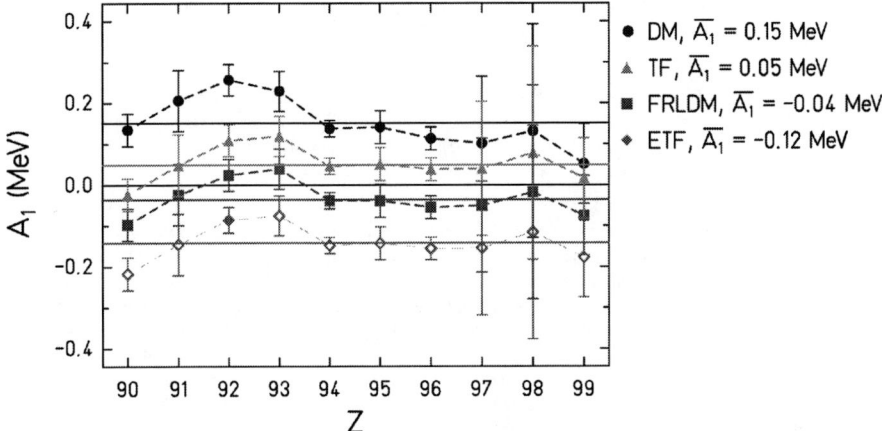

FIGURE 4. Slopes of δU_{sad} as a function of the neutron excess are shown as a function of the nuclear charge number Z obtained for the droplet model (points), the Thomas-Fermi model (triangles), FRLDM (squares) and the extended Thomas-Fermi model (rhomboids). The full lines indicate the average values of slopes. Average values are also written in the figure.

In conclusion, we have studied four different macroscopic models in order to find, which ones are most adapted for calculating the saddle-point masses of nuclei far from stability. The results of this study show that the preferential models should be the finite-range liquid-drop model [6,7] and the Thomas-Fermi model [8,9]. Severe doubts in the consistency of the droplet model [4] were seen, rising also the question of the applicability of the Howard-Möllers fission-barrier tables [5].

REFERENCES

1. J.M. Pearson and S. Goriely, *Nucl. Phys. A in print.*
2. S. Bjørnholm and J.E. Lynn, *Rev. Mod. Phys. 52 (1980) 725.*
3. M. Dahlinger, D. Vermeulen and K.-H. Schmidt, *Nucl. Phys. A376 (1982) 94.*
4. W.D. Myers, „*Droplet Model of Atomic Nuclei"*, 1977 IFI/Plenum, ISBN 0-306-65170-X.
5. W.M. Howard and P. Möller, *At. Data Nucl. Data Tables 25 (1980) 219.*
6. A. Sierk, *Phys.Rev. C33 (1986) 2039.*
7. P. Möller, J.R. Nix, W.D. Myers and W.J. Swiatecki, *At Data Nucl. Data Tables 59 (1995) 185.*
8. W.D. Myers and W.J. Swiatecki, *Phys. Rev. C 60 (1999) 014606-1.*
9. W.D. Myers and W.J. Swiatecki, *Nucl. Phys. A 601(1996) 141.*
10. A. Mamdouh, J.M. Pearson, M. Rayet and F. Tondeur, *Nucl. Phys. A 679 (2001) 337.*

Determination of the ^{243,246,248}Cm thermal neutron induced fission cross sections

O. Serot*, C. Wagemans†, S. Vermote†, J. Heyse†,**, T. Soldner‡ and P. Geltenbort‡

*CEA Cadarache, DEN/DER/SPRC/LEPh, F-13108 St. Paul-Lez-Durance, France
†Dept. of Subatomic and Radiation Physics, University of Gent, B-9000 Gent, Belgium
**EC-JRC-IRMM, Retieseweg 111, 2440 Geel, Belgium
‡Institut Laue Langevin, F-38042 Grenoble Cedex 9, France

Abstract. The minor actinide waste produced in nuclear power plants contains various Cm-isotopes, and transmutation scenarios require improved fission cross section data. The available thermal neutron induced fission cross section data for ^{243}Cm, ^{246}Cm and ^{248}Cm are not very accurate, so new cross section measurements have been performed at the high flux reactor of the ILL in Grenoble (France) under better experimental conditions (highly enriched samples, very intense and clean neutron beam). The measurements were performed at a neutron energy of 5.38 meV, yielding fission cross section values of (1240±28)b for ^{243}Cm, (25±47)mb for ^{246}Cm and (685±84)mb for ^{248}Cm. From these results, thermal fission cross section values of (572±14)b; (12±25)mb and (316±43)mb have been deduced for ^{243}Cm, ^{246}Cm and ^{248}Cm, respectively.

Keywords: NUCLEAR REACTIONS 243Cm(n,f) 246Cm(n,f) 248Cm(n,f), E=5.38 meV; measured fission yield; deduced reaction cross section; calculated fission cross section at 25.3 meV
PACS: 25.85.Ec; 28.20.Fc

INTRODUCTION

Amongst the minor actinides produced in nuclear power plants, various Cm isotopes could be candidates for transmutation. These transmutation studies require good knowledge of the Cm fission and capture cross sections. However, the thermal fission cross section data which can be found in the European (JEFF), Japanese (JENDL) and American (ENDF/B) neutron libraries for Cm isotopes show discrepancies. Most of these evaluations are based on old measurements (more than 30 years) and therefore new measurements have been recommended. In order to try to improve the situation, at least two conditions must be fullfilled: a highly enriched sample combined with an 'excellent' neutron beam, i.e. a very high and stable neutron flux, without any gamma's or fast neutrons. The fact that we could have a sufficient amount of highly enriched ^{243}Cm, ^{246}Cm and ^{248}Cm material with a suited neutron beam at the high flux reactor of the Institut Laue Langevin (ILL) in Grenoble, enabled us to perform the three measurements reported in the present paper.

TABLE 1. Isotopic composition of the ^{243}Cm sample.

	Number of atoms	At. %
^{243}Cm	**4.42 $\times 10^{15}$**	**97.110**
^{244}Cm	6.70 $\times 10^{13}$	1.473
^{245}Cm	2.95 $\times 10^{12}$	0.065
^{239}Pu	6.15 $\times 10^{13}$	1.352

NEUTRON BEAM

All three measurements were performed at the PF1B cold neutron guide at the high-flux reactor of the ILL. The neutron flux at the exit of the guide is 1.3×10^{10} n/s cm^2. The spectrum of the neutrons and its dependance on the horizontal position was measured by the ILL team and is reported on the ILL web site [1]. The average neutron wavelength is $\bar{\lambda}$=3.9 Å, which corresponds to a neutron energy of $E_{\bar{\lambda}}$=5.38 meV. Since the neutron guide is 76 m long and bent, essentially all gammas and fast neutrons coming from the reactor are removed, which is very important mainly for the determination of the ^{246}Cm and ^{248}Cm subthreshold fission cross sections.

^{243}Cm FISSION CROSS SECTION

Sample

We used a 2.0 μg Cm-oxide sample which was deposited on a 30 μm thick Al backing. The diameter of the sample was 15 mm and its isotopic composition in April 2005 (date of the experiment) is given in Tab.1. Due to the alpha decay of ^{243}Cm, ^{239}Pu nuclei are created. Since the ^{239}Pu thermal fission cross section is very high (747.7 b), ^{239}Pu nuclei are undesirable! So, a Cm - Pu separation was performed in September 2004, strongly reducing the ^{239}Pu contribution.

Experimental setup

The sample was mounted in the center of a vacuum chamber at a 45° angle compared to the direction of the incident neutron beam. A 450 mm^2 large Si-Au surface barrier detector was positioned outside the collimated neutron beam for the detection of the fission fragments. Signals were amplified, digitized and stored in a PC. The flux measurement was done by replacing the ^{243}Cm-sample by a ^{235}U-sample having exactly the same dimensions. The U-sample has an enrichment of 99.97 % in ^{235}U and a mass of 16.5 μg. Unfortunately, the distance between the sample and the detector could not be maintained for both measurements, modifying the detection geometry. As we will see later, this geometrical effect was taken into account by determining the detection efficiency for both configurations from Monte Carlo calculations.

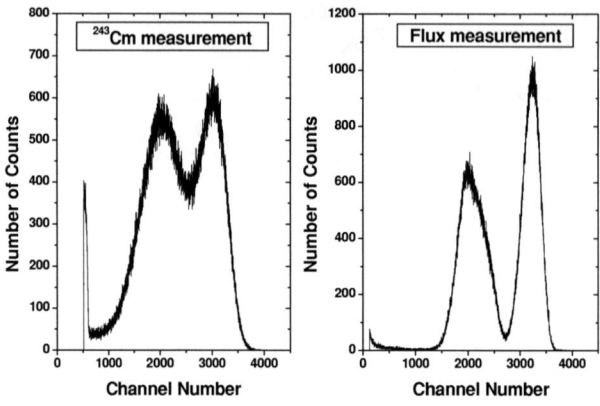

FIGURE 1. Pulse height spectra measured with the ^{243}Cm (left) and ^{235}U (right) samples.

Measurements and results

A clean detection of the fission fragments could be achieved as shown in Fig.1, where the ^{243}Cm(n,f) and ^{235}U(n,f) pulse height spectra are plotted respectively on the left and right part of the figure. We have verified by closing the neutron beam that the background and the spontaneous fission contributions were negligible for both measurements. The ^{243}Cm measurement can be briefly summarized as follows:

- the measuring time is 3470 s;
- the dead time is negligible;
- the fission counting rate obtained by integrating the pulse-height spectrum is: Y_{nf}^{243Cm}=(140.6 $\pm$ 1.4) fissions/s (the uncertainty is a combination of the statistical and systematic errors);
- the Monte Carlo calculation of the detection efficiency gives: $\varepsilon_{Cm}=\Omega_{Cm}/4\pi$=(1.3313 $\pm$ 0.0003) %.

For the flux measurement (with the ^{235}U sample):

- the measuring time is 300 s,
- the fission counting rate obtained by integrating the pulse-height spectrum and by adding a 0.4% dead time correction is: Y_{nf}^{235U}=(1409.8 $\pm$ 4.7) fissions/s,
- the Monte Carlo calculation of the detection efficiency gives: $\varepsilon_U=\Omega_U/4\pi$=(1.4063 $\pm$ 0.0003) %.

TABLE 2. Fission cross sections at $E_{\bar{\lambda}}$=5.38 meV and at thermal energy (E_0=25.3 meV) for all nuclei involved in the measurements reported in this paper.

Target	$\sigma_f(E_{\bar{\lambda}})$	$\sigma_f(E_0)$
^{244}Cm *	2.78 ± 0.57	1.03 ± 0.20
^{245}Cm †	5267 ± 211	2141.5 ± 64
^{247}Cm *	189 ± 12	81.8 ± 4.4
^{239}Pu *	1582 ± 20	747.7 ± 1.9
^{235}U *	1325 ± 16	584.90 ± 1.11

* Data from JEFF3.0
† Data from JENDL3.3

Assuming that the shape of all cross sections can be well described by a 1/v behaviour around $E_{\bar{\lambda}}$, the fission counting rate measured with the ^{243}Cm sample is simply given by:

$$Y_{nf}^{243Cm} = \varepsilon_{Cm}\Phi \times \sum_i N_i \sigma_{nf}^i(E_{\bar{\lambda}}) \quad (1)$$

where the sum is performed over all nuclei i in our sample (i=^{243}Cm, ^{244}Cm, ^{245}Cm and ^{239}Pu). $\sigma_{nf}^i(E_{\bar{\lambda}})$ corresponds to the fission cross section at $E_{\bar{\lambda}}$=5.38 meV. These cross sections were taken from the JEFF3.0 library (except for ^{245}Cm data, where the JENDL3.3 library was used) and are reported in Tab.2. Φ is the total neutron flux, and N_i represents the number of atoms (see Tab.1).

Similarly, the fission counting rate measured with the ^{235}U sample can be expressed as:

$$Y_{nf}^{235U} = \varepsilon_U \Phi \times N_{235U} \sigma_{nf}^{235U}(E_{\bar{\lambda}}) \quad (2)$$

Combining equations (1) and (2) enables the determination of the ^{243}Cm neutron induced fission cross section at $E_{\bar{\lambda}}$=5.38 meV. We have obtained:

$$\sigma_{nf}^{243Cm}(E_{\bar{\lambda}}) = (1240 \pm 28)b \quad (3)$$

In order to compare our results to previous results obtained with thermal neutron beams, we have calculated the thermal fission cross section assuming a 1/v behaviour between $E_{\bar{\lambda}}$ and E_0=25.3 meV:

$$\sigma_{nf}^i(E_0) = \sigma_{nf}^i(E_{\bar{\lambda}}) \times \frac{\sqrt{E_{\bar{\lambda}}}}{\sqrt{E_0}} \quad (4)$$

In this way, we obtained:

$$\sigma_{nf}^{243Cm}(E_0) = (572 \pm 25)b \quad (5)$$

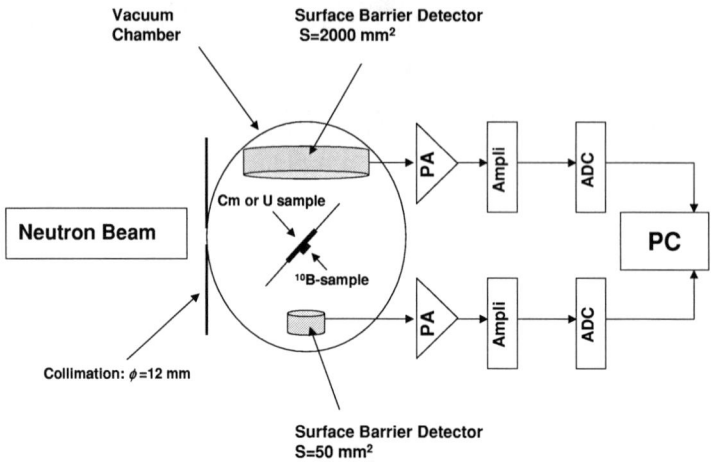

FIGURE 2. Schematic layout of the detection setup used for the ^{246}Cm and ^{248}Cm measurements. PA= Pre-Amplifier, ADC= Analog Digital Converter.

The uncertainty given at thermal energy has been increased compared with the one given at $E_{\bar{\lambda}}$ to account for the 1/v assumption. Within the error bars, our result is in agreement with the measurement performed by Bemis [2] who found (609.6 ± 26.9 b), somewhat lower than the values reported by Hulet [3] and Zhuravlev [4] which are respectively: (690 ± 50) b and (672 ± 60).

^{246}Cm AND ^{248}Cm FISSION CROSS SECTIONS

^{246}Cm and ^{248}Cm Samples

For both samples, Cm-oxide was deposited on a 30 μm thick Al backing. The diameter of the two samples is 15 mm and their masses are 46.4 μg for ^{246}Cm and 35.3 μg for ^{248}Cm. Their isotopic compositions are mentioned in Tab.3.

Experimental setup

For both ^{246}Cm and ^{248}Cm measurements, the same experimental setup was used (see Fig.2). This experimental setup is slightly different from the one used for the ^{243}Cm measurement. Indeed, due to the low fission counting rate expected for these

TABLE 3. Isotopic composition of the ^{246}Cm and ^{248}Cm samples.

	^{246}Cm - sample		^{248}Cm - sample	
	Number of atoms	At. %	Number of atoms	At. %
^{243}Cm	6.85×10^{12}	0.006		
^{244}Cm	6.81×10^{13}	0.060	6.98×10^{14}	0.817
^{245}Cm	9.46×10^{12}	0.008	1.73×10^{13}	0.020
^{246}Cm	$\mathbf{1.13 \times 10^{17}}$	**99.887**	1.13×10^{15}	1.323
^{247}Cm	2.83×10^{13}	0.025	1.98×10^{14}	0.232
^{248}Cm	1.58×10^{13}	0.014	$\mathbf{8.34 \times 10^{16}}$	**97.608**

subthreshold fission reactions, a 2000 mm^2 large surface barrier detector was used. The disadvantage of such a large area detector is that a non negligible dead time occurs during the flux measurement with the ^{235}U sample. In order to determine this dead time precisely, a ^{10}B sample was added back to back with the Cm (or U) sample. A 50 mm^2 large surface barrier detector facing the ^{10}B target was used to detect α and Li nuclei coming from the ^{10}B(n,α)^{7}Li reaction. Comparing the (n,α) counting rates obtained with the Cm and the U samples enables us to correct for the dead time and also for possible small neutron flux fluctuations. Lastly, for both ^{246}Cm and ^{248}Cm experiments, the flux measurement with the ^{235}U sample mentioned before could be performed strictly maintaining the same detection geometry.

^{246}Cm Measurement

The measurement with the ^{246}Cm target can be summarised as follows:

- the measuring time is 14.86 h;
- the dead time is negligible;
- the measured fission counting rate obtained with the neutron beam open is: $Y_{ON}^{246Cm}=(18.61 \pm 0.19)$ fissions/s;
- the contribution from spontaneous fissions is measured by closing the neutron beam during 2.16 h, yielding: $Y_{OFF}^{246Cm}=(8.97 \pm 0.09)$ fissions/s;
- so, the fission counting rate from neutron induced fission reactions is: $Y_{nf}^{246Cm}=Y_{ON}^{246Cm}-Y_{OFF}^{246Cm}=(9.64 \pm 0.28)$ fissions/s;
- the quoted uncertainties are a combination of the statistical and systematic errors.

For the flux measurement (with ^{235}U target):

- the measuring time is 300 s;
- a (6.80 ± 0.14) % dead time was determined;
- the measured fission counting rate is (after correction for the dead time): $Y_{nf}^{235U}=(7687 \pm 230)$ fissions/s (the uncertainty is a combination of the statistical and systematic errors).

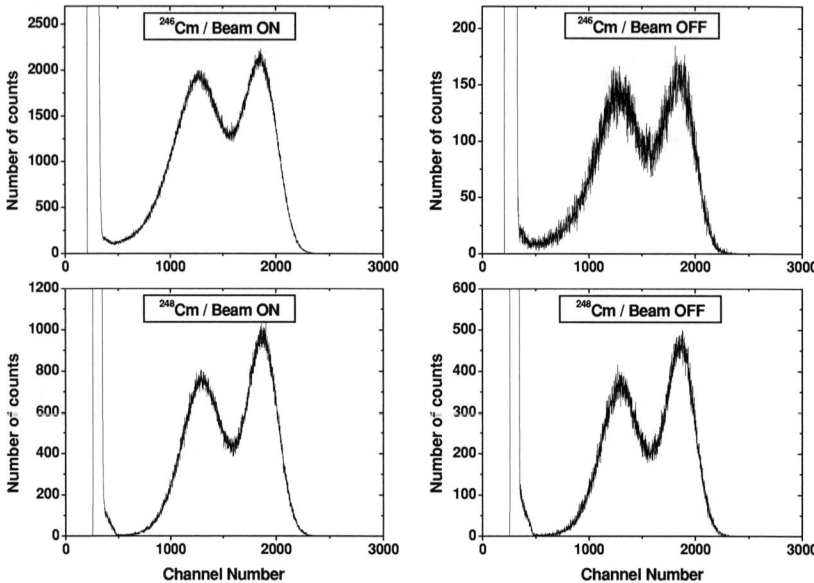

FIGURE 3. Pulse height spectra measured with the ^{246}Cm (top) and ^{248}Cm (bottom) samples. For each sample, spectra obtained with the beam open (left part) and closed (right part) are plotted.

The fission cross section at $E_{\bar{\lambda}}=5.38$ meV and at thermal energy (E_0=25.3 meV) was determined using the same procedure as the one described in the previous section. We obtained:

$$\begin{cases} \sigma_{nf}^{246Cm}(E_{\bar{\lambda}}) = (25 \pm 47) mb \\ \sigma_{nf}^{246Cm}(E_0) = (12 \pm 25) mb \end{cases} \quad (6)$$

The present thermal fission cross section value is much smaller than data reported by Benjamin [5]: (170 ± 100) mb and Zhuravlev [6]: (140 ± 50) mb.

^{248}Cm Measurement

The main characteristics of the ^{248}Cm measurement are:
- the measuring time is 1.48 h;

- the dead time is negligible;
- the fission counting rate obtained with the neutron beam open is: Y_{ON}^{248Cm}=(67.7 ± 0.1) fissions/s;
- the contribution from spontaneous fission events (measured by closing the neutron beam during 1.17 h) is: Y_{OFF}^{248Cm}=(40.0 ± 0.1) fissions/s;
- so, the fission counting rate from neutron induced fission reactions is: $Y_{nf}^{248Cm} = Y_{ON}^{248Cm} - Y_{OFF}^{248Cm}$=(27.7 ± 0.2) fissions/s;
- the quoted uncertainties are a combination of the statistical and systematic errors.

Concerning the flux measurement (^{235}U target):

- the measuring time is 120 s;
- the dead time is: (7.77 ± 0.15) %;
- the measured fission counting rate, after correction for the dead time is: Y_{nf}^{235U}=(7843 ± 235) fissions/s (the uncertainty is a combination of the statistical and systematic errors).

Again, the determination of the fission cross sections at $E_{\bar{\lambda}}$=5.38 meV and at thermal energy (E_0=25.3 meV) was done using the same procedure as the one described for the ^{243}Cm measurement. We have obtained:

$$\begin{cases} \sigma_{nf}^{248Cm}(E_{\bar{\lambda}}) = (685 \pm 84) mb \\ \sigma_{nf}^{248Cm}(E_0) = (316 \pm 43) mb \end{cases} \quad (7)$$

The thermal fission cross section is in agreement with the ones reported by Benjamin [5] and Zhuravlev [6] who found respectively (340 ± 70) mb and (390 ± 70) mb.

CONCLUSION

We have measured the neutron induced fission cross sections for ^{243,246,248}Cm at $E_{\bar{\lambda}}$=5.38 meV. From these results, we could also deduce fission cross sections at thermal energy. For ^{243}Cm, the present work shows a good agreement with the value reported by Bemis [2]. For ^{246}Cm, we find a thermal fission cross section much lower than the values obtained by both Benjamin [5] and Zhuravlev [6], while for ^{248}Cm, we find a good agreement with these authors.

REFERENCES

1. http: www.ill.fr
2. Bemis C.E., et al., *Nucl. Sci. Eng.* **63**, 413-417 (1977)
3. Hulet E.K., et al., *Phys. Rev.* **107**, 1294 (1957)
4. Zhuravlev K. and Kroshkin N.I., *At. En.* **47**, 55 (1979)
5. Benjamin R.W., MacMurdo K.W., Spencer J.D., *Nucl. Sci. Eng.* **47**, 203-208 (1972)
6. Zhuravlev K. et al., *At. En.* **39**, 285 (1975)

Neutron Cross Section Data for Pd-105, Ag-109, Xe-131, and Cs-133

Y.D. Lee and Y.O. Lee

Korea Atomic Energy Research Institute, P.O. Box105, Yusung, Daejon, Korea 305-600

Abstract. The neutron induced nuclear cross section data were calculated and evaluated from an unresolved energy to 20 MeV. The energy dependent potential parameters were extracted based on the recent experimental data and applied up to 20 MeV. A spherical optical model and a statistical model in an equilibrium energy, and a multistep direct and a multistep compound model in a pre-equilibrium energy were used in the calculation. The direct capture model was introduced for the fast neutron capture. The theoretically calculated cross sections were compared with the experimental data and the evaluated files. The evaluated cross section results were compiled in an ENDF-6 format and finally merged with the resonance part. The created library involves the data from the thermal to 20 MeV.

Keywords: Neutron, Cross Section, Evaluation
PACS: 28.20

INTODUCTION

The neutron cross section data evaluation for the selected fission products which mainly influence the reactivity in a fission reactor has been performed under a joint work with National Nuclear Data Center (NNDC) of Brookhaven National Laboratory (BNL). The cross section data file is basically divided into two energy ranges: thermal to resonance and upper resonance to 20 MeV. For the resonance energy region, the evaluation results[1] were already adopted in ENDF/B-VI.8. In this paper, the fast energy region results are evaluated from an unresolved energy to 20 MeV.

Neutron induced nuclear reaction data for fission products are important for predicting the burnup performance in a fission reactor, criticality for a spent fuel storage design, advanced fuel performance and a radiation damage estimation of structural material. Neutron capture cross sections of fission isotopes in several keV regions are significant in a fission reactor concerning the neutron absorption loss.

Pd-105, Ag-109, Xe-131 and Cs-133 are stable isotopes and have 22.33 %, 48.16 %, 21.2 % and 100 % in a natural elemental abundance, respectively. ENDF/B-VI on Pd-105 was modified by P.G. Young (Los Alamos National Laboratory) for the fast energy region in 1996, with some corrections by experimental data. For Ag-109, Xe-131 and Cs-133, ENDF/V evaluations were done in 1983, 1978 and 1978, respectively. They were converted to ENDF-6 format by the National Nuclear Data Center in 1990 and for the ENDF/B-VI in the fast energy region.

The calculated cross sections were graphically compared with the experimental data and the evaluated files (ENDF/B-VI, JENDL-3.2, JEF-2.2, BROND-2 and CENDL-2).

Finally, the results were merged with the resonance results[1]. The new data library was submitted to the ENDF/B-VII.

NUCLEAR MODELS

The evaluation consists of an optical model potential search[2,3] followed by a complete nuclear reaction model calculation and a validation with the experimental data[4]. Nuclear reaction cross sections were calculated using the recently released Empire code[5]. The potential as a function of the incident neutron energy was searched in a spherical optical model based on the experimental data. The Woods-Saxon well is used for the real part potential in the optical model. For the imaginary part potential, the derivative Woods-Saxon shape is used. Thomas form is taken in the optical model potential for the spin-orbit coupling:

The main utilities in Empire include the masses, level densities and discrete levels, decay schemes, deformation parameters, γ-ray strength functions, RIPL, ENDF-6 formatting and the plotting capabilities. The major modules are: Optical model, Multi-step Direct and compound, Pre-equilibrium exiton (DEGAS) and a Monte Carlo hybrid simulation (HMS) and a fully featured Hauser-Feshbach including a width fluctuation correction. The direct capture model was inserted to enhance the capture cross section.

RESULTS AND DISCUSSIONS

The calculation of several selected cross sections is presented here. The current fast energy evaluation and the resonance connected full data are nominated as ENDF/B-VII in the figures. The natural element experimental data[6] was used in the optical model potential parameter search for Pd-105. The capture cross section calculated by Empire using the optical model is shown in Figure 1. The calculation and the evaluated files are in good agreement with the experimental data[7,8]. The calculation shows a smooth connection with the resonance. However, the calculation shows a different shape from the other evaluations in the pre-equilibrium energy region. Figure 2 shows the capture cross section for Ag-109. The calculation and the ENDF/B-VI

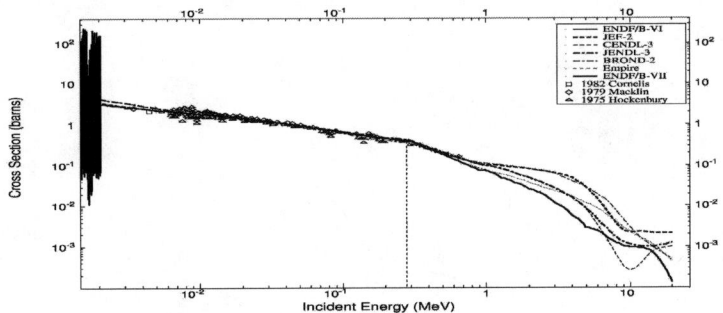

FIGURE 1. (n, γ) cross section of Pd-105.

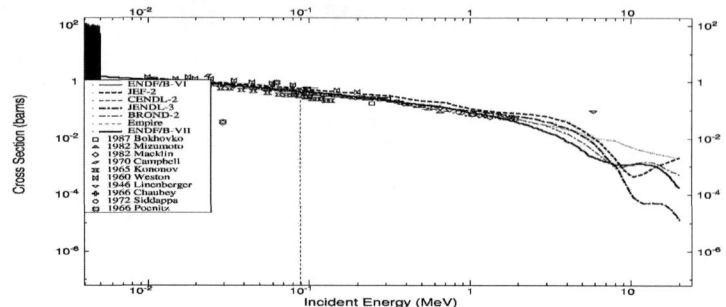

FIGURE 2. (n, γ) cross section of Ag-109.

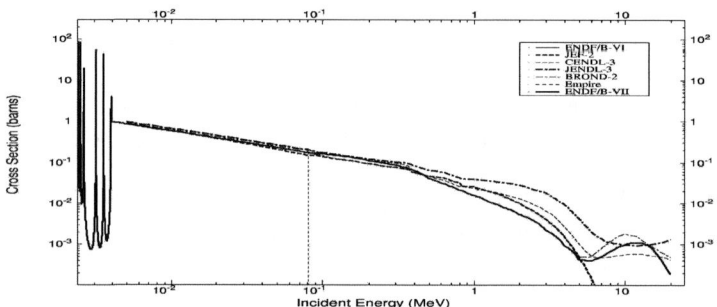

FIGURE 3. (n, γ) cross section of Xe-131.

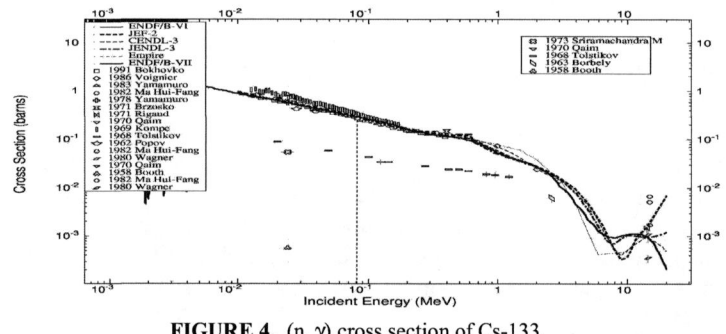

FIGURE 4. (n, γ) cross section of Cs-133.

FIGURE 5. (n, p) cross section of Ag-109.

agree well with the experimental data[9]. However, the calculation shows a direct capture contribution in the pre-equilibrium region, on the other hand, the ENDF/B-VI has no such feature. Figure 3 is the capture cross section for Xe-131. There is no experimental data. The calculation and other evaluations are together until 500 keV. ENDF/B-VI decreases continuously after 6 MeV. Figure 4 shows the capture cross section of Cs-133 for the calculation and the evaluated files. The calculation and the ENDF/B-VI agree well with the experimental data[10]. Figure 5 is the (n, p) cross section. The calculation agrees well with the reference experimental data[11].

CONCLUSION

Empire was successful in producing the reaction cross sections. All the evaluated cross sections, not presented here, are summarized in the final file. The total, capture and other calculated threshold cross sections are in good agreement with the experimental data. At the pre-equilibrium energy region, the calculated capture cross section prominently shows the fast neutron direct capture phenomena. The evaluated results represent an improvement over the current ENDF/B-VI. The connection with the resonance region for the capture cross section was smooth and continuous. The produced neutron data library is at the preliminary stage of ENDF/B-VII.

ACKNOWLEDGEMENTS

This work is performed under the auspices of Korea Ministry of Science and Technology as a long-term R&D project.

REFERENCES

1. Oh, S. Y. and Chang, J. H., *Neutron Cross Section Evaluations of Fission Products below the Fast Energy Region*, Daejon: Korea Atomic Energy Research Institute (KAERI/TR-1511/2000), 2000.
2. Lee, Y. D., *ABRXPL development for parameter decision of spherical optical model potential*, Daejon: Korea Atomic Energy Research Institute (KAERI, NDL-14), 1999.
3. Lawson, R. D., "ABAREX_A Neutron Spherical Optical-Statistical Model Code" in *Workshop on Computation and Analysis of Nuclear Data Relevant to Nuclear Energy and Safety*, edited by M. K. Mehta et al., Trieste: International Center for Theoretical Physics, 1992, pp. 447-516.
4. Lee, Y. D. and Chang, J. H., *Nuclear Engineering and Technology* **33**, 370-381 (2002).
5. Herman, M., *EMPIRE-2: Statistical Model Code for Nuclear Reaction Calculations*, Vienna: IAEA, 2002.
6. Poenitz, P. and Whalen, J. F., *Neutron Total Cross Section Measurements in the Energy Region from 47 keV to 20 MeV*, Chicago: ANL (ANL-NDM-80), 1983.
7. Macklin, L., Halperin, J., and Winters, R. R., *Nucl. Sci. Eng.* **71**, 182-195 (1979).
8. Cornelis, E., "Average Capture Cross Section of The Fission Product Nuclei Pd-104 Pd-105 Pd-106 Pd-108 Pd-110" ANTWER Conference, 1982, pp. 222-230.
9. Bokhovko, M. V., *J. YK* **2**, 21-30 (1987).
10. Bokhovko, M. V., *Neutron Radiation Cross-Section, Neutron Transmission And Average Resonance Parameters For Some Fission Product Nuclei*, FEI-2168-91,1991.
11. Gupta J. P., *J. PRM* **24**, 637 (1985).

What can we learn from fission times?

M. Morjean

GANIL, CEA-DSM and IN2P3-CNRS, B.P. 55027, F-14076 Caen Cedex, France

Abstract. Fission times have been measured by the blocking technique in single crystal for uranium nuclei and for super-heavy elements with Z = 120. The fission times measured for uranium nuclei can be reproduced by statistical calculations following the Bohr and Wheeler approach only if a friction coefficient increasing with temperature is considered. In the super-heavy element domain, a discrimination between the fast quasi-fission process and the slow fusion-fission one has been achieved from reaction time measurements. A minimum cross-section $\sigma = 22$ mb for formation of Z = 120 compound nuclei in ^{238}U+Ni reactions at 6.62 MeV/A has been inferred.

Keywords: Fission, blocking technique in single crystals, dissipation, fusion, super-heavy elements
PACS: 24.75.+i, 25.70.Jj, 61.85.+p

INTRODUCTION

A strong influence on fission dynamics of the nuclear dissipation has been long ago predicted [1, 2]. However, the magnitude of the nuclear dissipation involved as well as its possible evolution with excitation energy are still uncertain. In the first part of this paper, I shall show that the recently measured very long fission times for uranium nuclei heated up to about 3 MeV [3] provide us with a clear evidence for an increasing nuclear dissipation with temperature.

The fission time is also obviously quite sensitive to the fission barrier height. In the super-heavy element (SHE) domain, very high fission barriers are predicted [4, 5, 6, 7]. However, due to shell effect damping with temperature, the fission barriers for the nuclei actually formed are much lower and fission becomes the very dominant exit channel. In the second part of this paper, I shall present recent experimental results [8, 9] showing that SHE fission barriers are high enough to make the fission times accessible by direct measurements even at rather high excitation energies. The discrimination between fusion-fission events, corresponding to a very slow process, and quasi-fission events, corresponding to a fast process, becomes thus possible and the fusion cross-sections can be measured.

FISSION TIMES AND NUCLEAR DISSIPATION

For excited highly fissile nuclei, the retarding influence of nuclear friction [10] gives rise to very broad fission time distributions, with long lifetime components corresponding to nuclei fissioning at relatively low residual excitation energies, after cooling by particle emission. Due to these long lifetime components, the fission time (the average value of the fission time distribution) is quite sensitive to nuclear dissipation. Fig. 1 presents the fission time calculated [11] as a function of the initial excitation energy E* for ^{235}U

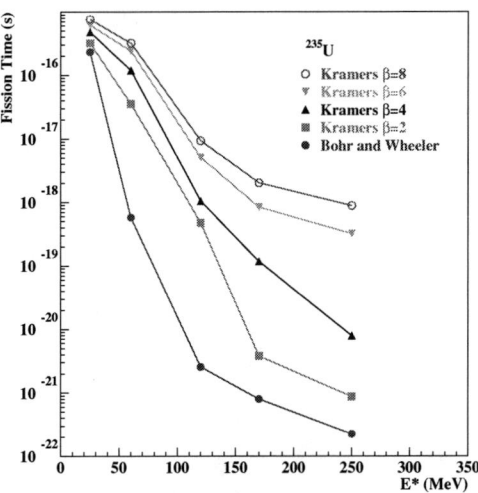

FIGURE 1. Fission times calculated as a function of the initial excitation energy. The lines are drawn to guide the eye. The β values are in units 10^{21}s^{-1}.

nuclei by an updated version of the Monte Carlo code SIMDEC [12]. The fission widths have been calculated either from the pure Bohr and Wheeler model [13], without any dissipation, or considering a width reduction, as predicted by Kramers [1], with various reduced friction parameters β. For each β value and excitation energy, the transient evolution of the fission width has been determined from a numerical resolution of a one dimension Langevin equation, including evaporation, and this transient evolution has been taken into account in SIMDEC. For $E^* = 250$ MeV, the fission time is increased by about 4 orders of magnitude between the calculation with Bohr and Wheeler approach and the one with $\beta = 8 \times 10^{21}\text{s}^{-1}$, confirming thus the very high fission time sensitivity on dissipation.

Since the first evidences for the influence of dissipation on fission dynamics [14], the fission times are often inferred from statistical evaporation calculations reproducing data on pre-scission neutron or γ-ray multiplicities. However, direct measurements, free of any nuclear model and parameter, lead to much longer times [3, 15, 16, 17] than those inferred from the neutron or γ clock. This discrepancy results from the lack of sensitivity of the nuclear clocks to the long lifetimes components [18, 11], as shown by Fig. 2 that presents, for ^{235}U excited at $E^* =120$ MeV, time distributions calculated by SIMDEC for fission (upper panel), pre-scission neutron emission (middle panel) and γ-ray emission (lower panel). This figure shows that pre-scission neutron emission becomes negligible after about 10^{-17}s, whereas fission is still quite probable at much longer times. In contrast, γ-ray emission is spread on the same time scale than fission, but the shadowed area in the lower panel indicates the time domain actually covered by γ-ray emission when a realistic experimental gate is applied on the measured γ energies. Considering

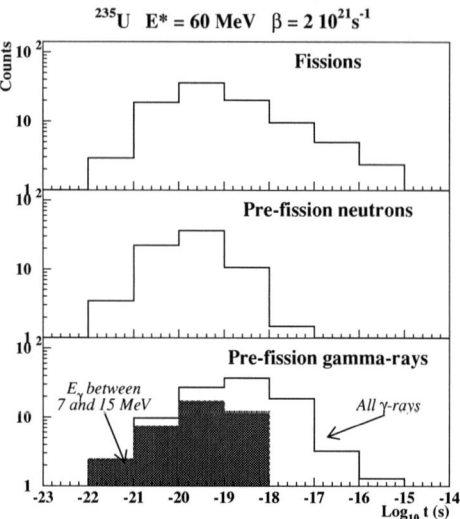

FIGURE 2. Upper panel: fission time distribution calculated by SIMDEC; middle panel: pre-fission neutron emission time distribution; lower panel: pre-fission GDR γ-ray emission time distribution (the shadowed area indicates the time distribution for γ-rays with energy between 7 and 15 MeV)

this experimental limitation, pre-scission γ-ray detection becomes also negligible at very long times. Therefore, these calculations show that the average value of the fission time distribution (corresponding to the fission time) inferred from either the neutron or γ clock are quite underestimated when long lifetime components are present.

The fission times of uranium-like nuclei have been measured on a broad range of excitation energy [3] at GANIL by the blocking technique in single crystals, a nuclear model free technique [19]. For excitation energies increasing between 60 MeV and 250 MeV, a roughly exponential decrease of the fission times between 5×10^{-18} and 3×10^{-19} s has been observed. Such a fission time evolution can only be taken into account by the calculations shown in Fig. 1 if a significant variation of the friction parameter β is considered. However, the comparison between data and calculations can be biased if the fissioning nuclei are experimentally identified by a selection performed on $Z_1 + Z_2$ (the sum of the measured atomic numbers of the coincident fission fragments) [18]: light charged particle emission, despite a weak probability for uranium-like nuclei, affects significantly the measured times. Nevertheless, the data of [3] have been obtained for $Z_1 + Z_2 = 92 \pm 5$, the width of the bin reflecting the experimental resolution. The very dominant projectile-like nuclei that undergo fission in this bin are uranium nuclei but projectile-like fragments fissioning after charged particle emission are also present in the bin due to its width. The data of ref. [3] can thus be directly compared to calculations performed for uranium nuclei. Fig. 3 presents for uranium nuclei the inferred β evolution with the initial temperature of the fissioning nuclei. As predicted by microscopic calculations [20, 21], an increase of nuclear friction is observed for increasing temper-

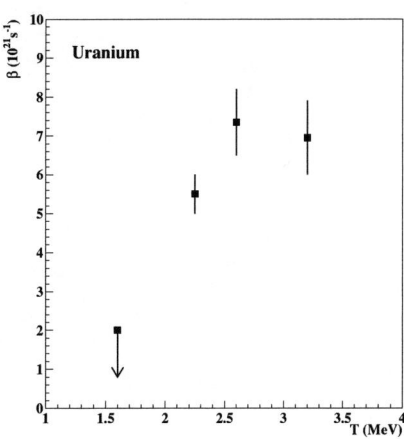

FIGURE 3. Friction parameter β inferred from the measured fission times of ref. [3].

atures. The present β evolution with temperature is weaker than the one inferred from pre-scission particle and γ emission [22]. However, in contrast to pre-scission multiplicities, the fission times are only sensitive to friction before the saddle point, at relatively small deformation, since they are much longer than the times associated with the saddle to scission path.

FISSION TIMES OF SUPER-HEAVY ELEMENTS

In order to determine if SHE fission times can actually be measured by the blocking technique in single crystals, an experiment has been performed at GANIL to form compound nuclei with Z=120 by bombarding a Ni single crystal with a ^{238}U beam at 6.62 MeV/A. Considering the rather high CN excitation energy, $E^* \approx 80$ MeV, the shell effects should be strongly damped and the CN fission barrier should thus be very low. Nevertheless, long lifetime components, accessible by the blocking technique, can arise from a progressive shell effect restoration due to decreases in temperature by fast pre-fission neutron emission. Evidence for very long reaction times, only compatible with compound nuclei formation, can thus be obtained from the detection of long lifetime components and a discrimination between the fast and highly anisotropic quasi-fission process and the slow fusion-fission process becomes possible.

Taking advantage of the reverse kinematics, the fission-like fragments were identified in atomic number and in energy (a careful atomic number and energy calibration has been achieved using beams of Kr, Xe, Pb, U at various energies and the corresponding Ni, Cu, Au target recoils). The blocking effects were measured at 11° (for projectile elastic scattering) and 20° (for fission-like events) by 3 specially designed telescopes, consisting of low threshold ionization chambers followed by 2-dimension position sensitive

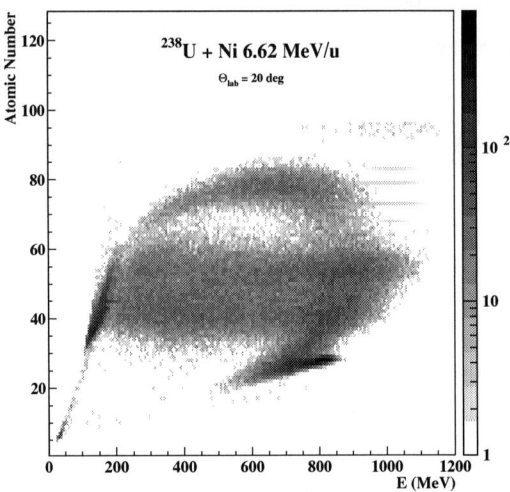

FIGURE 4. Atomic number versus kinetic energy for fragments detected at 20°.

silicon detectors, providing us with the atomic number, energy and emission angle (with a resolution better than 0.05°) of heavy fragments ($Z_1 \geq 6$). All the other charged reaction products (light charged particles, intermediate mass fragments, heavy fragments) were detected on a solid angle close to 4π by INDRA [23], allowing thus a control on the mechanisms involved. A complete description of the experimental set-up can be found in [8, 9].

Fig. 4 presents the correlation between the atomic number Z_1 and the energy E_1 measured by one of the telescopes located at 20°. A peak corresponding to target quasi-elastic scattering can be easily seen at $Z_1 \approx 28$ and $E_1 \approx 820$ MeV. Quasi-elastic scattering is a very fast process: the projectile- and target-like fragments separate within the thermal vibration domain of the Ni single crystal atoms. The blocking dip associated with these events (shown in Fig. 5-a) can thus be used as a reference for very short times. Below the quasi-elastic peak down to $Z_1 \approx 20$, a tail can be seen in fig. 4 arising from deep-inelastic reactions. Above the quasi-elastic and deep-inelatic regions, a large domain extending up to $Z_1 \approx 60$ is dominantly populated by sequential fission of uranium-like nuclei, although other mechanisms (quasi-fission, CN fission...) are also present with weaker probabilities. A very low light charged particle multiplicity ($M_{lcp} = 3 \times 10^{-2}$) is measured in this region, pointing for the excited uranium-like nuclei that undergo sequential fission to low excitation energies associated with very long fission times [3, 15, 17]. Therefore, the blocking dip associated with these events, shown in Fig. 5-b, can be used as a test for the sensitivity of the experiment to long times. Above the sequential fission, up to $Z_1 \approx 85$, a distinct domain can be seen, for which Fig. 5-c and 5-d present the dips for $E_1 < 630$ MeV and $E_1 > 630$ MeV, respectively. The analysis of the charged products in coincidence with these fragments

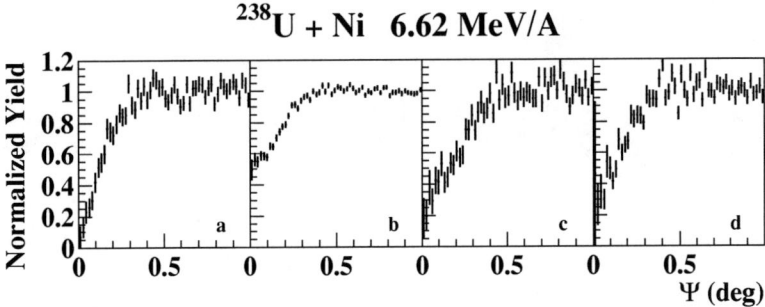

FIGURE 5. Blocking dips measured at 20° for: a) quasi-elastic target; b) dominant uranium sequential fission; c) low energy fission fragments with Z > 60; d) high energy fission fragments with Z > 60

indicates that: *i)* only two heavy fragments are present; *ii)* the sum of their atomic numbers have a gaussian shape distribution centered at $Z_1 + Z_2 = 120$ with a FWHM = 4, corresponding to the experimental resolution; *iii)* the intermediate mass fragment multiplicity is $\approx 4 \times 10^{-3}$, excluding therefore any incomplete fusion process; *iv)* the light charged particle multiplicity (Z = 1 or 2) is $\approx 7 \times 10^{-2}$, pointing to a sizeable excitation energy; *v)* the sum of the center-of-mass kinetic energies is in agreement with Viola systematics [24] (at most lower by 10%, depending on the asymmetry considered). All these characteristics correspond to what is expected either for complete fusion followed by fission or quasi-fission reactions. The discrimination between these two processes will therefore be achieved from the reaction times.

Fig. 5 shows striking differences between the measured blocking dips. However, the dip shapes depend on the atomic numbers and energies of the detected fragments and it is only through full simulations taking into account all the single crystal characteristics (including intrinsic defaults and damage due to irradiation) that reaction times can be reached from the shapes. Fortunately, a quite direct experimental proof of long times can be found in the χ_{min} value (the yield measured at $\psi = 0°$). Extensive experimental and theoretical works [19] on channeling and blocking processes have shown that, for a given single crystal, χ_{min} variations can only arise from time effects. The lowest possible χ_{min} value is obtained when the interacting system splits within the thermal vibration domain of the crystal. It obviously corresponds in the present experiment to $\chi_{min} \approx 0.1$, as measured for quasi-elastic scattering (Fig. 5-a). The other dips in fig. 5 have significantly higher χ_{min} values: ≈ 0.5 in Fig. 5-b for sequential fission, pointing as expected to very long times, and $\chi_{min} > 0.2$ in Fig. 5-c and -d. The time needed by a composite system recoiling with the center-of-mass velocity to move away from the well known thermal vibration domain of Ni atoms is $t_{min} = 7 \times 10^{-19}$s. Therefore, the high χ_{min} value observed in Fig. 5-c and -d is a direct evidence for reaction times $t_{reac} > 7 \times 10^{-19}$s, only compatible with a fusion process followed by fission.

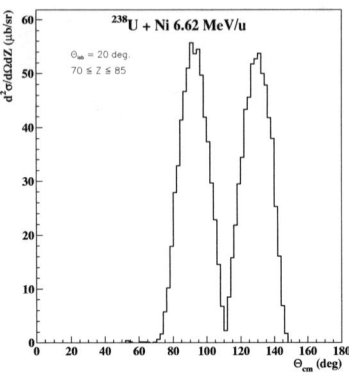

FIGURE 6. Angular distribution.

A sizeable part of the events with $Z_1 \geq 60$ arises thus from fission of compound nuclei with Z=120. In order to determine the amount of quasi-fission events in this region, the angular distributions in the center-of-mass frame have been studied. Fig. 6 presents this distribution for $70 \leq Z_1 \leq 85$. Two components corresponding to forward and backward emission in the center-of-mass frame can be seen with similar cross-sections. As shown in [25, 26, 27, 28], quasi-fission is a fast, non-equilibrium process, characterized by highly anisotropic angular distributions and strongly backward peaked emission for asymmetric splittings [25, 26, 28], whereas in the present experiment, even selecting very large fission asymmetries with $Z_1 \geq 80$, backward and forward emission keep similar weights. Taking into account all possible experimental errors (on energy calibration, atomic number identification, mass assumptions), the angular distributions are compatible with at most 20% of non-equilibrium events that represents thus the maximum possible amount of quasi-fission events. This low percentage is due to the center-of mass angles involved when heavy fragments with $Z_1 \geq 60$ are detected at 20° in the laboratory: quasi-fission has strongly decreased at these angles [25] and fusion-fission is dominant. An integration over 4π of the differential cross-sections $d^2\sigma/d\Omega_{cm}/dZ_1$ assuming isotropic emission for fragments with $60 \leq Z_1 \leq 85$ leads to a cross section $\sigma = 27$ mb. Considering the maximum possible mixture with quasi-fission, a minimum fusion cross-section $\sigma_{fusion}^{min} = 22$ mb can be inferred.

CONCLUSION

The fission times are very sensitive to the friction assumed in statistical calculations performed for fissile nuclei at high excitation energies. From a comparison between the fission times measured for uranium nuclei excited up to 250 MeV and statistical calculations considering reductions of the fission widths following the Kramers approach, an increase of the friction coefficient with temperature is found up to T$\approx$ 3 MeV, in agreement with predictions from microscopical calculations.

In the ^{238}U+Ni reaction at 6.62 MeV per nucleon, evidence for long lifetimes components, only compatible with fusion-fission processes, has been found for reactions leading to heavy fission-like fragments emitted at 20 degrees in the laboratory. A careful analysis shows that these fission-like events arise dominantly from the formation of compound nuclei with Z = 120 followed by fission. The shell effect damping at high excitation energy makes the compound nucleus fission barrier rather low whereas, in contrast, the long measured fission time points to high fission barriers in the fissioning nuclei. A shell effect restoration seems thus to arise from fast pre-fission neutron emission. These results give the opportunity to determine the position of the islands of stability for Z larger than 110 from the longest fission times measured at a given excitation energy.

ACKNOWLEDGMENTS

Many thanks are due to the authors of ref.[8] for their permission to present new results before publication.

REFERENCES

1. H.A. Kramers, *Physica (Utrecht)*, **7**, 84 (1940)
2. P. Grangé, L. Jun-Qing and H.A. Weidenmüller, *Phys. Rev. C*, **27**, 2063 (1983)
3. F. Goldenbaum et al., *Phys. Rev. Lett.*, **82**, 5012 (1999)
4. P. Möller et al., *At. Dat. And Nucl. Dat. Tab.*, **59**, 185 (1995)
5. J.F. Berger et al., *Nucl. Phys. A*, **685**, 1 (2001)
6. J.F. Berger, D. Hirata and M. Girod, *Acta Phys. Pol. B*, **34**, 1909 (2003)
7. T. Bürvenich et al., *Phys. Rev. C*, **69**, 014307 (2004)
8. D. Jacquet et al., *in preparation*.
9. A. Drouart et al., *Proc. of the International Symposium on Exotic Nuclei, Peterhof, 2004*, to be published in World Scientific.
10. Yu. A. Lazarev et al., *Phys. Rev. Lett.*, **70**, 1220 (1993)
11. S. Basnary, *Ph.D. Thesis, Univ. Caen*, (2002)
12. M. Ohta et al., *Tours Symposium on Nuclear Physics II*, edited by H. Utsunomiya et al., World Scientific, 1995, p. 480.
13. N. Bohr and J.A. Wheeler., *Phys. Rev.*, **56**, 426 (1939)
14. D. Hilscher and H. Rossner, *Ann. Phys. (Paris)*, **17**, 471 (1992)
15. J.U. Andersen et al., *Nucl. Phys. A*, **241**, 317 (1975)
16. J.U. Andersen et al., *Phys. Rev. Lett.*, **36**, 1539 (1976)
17. J.D. Molitoris et al., *Phys. Rev. Lett.*, **70**, 537 (1993)
18. I. Gontchar, M. Morjean and S. Basnary, *Europhys. Lett.*, **57**, 355 (2002)
19. D.S. Gemmell, *Rev. Mod. Phys.*, **46**, 129 (1974)
20. S. Yamaji et al., *Nucl. Phys.A*, **612**, 1 (1997)
21. H. Hofmann et al., *Phys. Rev. C*, **64**, 054316 (2001)
22. I. Diószegi et al., *Phys. Rev. C*, **61**, 024613 (2000)
23. J. Pouthas et al., *NIM A*, **357**, 41 (1995)
24. V.E. Viola, K. Kwiatkowski and M. Walker, *Phys. Rev. C*, **31**, 1550 (1985)
25. J. Töke et al., *Nucl. Phys.A*, **440**, 327 (1985)
26. R. Bock et al., *Nucl. Phys. A*, **388**, 334 (1982)
27. B.B. Back et al., *Phys. Rev. Lett.*, **50**, 818 (1983)
28. H. Keller et al., *Z. Phys. A*, **326**, 313 (1987)

FRAGMENT EXCITATION AND NEUTRON EMISSION

CURRENT STATUS OF THE SEARCH FOR SCISSION NEUTRONS IN FISSION AND ESTIMATION OF THEIR MAIN CHARACTERISTICS

G.A. Petrov

PNPI of RAS, Gatchina, Leningrad District, 188300 Russia

Abstract. A short review on the main prompt neutron characteristics in fission at low excitation-energy is presented. It is pointed out that the neutron spectra and their angular and energy distributions may be adequately described under the assumption of its evaporation from the fully accelerated fragments. But about (10 – 20)% of the neutrons may be emitted near the rupture point of the fissioning system. As such the "scission neutron" emission mechanism has to be closely connected with fission dynamics, the perspectives of further investigations are discussed.

INTRODUCTION

The question of the scission neutrons existence and values of their parameters have been the object of much attention during the whole long history of fission physics. But in spite of a lot of theoretical and experimental attempts to get definite answers, this question remains to be solved.

However, a neutron emitted near the rupture point of the fissioning nuclear system might be a very effective probe for some peculiarities of the fission dynamics.

The first theoretical analysis of the scission-neutron emission probability during the descent of the fissioning system from the saddle to the scission point had been done in the works [1,2]. As a result, it was concluded that the yield of neutrons about 0.4 n/fis is possible only if the descent time is less than $1.5 \cdot 10^{-21}$ sec.

It would appear reasonable that some neutrons might be emitted just at the moment of rupture (real scission neutrons) as, for example, the well-known light charged particles in ternary fission. In the work of Madler [3] the so-called "catapult" mechanism of fast particle emission had been proposed and analyzed in the framework of time-dependent mean-field theory. In this connection it should be noted that the author of [4] has estimated possible yields of scission neutrons considered as the third particle in ternary fission on the basis of a statistical approach. Using the interpolation formula, he obtained unexpectedly high yields for such scission neutrons - 0.55 n/fis for ^{235}U induced fission and 0.18 n/fis for the ^{252}Cf spontaneous fission. It is known as well that a few neutrons can be emitted as the products of some unstable ternary particles decay. But the yields of these neutrons are too small for discussions.

In any case, the vast of experimental information accumulated on the prompt fission neutrons lead to the conclusion that the main part of the neutrons is emitted much later from highly excited fission fragments, which had already obtained their final velocities in the Coulomb field.

To test other possible mechanisms of prompt neutron emission in low-energy fission, the experimentalists usually have at their disposal only neutron multiplicities, energy, and angular distributions of neutrons with respect to the fission axis in the laboratory coordinate system (LCS), and (n-n) angular correlations. That is why henceforth we shell consider shortly only those properties of fission neutrons that are important for the scission-neutron search and further investigations. Thereafter all specific experimental possibilities of the search for the so-called scission neutrons emitted near the rupture point of the fissioning nuclei will be discussed in more details. And finally, the main concepts of our new Project directed to finding solid evidences for the existence of scission neutron in the fission process will be presented.

THE MAIN FEATURES OF PROMPT NEUTRON EMISSION IN LOW ENERGY FISSION

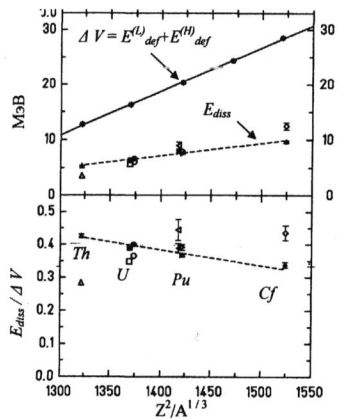

Fig.1. Dependences of dissipation and deformation energies at scission point on $Z^2/A^{1/3}$ /5/.

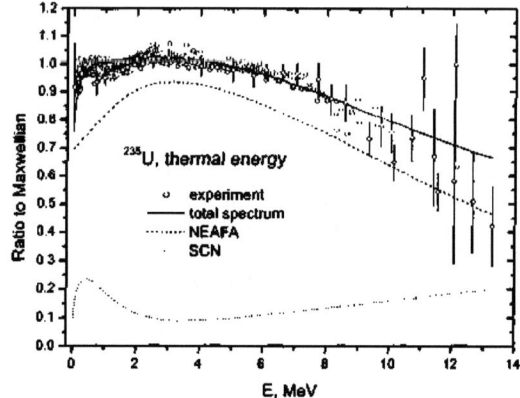

Fig.2. Comparison of experimental neutron spectrum in ^{235}U fission with the spectrum calculated in the NEFAF model with the scission neutron component (SCN)) /6/.

It is well known that prompt neutrons in the cases of ^{252}Cf spontaneous fission and thermal neutron induced fission of ^{235}U carry away about 82% and 70% of the total excitation energy respectively. In its turn this energy amounts correspondingly to about 17% and 12% of the total energy finally realized in the reaction in form of excitation and kinetic energies. As regards to these energies at the rupture point (that seems to be of interest for the problem of scission-neutron emission), up to now there are only circumstantial data or even only rough estimates (see Fig.1).

The shape of the fast-neutron spectrum observed in LCS is usually described by a Maxwellian distribution. But it was recently shown in the work /6/ for the case of ^{235}U fission that a much better description of numerous experimental data can be achieved if one uses a specific model of neutron evaporation from fully accelerated fragments (NEFAF) and includes scission-neutron component with the parameters shown in Fig.2.

The averaged prompt-neutron yields grow with the mass of the fissioning system (Fig.3.) as a result of increase of the total energy realized in fission and some changes in the fission dynamics with $Z^2/A^{1/3}$ (see Fig.1.). As one can see from the Fig.4, the variance of the prompt-neutron distribution remains practically invariable.

Prompt neutron yields have a very strong specific dependence on the masses of fission fragments (Fig.5) that is governed mainly by the excitation energies of the nucleus ejecting the prompt neutrons. An approximately linear dependence of the overall averaged neutron multiplicity on the total kinetic energies (TKE) of fragments readily follows from the equation Q = (TKE + E_{excit}) (see Fig.6).

What catches the eye in the Fig.7 is the distinctly pronounced angular anisotropy of prompt neutron emission as a function of $\mu = \cos\varphi$. It is evident from the rather simple calculations that these neutrons are emitted from the moving fragments, which completely or partially reached their maximum velocities.

If one proposes that all neutrons have been emitted isotropically in the center-of-mass system (CMS) and fission fragments at that time had been fully accelerated, then the ratio of integrals of experimental and calculated neutron spectra should remain constant for all values of $\mu = \cos\varphi$. As one can see from the Fig.8, it is not the case. Namely, much more neutrons are emitted at the angles close to 90^0 relative to the fission axis then it is predicted by theoretical calculations. This excess looks especially significant in the case of ^{235}U neutron- induced fission. This already shows that it is necessary to look for some additional mechanism of neutron emission in fission (e.g. scission neutrons) or for some possible systematical errors not taken into account in the data evaluation.

The (n-n) angular correlation experiments present some additional possibilities to test different models of fission neutron emission. In principal such measurements may be performed without registration of fission fragments and their characteristics. In spite of the fact that the experimental results are averaged over neutron energies and

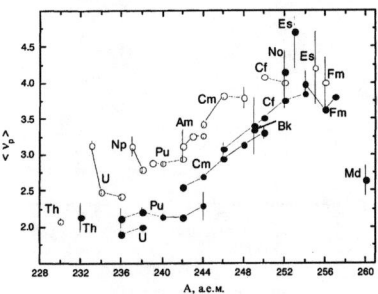

Fig.3. Average neutron multiplicities for different fissile nuclei /7/.

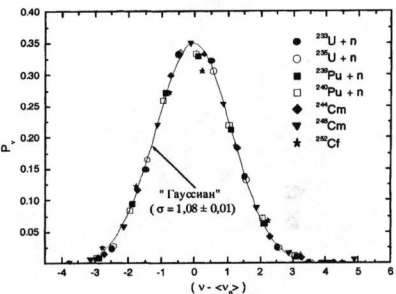

Fig.4. Relative probabilities of prompt neutron emission in different reactions /8/.

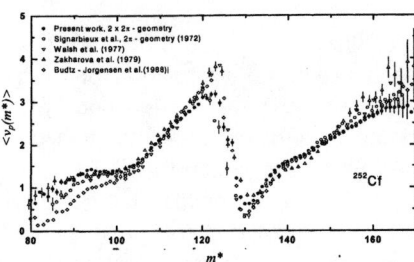

Fig.5. Neutron multiplicities as a function of ^{252}Cf fission fragment masses /9/.

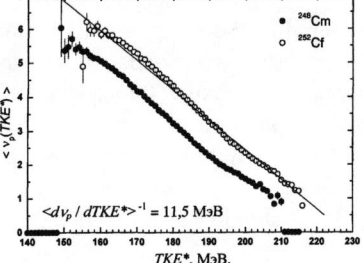

Fig.6. Average neutron multiplicity as a function of total kinetic energy of the fission fragments for Cm and Cf /9/.

fission fragment characteristics including flight directions, such measurements appear to be rather sensitive to the existence of scission neutrons. Two experiments of such type had been performed long ago for the ^{252}Cf spontaneous fission /15/ and for neutron-induced ^{235}U fission /16/.

Although very similar experimental methods were used in these both measurements, the resulting data turned out to be quite contradictory. Whereas in the case of ^{235}U-induced fission the (n-n) angular correlation data together with the (n-f) angular and energy distributions might be successfully described by simple NEFAF model with 20% isotropic scission-neutron mixture, the inclusion of the scission-neutron component in the case of ^{252}Cf spontaneous fission practically did not improve the theoretical description of similar experimental data. As a result, one needs to find some additional mechanism of polar neutron emission to achieve acceptable agreement between the performed calculations and the obtained experimental data.

It is clear that these puzzles may be resolved only with the help of new experiments and new theoretical approaches.

THE MAIN DIRECTIONS AND RESULTS OF THE SEARCH FOR SCISSION NEUTRONS IN LOW-ENERGY FISSION.

From a general point of view emission of scission neutrons in binary fission might

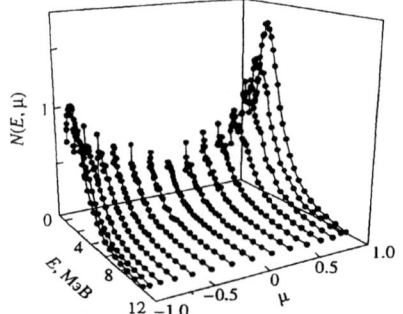

Fig.7. Energy and angular distributions of prompt neutrons in ^{252}Cf fission in LCS. $\mu = \cos\varphi$ stands for the LCS angle relative to the direction of the light fission fragment /10/.

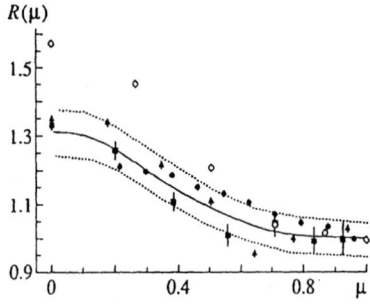

Fig.8. Ratio of the experimental neutron spectrum integrals for ^{252}Cf (•) /11-13/ and for ^{235}U (o) /14/ to the integrals calculated in NEFAF model /10/.

be considered in the same way as the arising of the light particles in ternary fission. But such a point of view does not give us any new and useful information, because the mechanism of the charged ternary particle emission is also poorly understood by now.

It was mentioned above that an appreciable quantity of pre-scission neutrons (about 0.4 n/fis) might be emitted only if the descent time is shorter then $2 \cdot 10^{-21}$sec /1/. As long as such a fast a descent is considered to be not acceptable, the pre-scission neutron yield should be negligibly small. But the times of this order of magnitudes seem to be reasonable for the duration of the rupture of the fissioning system or for the time of the neck sucking up or "ripple" propagation /3/. If this were the case, such scission neutrons would appear as the result of a non-adiabatic rupture process. But presently nobody can predict the main characteristics of these hypothetical neutrons or their specific distinctions from the ordinary neutrons evaporated from the highly excited fission fragments. Actually we have to search for the neutrons emitted from fissioning system being as a whole at rest. But the search for these neutrons superimposed on the background of prompt fission neutrons is not a simple task.

Up to now practically the only way had been used in order to find an answer concerning scission neutron existence and their properties - precise comparison of the experimental energy and angular distributions of fission neutrons emitted by the fragments of definite kinetic energy and masses with theoretical calculations performed under different suppositions about the mechanism and the conditions of neutron emission. The main information about scission neutrons accumulated as a result of different experimental works is presented in the Table I.

TABLE I. Experimental data on the scission neutron characteristics.

Nuclear reaction	Type of experiment	% of scission neutrons (SN)	Typical average SN energy, Mev	Reference
^{235}U + n$_{th}$	(n-f) correlation	15%	1.58	Skarsvag (1963)
^{235}U + n$_{th}$	– " –	10%	3.2	Kapoor (1963)
^{235}U + n$_{th}$	(n-n) correlation	20%	—	Franklyn (1978)
^{235}U + n$_{th}$	(n-f) correlation	(10 ± 2)%	—	Samant (1995)
^{235}U + n$_{th}$	Compilation	14%	0.98 and 2.74	Kornilov (2001)
^{235}U + n$_{th}$	Total spectra	15%	—	Kornilov (2002)
^{252}Cf sp. fission	(n-f) correlation	10%	2.6	Bowman (1962)
^{252}Cf sp. fission	– " –	20%	> 1.5	Piksaikin (1977)
^{252}Cf sp. fission	– " –	(13.2 ± 3.1)%	~ 1	Riehs (1981)
^{252}Cf sp. fission	– " –	~10%	(1.5 ±0.3)	Seregina (1985)
^{252}Cf sp. fission	– " –	1.1%	0.39	Budth-Jorgenson (1988)
^{252}Cf sp. fission	– " –	No SN	—	Marten (1989)
^{252}Cf sp. fission	– " –	< 1%	—	Blinov (1989)
^{252}Cf sp. fission	Compilation	10%	0.9 and 3.0	Kornilov (2001)
^{252}Cf sp. fission	(n-n) correlation	10%	—	Pringle (1975)

As seen from Table I, all experimental works on ^{235}U fission present strong indications for the evident existence of an additional source of neutrons that is not connected with evaporation from the moving fission fragments. As to the case of ^{252}Cf spontaneous fission the situation looks not so evident, may be because the admixture of such neutrons here is smaller then in the case of ^{235}U neutron-induced fission. But practically in all cases the estimates of possible scission neutron yields and energies have no sufficiently justified statistical and systematical errors.

In addition to this information, new and very interesting results on the possible yield of pair scission neutrons has been received recently under multi-parametric investigations of the cascade gamma transitions in the pair of even- even fission fragments of ^{252}Cf spontaneous fission /23/.

Another possibility of direct estimation of the scission-neutron yields in low-energy fission appeared recently as a result of the first observation and the investigations of the so-called effect of the T-odd asymmetry of the light-particle emission in ternary fission induced by the cold polarized neutrons /24/. It was shown /25/ that such effect exists for different light ternary particles but only under the condition of their simultaneous emission with fission fragments. In our case it means that the effect has to be absent for the neutrons evaporated from the fission fragments.

A very interesting and promising possibility to study space-temporal properties of the particles emitted in nuclear reactions has been proposed recently in the works of Lyuboshitz et al. /26/. Such information may be obtained from the measurements of interference effect in the angular distribution of two particles emitted under very small relative emission-angle and linear-momentum difference.

MAIN FEATURES OF THE NEW PROJECT ON THE SEARCH FOR SCISSION NEUTRONS AND THE FIRST EXPERIMENTAL RESULTS.

As an object of investigation we selected thermal neutron-induced fission of ^{235}U. This choice was caused by the existing essential disagreements in the experimental data on general fission neutron characteristics in ^{235}U fission and by the fact that just here the maximum yield of scission neutrons has been theoretically predicted /4/ and observed in the subsequent experiments (see the Table I).

As the first and the most important aim of our work we consider the direct and independent evidence for the existence of scission neutrons in fission. We hope to do this by observation of a T-odd asymmetry effect for prompt neutrons emitted at the right angle to the plain formed by the directions of the light fragment linear momentum and longitudinal polarization of the cold neutrons that initiated ^{235}U fission.

The key features of our project compared to the large body of preceding researches referred to the search for scission neutron search are as follows:
- special directionality of the Project to the solution of this task,
- combined performance of different types of measurements sensitive to the existence of scission neutrons existence and their properties,
- systematic error analysis and execution of special control experiments to check the possible influence of different factors on the conclusions about scission neutron yields and their characteristics.

As the most effective and sensitive way to get reliable information about the yield, energy and angular distribution of scission neutrons, we use precise combined measurements of the double differential distributions of prompt neutrons emitted in fission with fixed masses and kinetic energies of the fragments, and neutron - neutron angular correlations with simultaneous measurements of both neutron energies. In these experiments the special attention is paid to provision of low background, firm discrimination against fission γ-quanta, steady control of the prompt neutron registration threshold and efficiency, elimination of neutron re-scattering at the set-up elements, and exclusion of detector's cross-talk.

With the aim to achieve a correct evaluation of the obtained experimental data, the broad program of theoretical calculations is incorporated. Prompt neutron energy and angular distributions are calculated with the Monte-Carlo method taking into account the cascade-neutron spectra, evaporated by the fragments with the large aligned angular momentum arising in the rupture point /27/. Besides, the yield and energy spectra of neutrons evaporated during the process of fragment acceleration in the Coulomb field will be calculated under different reasonable suppositions about neutron emission time. Scission neutron yields and theirs energy and angular distributions together with estimates of possible systematic errors will be extracted from the comparison of all experimental data with proper theoretical calculations.

As a part of this Program realization the first search measurements of the T-odd asymmetry coefficient of fission-neutron emission were performed at the WWR-M reactor of the PNPI in the ^{235}U and ^{233}U fission induced by thermal polarized neutrons /28/. The values of T-odd asymmetry coefficients looks as follows:
$$<D>(^{235}U) = -(9 \pm 5)\cdot 10^{-4} \text{ and } <D>(^{233}U) = -(3 \pm 7)\cdot 10^{-4}$$
Taking into account these values we can conclude that the yields of scission neutrons are not higher then (10 ÷15)% (90% c.l.). In 2005 we hope to improve the statistical accuracy of these measurements by (3 ÷ 5) times at the cold polarized neutron beam of the ILL high-flux reactor.

From the other side we could estimate the scission-neutron yield in the ^{235}U fission from the results of measurements of double-differential distributions of prompt fission neutrons performed in collaboration with the V.G. Khlopin Radium Institute at the WWR-M reactor (ISTC-554, 554-2). The experimental set-up consisted of 12 photo multipliers with stilben crystals without shielding and a back-to-back ionization chamber.

Average yield and energies of scission neutrons obtained by comparing the experimental neutron spectra with the theoretical ones (calculated in the framework of NEFAF model) turned out to be equal /29/:

Y_n/fission = 8% ± 2%(stat) ± 2%(sys), and the energies ~ 0.4 MeV and 2.1 Mev.

Much more careful measurements of energy and angular distributions of prompt neutrons in the ^{235}U thermal neutron-induced fission are carried out now at the reactor WWR-M under quite different environmental conditions /30/.

The experimental installation for these measurements consists of a low-pressure chamber with 8 pairs of MWPC detectors for registration of pair fission fragment velocities at the eight angles with respect to the axes of two neutron detectors, and a zero-time MWPC counter placed adjacent to the thin fissile target. Two neutron detectors with the angle 90° between them represent the stilben crystals (60 x 40) mm^2 and fast photo multipliers FEU-30 placed in the plane of 8 pairs of MWPCs and MWPC start –target assembly at 50 cm from the center.

New measurements of neutron – neutron angular distributions in thermal neutron-induced fission of ^{235}U and ^{252}Cf spontaneous fission are now in a preliminary. As mentioned above, the results of two previous investigations /15, 16/ of such type had come to contradictory conclusions. We will repeat these investigations under conditions of essential suppression of background of scattered neutrons and γ-radiation from the fission fragments and with additional measurements of fission-neutron energies.

New Monte-Carlo calculations of the (n-n) angular distributions will be performed taking into account the large angular momentum of the fission fragments, different types of cascade neutron evaporation spectra in CMS and possibility of neutron emission during fragment acceleration.

If the yield of scission neutrons will turn out to be not less then 10% we plan to explore the interference effect on the angular correlation for two neutrons emitted in fission with a small difference of their values of linear momentum and emission direction /26/. First experiments of such type had been performed already /31, 32/ and today's task consists in essentially increasing the accuracy.

The total execution of the Program will take about two years and will be realized at the WWR-M reactor of PNPI of RAN and partly at the High Flux Reactor in Grenoble. This Program is supported partly by INTAS Grant 03-51-6417 and the Grant RFBR 02-02-17051.

REFERENCES

1. R.W. Fuller. Phys. Rev. 126, 684 (1962).
2. Y. Bonch and Z. Fraenkel. Phys. Rev. 10 C , 893 (1974).
3. P. Madler. Z. Phys. A 321, 343 (1985)
4. G.V. Val'ski. Yad. Fiz. 67, 1288 (2004).
5. F.Rejmund, A.V, Ignatyuk et al. Nucl. Pys. A678, 215 (2000).
6. N.V. Kornilov, A.B. Kogalenko. INDC Nucl. Const. Iss.2 (2002).
7. V.V. Malinovski, V.G. Vorobyev et al. Questions of Nucl. Sci. and Tech. Ser. Nucl. Const. (Rus). Iss.5(54), 19 (1983)
8. N.E. Holden and M.S. Zuker. Proc. Int. Conf. "Nucl. Data for Basic and Appl. Science" Santa Fe, May 1985, 1631 (1986).
9. A.C. Vorobyev. PhD Thesis. PNPI of RAS (2004).
10. N.V. Kornilov, A.B. Kogalenko et al. Yad. Fiz. (Rus) 64, 1 (2001).
11. H.R. Bowman, J.C.D. Milton et al. Phys. Rev. 126, 2120 (1962).
12. E.A. Seregina. PhD Thesis (1985). Yad.Fiz. (Rus) 42,1337 (1985).
13. C.Budtz-Jorgensen and H.-H. Knitter.Nucl. Phys. A490, 307 (1988).
14. K. Skarsvag and K. Bergheim. Nucl. Phys. 45, 72 (1963).
15. J.S. Pringle and F.D. Brooks. Phys. Rev. Lettr. 35, 1563 (1975).
16. C.B. Franklyn, et al. Phys, Lett. 78B, 564 (1978). Proc. Int. Conf. "Nucl. Data for Basic and Appl. Sci.", Santa Fe, May 1985, v.1, p. 323 (1986).
17. S.S. Kapoor, R. Ramanna et al. Phys. Rev. 131, 283 (1963).
18. V.M. Piksaikin, P.P. Dyachenko et al. Yad. Fiz. (Rus) 25, 723 (1977).
19. P. Riehs. Acta Phys. Austriaca 53, 271 (1981).
20. H. Marten, D. Richter et al. Proc.of IAEA Consulting Meeting, Vienna, (INDC/NDC) –220, 161 and 245 (1989).
21. M. Blinov, O. Batenkov. Proc.of IAEA Consulting Meeting, Vienna, (INDC/NDC) –220, 207 (1989). At. Energy, 64, 429 (1988).
22. M.S. Samant, R.P. Anand et al. Phys. Rev. C51, 3127, (1995).
23. J.R. Hwang, A.V. Ramayya et al. Phys.Rev. C 60, 044616-1, (1999).
24. V. Bunakov, F. Goennenwein et al. ILL Report ILL01 BU03T, ILL, Grenoble (2001).
25. A.L. Barabanov, V.E. Bunakov et al. Yad. Fiz. 66, (2003)
26. R. Lednicky and V. Lyuboshits. Sov. J. Nucl. Phys. 35(5), 770, (1982).
27. I.S. Guseva and G.A. Petrov. Proc. Intern. Seminar ISINN-12, Dubna, JINR, (2004).
28. A.M. Gagarski, F. Goennenwein et al. Proc. Int. Seminar ISINN-12, Dubna, JINR (2004)
29. A.S. Vorobyev, V.A. Kalinin et al. Preprint PNPI of RAS, PNPI-2004, No. 2591 (2004).
30. G.V. Val'ski, A.M. Gagarski et al. Preprint PNPI-2003, No.2546 (2003).
31. M.M. Danilov, Yu..D. Katarzhanov et al. Phys. Atom. Nucl. 58, 349 (1995).
32. Yu.D. Katarzhanov, V.G. Nedopekin et al. Phys. Atom. Nucl., 62, 170 (1999).

Influence of fission fragment excitation energy on prompt fission neutron observables

S. Lemaire*, P. Talou*, T. Kawano*, M. B. Chadwick[†] and D. G. Madland*

*Theoretical Division, Nuclear Physics group, Los Alamos National Laboratory, Los Alamos, NM, 87545
[†]PADNWP, Los Alamos National Laboratory, Los Alamos, NM, 87545

Abstract.
We have implemented a Monte-Carlo simulation of the statistical decay of fission fragments by sequential neutron emission. In this presentation, we report on some numerical results obtained for the spontaneous fission of ^{252}Cf and neutron-induced fission of ^{235}U at 0.53 MeV neutron energy, and compare them to the results of the Los Alamos model for the calculation of the average number of prompt fission neutrons $\bar{\nu}$ and the prompt fission neutrons spectrum $N(E)$. Within this approach, we also calculate neutron multiplicity distributions $P(\nu)$ as well as neutron-neutron correlations such as the full matrix $\bar{\nu}(A, TKE)$. Two assumptions for partitioning the total available excitation energy among the light and heavy fragments are considered. The influence of the fission fragments excitation energy for low total kinetic energies on prompt neutron energy spectrum, multiplicity distributions and $\bar{\nu}(A, TKE)$ is discussed.

Keywords: neutron induced fission, spontaneous fission, fragment excitation energy, neutron multiplicity, neutron energy, total excitation energy, fission, prompt fission neutron spectrum, correlations
PACS: 25.85.Ca, 25.85.E, 21.10.Gv, 21.60.Ka, 24.10.Pa, 24.60.Dr, 24.75.+i, 25.40.-h

INTRODUCTION

For years, the Los Alamos model [1] has been successfully used to predict, with few parameters fitted to experimental data, the neutron energy spectrum $N(E)$ and the average number of prompt fission neutrons $\bar{\nu}$ as a function of the fissioning nucleus and its excitation energy. However, the Los Alamos model can only calculate physical quantities averaged over many components, such as the fission fragments (FF) mass yields and the whole nuclear decay chain of a given FF.

From a fundamental point of view, the knowledge of more specific informations on prompt fission neutrons will help improve our knowledge of the fission process. To go a step further than the Los Alamos model, we have developed a Monte-carlo approach to the FF statistical decay based on sequential neutron emission, which allows us to investigate quantities such as neutron multiplicity distributions $P(\nu)$ as well as neutron-neutron correlations such as the full matrix $\bar{\nu}(A, TKE)$.

In this proceeding, we will present, in a first part, the theoretical framework and input parameters used in our Monte-Carlo code. Then numerical results are provided for two different cases, spontaneous fission of ^{252}Cf and neutron-induced fission of ^{235}U (at 0.53 MeV neutron energy), and compared with experimental data. Finally, potential improvement to this model will be proposed.

THEORETICAL APPROACH

In this work, we have implemented a Monte-Carlo simulation of the statistical neutron emission from FF. A Monte-Carlo approach allows us to follow each fragment and subsequent neutrons throughout the decay chain so that we can investigate: neutron energy spectra $N(E)$, neutron multiplicity distributions $P(\nu)$ as well as correlations such as the full matrix $\bar{\nu}(A,TKE)$.

Methodology

We first sample the FF mass and charge distributions, and pick a pair of light and heavy nuclei. The FF mass and charge distributions is given by $Y(A,Z) = Y(A)_{exp} \times P(Z)$, where $Y(A)_{exp}$ represents an experimental pre-neutron FF mass distribution. The charge distribution $P(Z)$ is assumed Gaussian in shape.

The total excitation energy available for the pair $(A,Z)_L$ (light), $(A,Z)_H$ (heavy) reads

$$TXE(A_L,A_H,Z_L,Z_H) = E_r^*(A_L,A_H,Z_L,Z_H) + B_n(A_c,Z_c) + E_n - TKE(A_L,A_H), \tag{1}$$

where $E_r^*(A_L,A_H,Z_L,Z_H)$ is the energy release in the fission process, which is given, in the case of binary fission, by the difference between the compound nucleus and the FF masses. $B_n(A_c,Z_c)$ and E_n are the separation and kinetic energies of the neutron inducing fission (in the case of spontaneous fission, both $B_n(A_c,Z_c)$ and E_n terms in Eq. (1) vanish). $TKE(A_L,A_H)$ is the total FF kinetic energy obtained from the distribution of TKE which is assumed to be Gaussian with a mean value and width taken from experiment.

In the present study, we have considered three hypotheses aimed at understanding what is total excitation energy available to the FF for neutron emissions and how this energy gets partitioned among the light and heavy fragments:

- Partitioning (H1) so that both light and heavy fragments share the same temperature at the instant of scission (hypothesis identical to the one made in the Los Alamos model [1]). From this condition, it follows that the initial excitation energy of a given FF is:

$$E_{L,H}^* = TXE \frac{1}{1 + \frac{a_{H,L}}{a_{L,H}}}, \tag{2}$$

where L and H refer to the light and heavy system, respectively.

- Partitioning (H2) using the experimental $\bar{\nu}(A)$ to infer the initial excitation of each fragment. This condition writes as follow:

$$E_{L,H}^* = TXE \frac{\bar{\nu}_{exp}(A_{L,H})\langle \eta \rangle_{L,H}}{\sum_{i=L,H}(\bar{\nu}_{exp}(A_i)\langle \eta \rangle_i)}, \tag{3}$$

where $\langle \eta \rangle_{L,H}$ is equal to the average energy removed per emitted neutron

$$\langle \eta \rangle_{L,H} = \langle \varepsilon \rangle_{exp}^{L,H} + \frac{1}{2}B_{2n}(A_{L,H},Z_{L,H}), \qquad (4)$$

where B_{2n} is the two neutron separation energy and $\langle \varepsilon \rangle_{exp}^{L,H}$ the average neutron kinetic energy for a given initial FF.

- Rescaling (H3) using the experimental $\bar{\nu}_{TKE}$ and $\overline{E}_\gamma(TKE)$ to infer the TXE available for neutron emission and then partitioning using hypothesis (H2),

$$TXE(A_L,A_H,Z_L,Z_H) = \overline{\nu}(TKE) \times (\langle \varepsilon \rangle + \langle B_n \rangle) + \overline{E}_\gamma(TKE), \qquad (5)$$

where $\langle \varepsilon \rangle$ and $\langle B_n \rangle$ are the average center of mass neutron energy and binding energy taken equal to 1.265 MeV and 5.5 MeV respectively.

Within the Fermi-gas model, the initial FF excitation energy $E^*_{L,H}$ is simply related to the nuclear temperature $T_{L,H}$. The probability for the FF to emit a neutron at a given kinetic energy is obtained by sampling over the Weisskopf spectrum at this particular temperature [2]:

$$\phi(A,Z,\varepsilon,T) = \frac{\varepsilon}{T^2_{A-1,Z}} e^{\frac{-\varepsilon}{T_{A-1,Z}}}, \qquad (6)$$

where $T_{A-1,Z}$ is the nuclear temperature of the residual nucleus given by

$$T_{A-1,Z} = \sqrt{\frac{E^*(A,Z) - B_n(A,Z)}{a_{A-1,Z}}}, \qquad (7)$$

with $a_{A-1,Z}$ the level density parameter of the nucleus.

The emission of a neutron of energy ε from the FF at the excitation energy E^* produces a residual nucleus with the excitation energy

$$E^*(A-1,Z) = E^*(A,Z) - \varepsilon - B_n(A,Z). \qquad (8)$$

The sequential neutron emission ends when the excitation energy of the residual nucleus is less than the sum of its neutron separation energy and pairing energy.

The transformation of the center-of-mass spectrum to the laboratory spectrum is done by assuming that neutrons are emitted isotropically in the center-of-mass frame of a FF. So, sampling over the angle of emission of the neutron $\theta_n \in [0,\pi]$ for each nucleus (A, Z), we infer the neutron energy in the laboratory frame, taking into account the recoil energy of the residual nucleus.

Input Parameters

In the present calculation, we sample over the pre-neutron fragments yields $Y(A)_{exp}$, i.e., before neutron evaporation, as reconstructed from the experimentally measured fission products mass distribution. In particular, we use the data by Hambsch [3] in the case of ^{252}Cf(sf), and the data by Schmitt [7] in the case of the neutron-induced fission (at 0.53 MeV) on ^{235}U.

We used 255 fragments to represent the $Y(A,Z)$ for the neutron induced n(0.53 MeV)+^{235}U reaction. In particular, we considered 85 equispaced fragment masses (between $76 \leq A \leq 160$) with 3 isobars per fragment mass, around the most probable charge Z_p. In the case of spontaneous fission of ^{252}Cf, we used 315 FF between $74 \leq A \leq 178$ with 105 fragment masses.

Nuclear masses are used to calculate the energy release for a given pair of FF and were taken from the data tables by Audi, Wapstra, Thibault [4].

In our calculation, the level density parameter is:

$$a(A,Z,U) = a^* \left\{ 1 + \frac{\delta W(A,Z)}{U} \left(1 - e^{-\gamma U}\right) \right\}, \qquad (9)$$

where $U = E^* - \Delta(A,Z)$, $\gamma = 0.05$, a^* is the asymptotic level density parameter [5]. The pairing Δ and shell correction δW energies for the FF were taken from the nuclear mass formula of Koura et al.[6].

The total kinetic energy is used to assess the total FF excitation energy distribution. It is assumed to be approximately Gaussian in shape with an average value and width taken from the experiment (Ref. [3] for spontaneous fission of ^{252}Cf and Ref. [7] for the neutron induced n(0.53 MeV)+^{235}U reaction).

For sake of simplicity, we have assumed no mass, charge or energy dependence of the cross section for the inverse process of compound nucleus formation. This approximation will be reviewed later on.

We have used the average number of emitted neutrons $\bar{\nu}_{TKE}$ and $\bar{\nu}(A)$ as a way of rescaling and partitioning the total excitation energy distribution between the light and heavy fragment. For the spontaneous fission of ^{252}Cf we used data from Refs. [8]. For the neutron induced n(0.53 MeV)+^{235}U reaction, we used data from Ref. [9]. The (H3) hypothesis requires the knowledge of the average total energy removed by γ-rays, $\bar{E}_\gamma(TKE)$. For the spontaneous fission of ^{252}Cf we used data from Refs. [10]. For the neutron induced n(0.53 MeV)+^{235}U reaction, we used data from Ref. [11].

RESULTS AND DISCUSSION

Our Monte-Carlo simulations were done using 10^9 events for both spontaneous fission of ^{252}Cf and neutron induced n(0.53 MeV)+^{235}U reactions. Numerical results were obtained for various prompt fission neutron observables for the three hypotheses considered, (H1), (H2) and (H3).

We report in Fig. 1 the results obtained under the three, (H1), (H2) and (H3) hypotheses with experimental data from Nishio et al. [9] on the total average number of emitted neutrons as a function of the total FF kinetic energy for ^{235}U. The fact that in our approach the total excitation energy increase with decreasing TKE (see Eq.(1)) is responsible for the increase of $\bar{\nu}_{TKE}$ in the case of (H1) and (H2) hypotheses. In addition, since the total excitation energy TXE available to the FF does not depend on the partitioning, similar results are obtained for the calculated $\bar{\nu}_{TKE}$ under both (H1) and

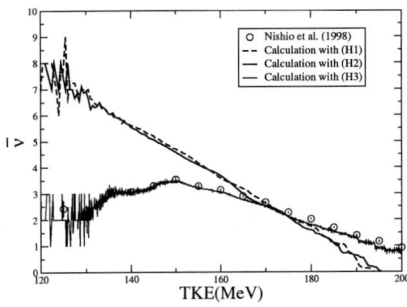 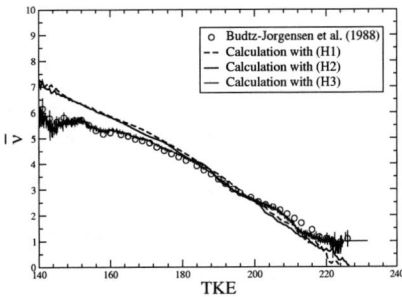

FIGURE 1. Sum of the neutron multiplicities from both FF, $\bar{\nu}_L + \bar{\nu}_H$ plotted as a function of TKE for n(0.53 MeV)+^{235}U reaction (left) and ^{252}Cf (sf) (right). The open circles are experimental data from [9].

(H2) assumptions (Fig. 1). Our calculations under (H1) and (H2) hypotheses deviate from experimental results by overpredicting $\bar{\nu}$ for low TKE (below about 160 MeV). Some deviations also appear for higher TKE (above about 180 MeV) where we predict too many prompt neutrons as compared to experimental data. A dramatic deviation between calculation and experiment on $\bar{\nu}$ is observed for low TKE, that would indicate the presence of additional opened channels. To understand the effect of the low TKE part of the total excitation energy available to FF on prompt fission neutron observables, we rescaled TXE in an (H3) hypothesis using experimental data available on $\bar{\nu}_{TKE}$. Of course, we expect the results for $\bar{\nu}(TKE)$ under (H3) assumption to be in better agreement with experiment as compared to the others results since these experimental data have been used to rescale TXE (see Fig. 1). The same observations but less pronounced are observed for the spontaneous fission of ^{252}Cf.

As pointed out earlier, when looking at $\bar{\nu}(TKE)$, we cannot distinguish between the (H1) and (H2) hypotheses. However, one observable that would be sensitive to the partitioning of TXE is the distribution $\bar{\nu}(A,TKE)$. Both measurements and calculations are compared on Fig. 2 for ^{235}U. Figure 2 shows some cuts of $\bar{\nu}(A)$ versus TKE (for the following specific total kinetic energies 140, 145, 150, 155, 160 and 165 MeV). The comparison of our results with data on Fig. 2 clearly show different behaviours under (H1) and (H2) assumptions. The (H2) calculation is in better agreement with experimental points. However, some large deviations are observed at low TKE (140, 145, 150 MeV), reflecting the observation made earlier on $\bar{\nu}(TKE)$. In the particular region of total kinetic energy peak (TKE~ 165MeV), our calculation under the (H2) assumption is in very good agreement with experiment (Fig. 2). One important feature observed after rescaling TXE in the (H3) hypothesis, is that the agreement with experimental data is restored at low TKE while keeping a fairly good agreement for higher TKE.

For the neutron-induced reaction on ^{235}U, the energy spectrum in the laboratory frame is shown in Fig. 3. Also shown for comparison are the results obtained with the Los Alamos model for the same reaction using the optical model potential of Becchetti and Greenlees for the average fragment of each peak. Experimental data points are taken

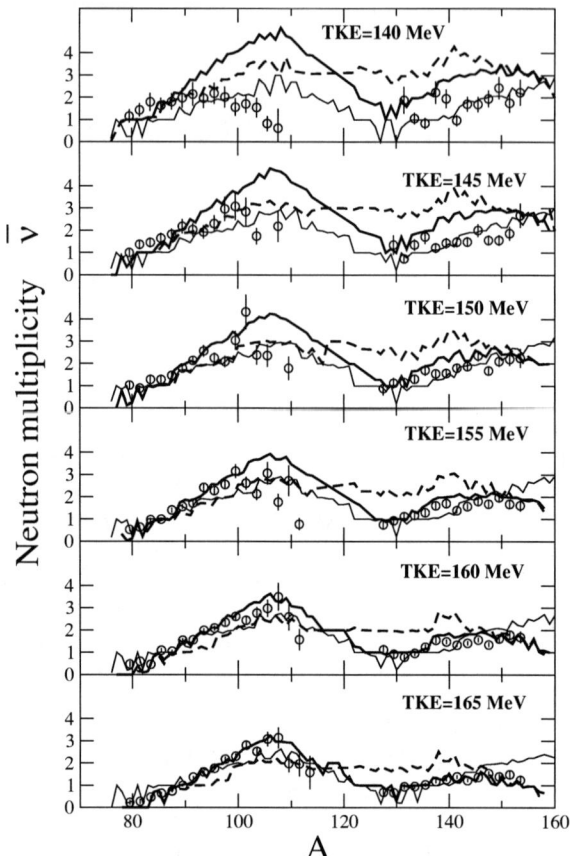

FIGURE 2. Average neutron multiplicity versus FF mass for specific 5 MeV TKE bins for n(0.53 MeV)+^{235}U reaction. The open circles are experimental data from [9]. The dashed, thick and thin lines are obtained under (H1), (H2) and (H3) hypotheses respectively.

from Johansson and Holmqvist [12]. The spectrum obtained by assuming equal nuclear temperatures in both FF at scission is not represented in Fig. 3 but is found too soft compared to the result under (H3) hypothesis. This is illustrated by an average laboratory neutron energy of $\langle E \rangle \approx 1.977$ MeV in the (H1) case and $\langle E \rangle \approx 1.988$ MeV for the assumption (H3). The calculated spectrum obtained with the (H2) hypothesis of splitting the TXE according to $\bar{\nu}_{exp}(A)$ exhibits a much too hard spectrum with $\langle E \rangle \approx 2.089$ MeV (the Los Alamos model gives $\langle E \rangle \approx 2.046$ MeV). It is interesting to see in Fig. 3 that the decrease of TXE at low TKE in the (H3) case restores in part the agreement with the neutron energy spectrum affecting especially the high energy tail of that spectrum. In particular, in the case of (H3) hypothesis, the spectrum obtained is no longer too hard,

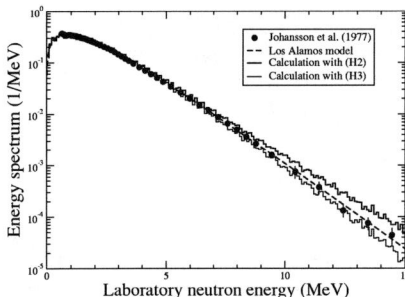 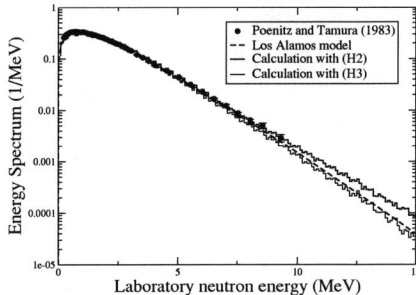

FIGURE 3. Neutron energy spectrum for n(0.53 MeV)+^{235}U reaction (left) and ^{252}Cf (sf) (right) in the laboratory frame. The dashed line is result of the Los Alamos model calculation using the optical model potential of Becchetti and Greenlees for the inverse process of compound nucleus formation. The experimental points are from Johansson and Holmqvist [12].

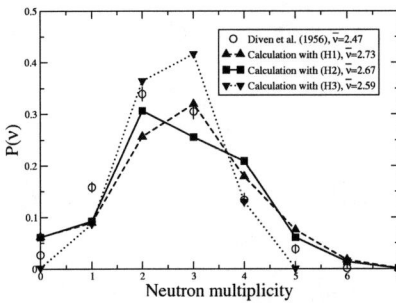 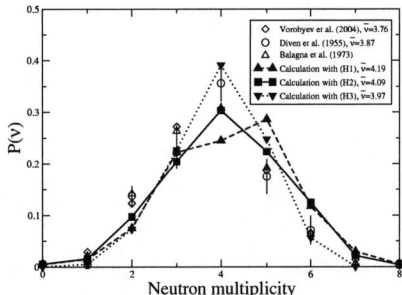

FIGURE 4. Neutron multiplicity distribution for n(0.53 MeV)+^{235}U reaction (left) and ^{252}Cf (sf) (right). Open square symbols are from our Monte-Carlo calculation assuming partitioning of FF total excitation energy as a function of $\bar{v}_{exp}(A)$ (H2 hypothesis), triangles up are the result obtained under the assumption of an equal temperature of complementary FF (H1 hypothesis), triangles down are the result obtained under the assumption (H3). The open circles are experimental data from Diven et al. [13].

as in the case of the (H2) assumption.

The same conclusion can be drawn for ^{252}Cf (*sf*).

The prompt neutron multiplicity distribution $P(v)$ can be inferred from our Monte-Carlo calculations, while most other approaches can only assess the average value of this distribution, $\bar{v}$. To the best of our knowledge, only a limited number of experimental data exist for $P(v)$. Our numerical results are compared with the experimental distribution by Diven et al. [13] in Fig. 4 for ^{235}U. In both calculated cases (H1) and (H2), the average $\bar{v}$ of the distribution is about 10% higher than the experimental value. Interestingly, the (H3) assumptions restore also in part the agreement with experimental distribution

$P(v)$. We found, in the case of neutron induced fission of ^{235}U, $\bar{v} \approx 2.59$ very close to the experimental value 2.47 from Diven et al. [13]. In the case of spontaneous fission of ^{252}Cf, we found $\bar{v} \approx 3.97$ in good agreement with 3.7661 from Boldeman [14].

CONCLUSION

We have developed a powerful tool aimed at exploring neutron evaporation from fission fragments. We have used a Monte-Carlo technique that allows to investigate all correlations in the reaction. In particular, we have been able to extract quantities such as neutron multiplicity distribution as well as correlation between the number of emitted neutrons, the fission fragment masses and their total kinetic energy $\bar{v}(A, TKE)$.

Several hypotheses have been used to estimate the total fission fragments excitation energy available for neutron evaporation and how this energy gets partitioned between the light and heavy fragments. We found that the low total kinetic energy part of the fission fragment excitation energy distribution could explain in part the discrepancies obtained with experimental data on neutron energy spectrum, neutron multiplicity distribution as well as in the correlations. One of the questions we have to answer now is related to the origin of that excitation energy which is not converted into prompt fission neutrons.

REFERENCES

1. D. G. Madland, J. R. Nix, Nucl. Sci. Eng. **81** (1982) 213.
2. V. Weisskopf, Phys. Rev. **52** (1937) 295.
3. F. J. Hambsch, S. Oberstedt, Nucl. Phys. **A617** (1997) 347.
4. G. Audi, A. H. Wapstra, C. Thibault, Nucl. Phys. **A729** (2003) 337.
5. A. V. Ignatyuk, K. K. Istekov, G. N. Smirenkin, Sov. J. Nucl. Phys., **29**, 450 (1979).
6. H. Koura, M. Uno, T. Tachibana, M. Yamada, Nucl. Phys., **A674**, 47 (2000).
7. H. W. Schmitt, J. H. Neiler, F. J. Walter, Phys. Rev. **141** (1966) 1146.
8. C. Budtz-Jørgensen, H. H. Knitter, Nucl. Phys. **A490** (1988) 307.
9. K. Nishio, Y. Nakagome, H. Yamamoto, I. Kimura, Nucl. Phys. **A632** (1998) 540.
10. H. Nifenecker, C. Signarbieux, M. Ribrag, J. Poitou and J. Matuszek, Nucl. Phys. **A189** (1972) 285.
11. F. Pleasonton, R. L. Ferguson and H. Schmitt, Phys. Rev. **C6** (1972) 1023.
12. P. I. Johansson and B. Holmqvist, Nucl. Sci. Eng. **62** (1977) 695.
13. B. C. Diven, H. C. Martin, R. F. Taschek and J. Terrell, Phys. Rev. **101** (1956) 1012.
14. J. W. Boldeman, in Proceedings of the NEANDC Meeting on Nuclear Data Standards for Nuclear Measurements, Uppsala University, Uppsala, Sweden, Oct. 1991, p. 108-109.

Distributions for Excitation Energy and Kinetic Energy in Nuclear Fission

Herbert R. Faust* and Zongyu Bao*

Institut Laue-Langevin

Abstract. New experiments on the kinetic energy distribution functions of fission fragments have shown, that the probability of excitation is a function of the Q-value of the specific mass/charge split, and a function of the nuclear charge of the fissioning actinide. Both, Q-value and nuclear charge Z_c enter linearly in the exponent of the excitation distribution function. It is straightforward to calculate from the functions for the excitation energy the remaining observable in fission: neutron evaporation, total excitation energy, and total and single kinetic energy. The calculations are done by a Monte-Carlo code in an event-by-event basis, and therefore supply at the same time all possible correlations between the observable. If a spin distribution function is assumed, the entry states can be constructed, and it is possible to calculate furthermore gamma emission, and the population of isomeric states in fission.

Keywords: fragment excitation energy, fragment kinetic energy, neutron emission, *REX*-model

INTRODUCTION

Only few attempts have been made in the past to directly calculate excitation energy distributions of fission fragments. In particular, in the early days of fission Viola [1] established a systematics of the dependence of the most probable value for the total fission fragment kinetic energy on the Coulomb-energy parameter, which was shown to have a more or less linear dependence. However, this systematic is based on an averaging over all mass and nuclear charge splits in a given system. A further attempt has been made by Brosa, who correlated mean neutron evaporation values with fragment deformation values at scission. These calcualtions were done particularly for spontaneous fission of ^{252}Cf, [2].

Recently a model has been developed where it was shown, that a simple distribution function for fragment excitation leads to quantitative agreement with neutron evaporation and kinetic energy distribution characteristics over mass ranges from $A = 80$ to $A = 160$ in thermal neutron induced fission of ^{233}U, [3]. New experiments on kinetic energy distributions in $^{245}Cm(n, f)$ show, that there is also quantitative agreement for kinetic energy distributions in this compound system. Moreover, a dependence of fragment excitation on the nuclear charge of the compound system was established.

THE MODEL DESCRIPTION

The *REX*-model assumes an excitation energy distribution for fission fragments of the form

$$P(E^*) = \frac{1}{N} \cdot \rho(E^*) \cdot \exp(-\frac{E^*}{\Xi}) \qquad (1)$$

Here $\rho(E^*)$ is the level density of the excited fragment at the excitation energy E^*, and the exponential contains the parameter Ξ. This parameter is dependent on the reaction mechanism in fission.

An intuitive picture of fission suggests dependencies of the parameter Ξ on the reaction Q-value, and on the time scale with which fission proceeds, e.g. on the value dQ/dt. This last term is dependent on the parameter Z_c, the nuclear charge of the compound nucleus, which is responsible for the driving force towards scission. The value of Q in fission is about 200 MeV, and the time-scale of the fission process is about $10^{-21} s$.

N is a normalization constant in eq. 1. For the level density dependence we use the Fermi-gas model, with inclusion of the nuclear pairing

$$\rho(E^*) = \frac{\sqrt{(\pi)}}{12} \frac{exp(2\sqrt{a(E^* - \delta)})}{a^{1/4}(E^* - \delta)^{5/4}} \qquad (2)$$

Here

$$\delta = \chi \cdot \frac{12}{\sqrt{A}} \qquad (3)$$

is the nuclear pairing term, and $\chi = 0, 1, 2$ for odd-odd, odd-even and even-even nulcei. a is the level denstity parameter, which is taken from experiment, [5], and A is the mass number of the fragment.

Equation 1 completely determines the excitation function of a fragment, if the level density parameter a and the value of Ξ are known. It has been shown that for $^{233}U(n,f)$ the value of Ξ is linearly dependent on the Q-value of the reaction, with $\Xi = \bar{f} \cdot Q$, and $\bar{f}$ a constant, which is determined from experiment. The Q-vlue is determined by the mass excesses Δ of the implied fragments

$$Q = \Delta(Ac, Zc) - [\Delta(f1) + \Delta(f2)] \qquad (4)$$

DETERMINATION OF $\bar{F}$ FOR OTHER COMPOUND SYSTEMS

We have fitted the parameter $\bar{f}$ to data from other compound systems, in particular to total kinetic energy values and mean fragment kinetic energies well known from experiment. The systematic survey established a linear depencence of the parameter $\bar{f}$ from the nuclear charge Z_c of the fissioning system. This dependence is shown in fig. 1. The dependency of the parameter of $\bar{f}$ from the nuclear charge of the compound system Z_c is determined to be

$$\bar{f} = \alpha \cdot Z_c + \beta \qquad (5)$$

The linear fit gives values of $\alpha = 3.29 \cdot 10^{-4}$ and $\beta = -0.0258$.
Up to now this linear relationship has been verified for actinides from Ac to Cm.

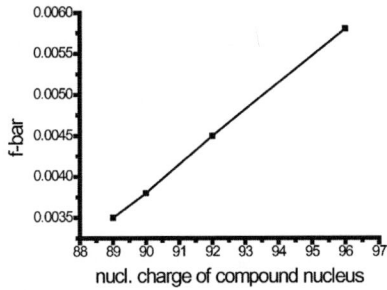

FIGURE 1. Dependence of the parameter $\bar{f}$ from the nuclear charge of the compound nucleus.

RELATIONSHIP OF THE PARAMETER Ξ WITH TEMPERATURE

Equation 1 is formally identical with the equation known from statistical thermodynamics [6]

$$P(E^*) = \frac{1}{N} \cdot \rho(E^*) \cdot exp(-\frac{E^*}{T}) \qquad (6)$$

In this formulation T is a temperature which is derived according to the statistical model. The parameter Ξ is identical to the parameter T used above. However, it is clear from the different dependencies, in our case from the Q-value and the compound nuclear charge Z_c, that the notion of temperature is different in both cases. If we write

$$T = \Xi \qquad (7)$$

than the meaning of the temperature is rather to be the parameter which determines the distribution in eq. 1, than a temperature derived from the statistical model. A conclusion may be, that the excitation energy distribution of 1, which is dependent on Q and Z_c, cannot be derived from the standard statistical model. This is not too surprising, since the failure of the statistical model to reproduce mass and nuclear charge distributions in fission is known since about 30 years. There is no reason to believe that energetics in fission should follow the standard statistical model, when the yield does not.

ENTRY STATES IN FISSION FRAGMENTS

The calculation of neutron and gamma emission from the excited fragments requires, in addition to the knowledge of the excitation functions, the probability function for the nuclear spin. Few is known about this function, and we assume a spin distribution function which is Gaussian, centered at around $5\hbar$, with $\sigma = 3\hbar$. Recent experiments on the population of high spin isomers in nuclear fission indicate, that this distribution function may be unappropriate in cases, where the spin seems to be generated to a substantial part by single particle spin, probably created during pair breaking in the process. Extensive tests have shown that neutron evaporation probabilities are little dependent on the

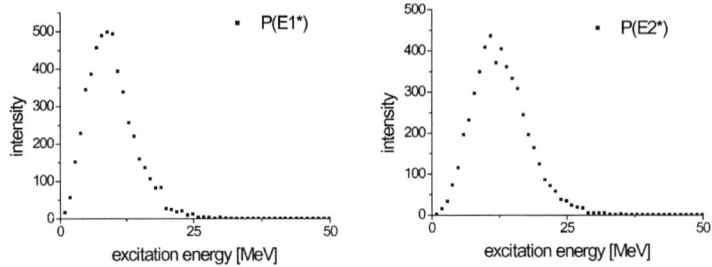

FIGURE 2. Excitation functions for the complementary fragments $A = 94$ and $A = 140$ in thermal neutron induced fission of ^{233}U.

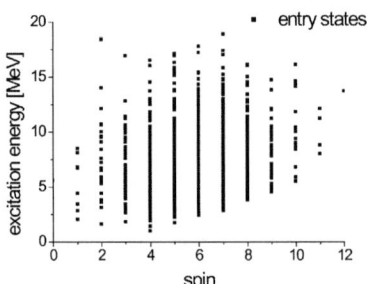

FIGURE 3. Entry states for $A = 94$ from thermal neutron nduced fission of ^{233}U.

details of the spin distribution function. However, if gamma decay is to be calculated by the statistical model, the knowledge of the spin distribution function is necessary.

We show in fig. 2 an example for the distribution of excitation energy in the fragmentation $(A,Z) = (94,38),(140,54)$ from thermal neutron induced fission of ^{233}U. If we input in addition the spin distribution function, we obtain the entry states shown in fig. 3. Most of the entry states are concentrated around the region $<E^*> = a \cdot \Xi^2$ and spin $<I> = 5\hbar$. At low excitation energies and high spins the yrast line is visible. From the entry states it is possible to calculate gamma and neutron decay characteristics with a standard statistical model code. We use here the statistical code *PACE II*. The results of the calculations are scetched in fig. 4. It is seen that neutron emission does not take away much angular momentum, and ends up in states near to the yrast-line. Then gamma decay proceeds rapidly along the yrast line to the ground state. We show in fig. 5 the neutron multiplicity summed over all fragments. Finally in fig. 6 we show calculated mean neutron emission as a function of fragment mass. These calculations agree reasonably well with the experimental findings.

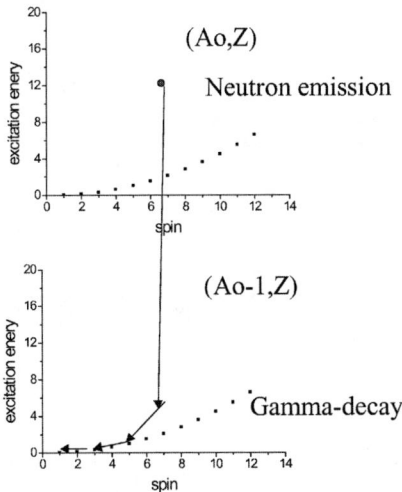

FIGURE 4. Principal paths of neutron and gamma decay in a fission fragment. The probability of neutron emission depends mainly on the excitation energy of the fragment, whereas the probability for gamma emission is maily determined by its spin.

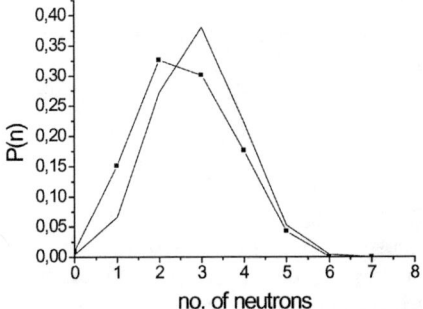

FIGURE 5. Probability for neutron emission summed over all masses in $^{233}U(n,f)$. Experimental values are given by points. The calculation is given by the solid line.

KINETIC ENERGY DISTRIBUTION FUNCTIONS

The distribution function for the total excitation energy TXE is constructed in the *REX* code [4] by randomly selecting events from the excitation functions of the two fragments. The distribution function for the total kinetic energy follows then by the energy conservation law, also in an event-by-event mode

$$TXE = E_1^* + E_2^* \tag{8}$$

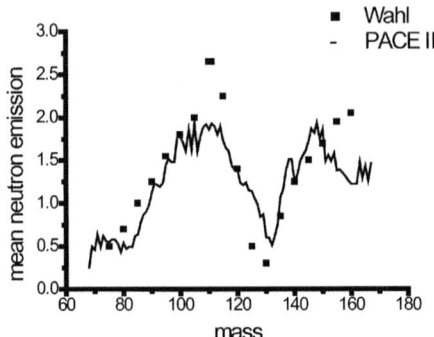

FIGURE 6. Mean neutron emission values over fragment mass in $^{233}U(n,f)$. Data points are from the compilation, the solid line represents the calculation.

$$TKE = Q - TXE \qquad (9)$$

The distribution function for TKE is in fact the mirror function of the TXE distribution, with the endpoint being at Q. The distribution functions for the single fragment kinetic energies are constructed from the TKE-distribution by observing the momentum conservation law. This also is done on an event-by-event basis.

$$E_1^{kin} = TKE/(1 + \frac{A_1}{A_2}) \qquad (10)$$

$$E_2^{kin} = TKE/(1 + \frac{A_2}{A_1}) \qquad (11)$$

Examples for kinetic energy distribution functions are shown in [3] for fragments from $^{233}U(n,f)$, and for $^{245}Cm(n,f)$ in the following contribution. All calculated distribution functions agree quantitatively with the experiment.

TKE-DISTRIBUTIONS FOR DIFFERENT COMPOUND SYSTEMS

If the compound system is formed in nuclear reactions, the excitation energy from the reaction process has to be considered. Because $\bar{f} \cdot Q$ can be formally considered as a temperature parameter, the exictation energy of the compund nucleus is to be added quadratically

$$\Xi = \sqrt{(\bar{f} \cdot Q)^2 + \frac{\varepsilon_c}{a_1 + a_2}} \qquad (12)$$

Here ε_c is the excitation of the compund nucleus, and a_1, a_2 are the level density parameter of fragment 1 and 2, respectively. The application of eq. 1 implies, that on the mean the excitation energy ε_c is distributed onto the fragments according to their level

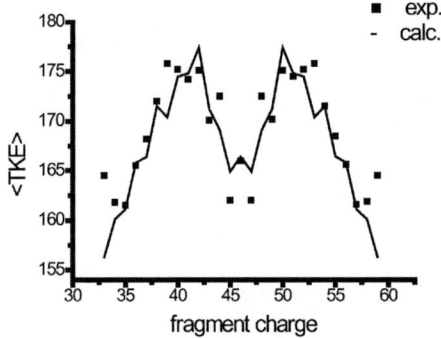

FIGURE 7. Mean total kinetic energies for the compound system ^{234}U at an excitation energy of about 12 MeV. Data points are from [7], the solid line represents the calculations.

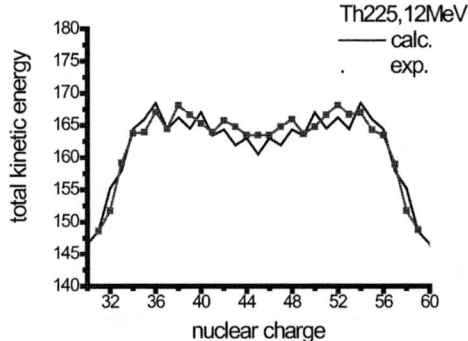

FIGURE 8. Mean total kinetic energy distributions for the system ^{225}Th. Data points are from [7], the calculations are given by the solid line.

density parameter.

We show in fig. 7 mean total kinetic energies for ^{234}U and in fig. 8 the values for ^{225}Th at an excitation energy of about 12 MeV, [7]. Also here calculations and experiment agree quantitatively.

FINE STRUCTURE ON THE MEAN KINETIC ENERGIES

It is known from experiment that the single fragment kinetic energies for even and odd fragmentations show a pronounced fine structure. Even fragmentations have in general higher kinetic energies by about 300 KeV than odd fragmentations. This effect is inherent contained in the *REX*-model. By calculating Ξ only part of the Q-value goes into fragment excitation. Most of the available Q-value is, however, distributed onto the fragments in form of kinetic energy. The result is, that the mean kinetic

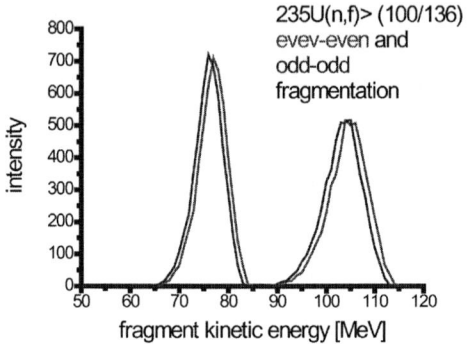

FIGURE 9. Kinetic energy distribution of the light and the heavy fragment of fragment masses $A = 100$ and $A = 136$ in fission of ^{236}U. Fragmentation in odd-odd fragments show a considerably lower mean kinetic energy than in even-even fragments.

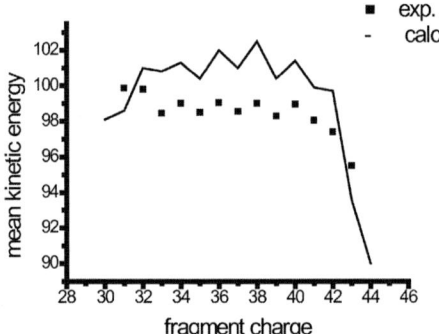

FIGURE 10. Mean single fragment kinetic energies as function of fragment charge in $^{233}U(n,f)$. Data points are from [8], the calculations are represented by the solid line.

energies of even-even fragmentations is in general higher by about 1 MeV than for odd-odd fragmentations, see fig. 9. Calculation and experiment are shown in fig. 10 for $^{233}U(n,f)$. Experimental results are from [8]. It is seen that the odd-even staggering of the mean kinetic energies is well reproduced.

KINETIC ENERGY DISTRIBUTIONS FOR FRAGMENTS IN EXCITED STATES

If a fragment has an isomeric state at high excitation energy, and if the kinetic energy disribution of this fragment is measured, the shape of the distributions changes characteristically. In the entry state diagram an isomer cuts out a region, which permits only excitation energies above the energy of the isomeric state, and the feeding by levels with

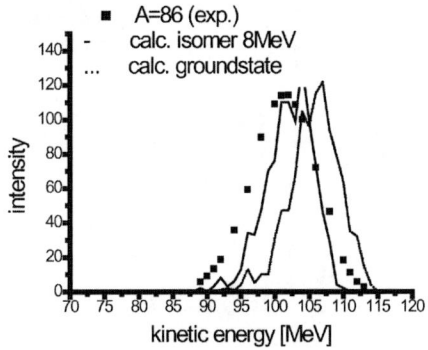

FIGURE 11. Kinetic energy distribution for mass $A = 86$. The experimental data are given by dots. The solid line far right shows the calculated distribution if no isomer is present. The solid line far left shows the calculated distribution, if an isomeric state at an excitation energy of $8\,MeV$ is assumes.

spin values above or around the spin of the isomeric state. The cut in the excitation energy modifies the kinetic energies of the light and the heavy fragment. An exampleis mass $A = 86$, where the kinetic energy distribution has been measured at *LOHENGRIN* for high ionic charge states. These ionic charge states are anomal, they are generated by conversion electron decay from the isomeric state [9]. Fig. 11 shows the experimental distribution, the calculated distribution without the presence of an isomer, and the kinetic energy distribution with the assumption that the energy of the isomeric state is located at about $8MeV$. This assumption leads to good agreement for the mean and the higher moments of the distribution.

VALUE OF HIGH SPIN FRACTION FOR ISOMERIC STATES

Our systematic investigation of neutron and gamma decay from excited fragments with the statistical code *PACE II* have shown, that the neutron multiplicity is not very much effected by the spin distribution function which is assumed for the fragments. On the contrary, the spin distribution function is a very sensitive input for prompt gamma emission characteristics. The number of gamma rays emitted essentially reflect the number of low multipolarity gama rays running down the yrast line.

In selecting a kinetic energy value in an experiment, automatically the associated permitted region for the entry states in terms of ecitation energy is selected. This is exemplified in fig. 12, where the allowed region of entry states for $A = 86$ is plotted in dependence of the kinetic energy chosen for the fragment. It is seen that, when a gate is set on low kinetic energies, the whole region in the yrast diagram is filled, whereas, if the gate is set on high kinetic energies, only a small region in the yrast plot is concerned. If an isomer is present in the fission product, then the feeding of the isomer is a characteristic function of the kinetic energy of the fragment chosen. In particular at high fragment kinetic energies the isomer can no longer be fed due to the energy conservation law, and the high spin fraction drops to zero. An example of this is seen in fig. 13, where the high

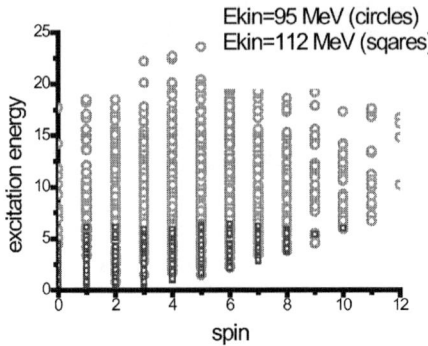

FIGURE 12. Allowed entry states for two kinetic energy values for the single fragment kinetic energy. For $E_{kin} = 95 MeV$ the whole region is allowed, whereas for $E_{kin} = 112 MeV$ only entry states with very low excitation energy are populated.

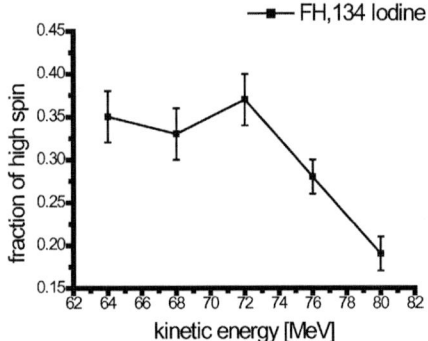

FIGURE 13. Fraction of high spin population in Iodine-134. The kinetic energies are not yet corrected for energy loss in the target and the cover-foil.

spin fraction for the 8- isomer in Iodine-134 was measured as a function of the kinetic energy.

CONCLUSIONS

We have shown that the energetics in fission is determined by a single parameter Ξ. This parameter is a linear function of the nuclear charge Z_c of the compound system, and the Q-value of the respective fragment split. From the parameter Ξ the distribution function for fragment excitation is derived, and with the functional dependence for the spin distribution the entry states for fission fragments can be constructed. With a Monte-Carlo code the distribution functions for TXE, TKE and the single fragment kinetic energies are derived by observing energy and momentum conservation laws.
The results of the model are in quantitative agreement with experimental data on neutron

evaporation and kinetic energy distribution functions.

REFERENCES

1. V. E. Viola, Jr., *Nucl. Data*, Sect. A1, 391(1966).
2. U. Brosa, *Phys .Rev. C*, 38, 1944 (1988).
3. H. R. Faust and Z. Bao, *Nucl Phys A*, 736, 55-76, (2004).
4. H. R. Faust, *REX*, a Fortran program to calculate fragment excitation and kinetic energies in nuclear fission, avaiable from the author.
5. C. Butz-Jøergenson and K. H. Knitter, *Nucl. Phys. A*, 490, 307 (1988).
6. W. Greiner, L. Neise and H. Stocker, Classical Theoretical Physics, Thermodynamics and Statistical Mechanics, Springer Verlag, 1997
7. K. H. Schmidt et al., *Nucl. Phys. A*, 665, 221 (2000).
8. U. Quade et al., *Nucl. Phys. A* 487, 1, (1988)
9. H. Wohlfahrt, Thesis, Technische Hochschule Darmstadt, 1976.

Kinetic Energy Distributions in Thermal Neutron Induced Fission of ^{245}Cm

Beatrice Weiss*, Herbert Faust† and Nicolas Bessolaz†

University Nice-Sophia Antipolis
†*Institut Laue-Langevin*

Abstract. The kinetic energy distribution functions of selected fragments in thermal neutron induced fission of ^{245}Cm have been measured at the *LOHENGRIN* spectrometer of the institut Laue-Langevin in Grenoble. The results are compared with the random excitation model *REX*. Quantitative agreement between experiment and calculations are obtained over a wide range of fragment masses.

Keywords: thermal neutron induced fission, fragment kinetic energies, *REX*-model

INTRODUCTION

The kinetics in the fission process can be studied by decay processes of the fragments, which depend characteristically on the entry states at the end of the fission process. Apart from the excitation energy distribution also the spin distribution has to be known to construct the states. From the entry states neutron evaporation and gamma emission can be calculated in a straighforward manner, using standard statistical model codes.//
A complementary method to investigate the excitation energy distribution of fragments consists in the measurement of the fragment kinetic energy distribution function. This function can be derived directly from the exitation energy distribution with the help of energy and momentum conservation laws. In the present experiment we compare single fragment kinetic energy distributions from $^{245}Cm(n,f)$ with calculated functions from the *REX*-model, [1].

EXPERIMENTAL SET-UP

We used the *LOHENGRIN* mass spectrometer at the Institut Laue-Langevin in Grenoble/France to measure kinetic energy distributions of selected fission fragments in thermal neutron induced fission of ^{245}Cm. The separator consists of a sequence of two magnetic and one electrostatic condensor field, which separate fission fragments according to their mass and kinetic energy. The detection of the fragments is done in a double anode ionisation chamber, which, for light fragments permits additionally the determination of the fragment charge. The target consisted of a layer of Cm_2O_3 with an enrichment of 98.84 % in ^{245}Cm. The target mass was 15.1 $\mu g/cm^2$ on a surface of 5x40 mm^2. The target was covered by a self supporting Ni-foil of 0.25 μm thickness to avoid loss of material due to the sputtering process during fission. The thermal neutron flux at the target postition was $5.3x10^{14} n/cm^2.s$. By limiting the exit slit length we chose

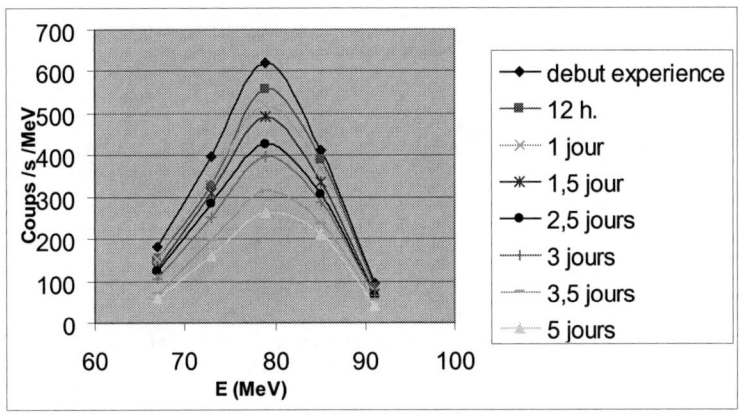

FIGURE 1. Burnup characteristic of the ^{245}Cm target.

an energy resolution of about $\Delta E/E = 5 \cdot 10^{-3}$. The mass resolving power was about $\Delta M/M = 1.3 \cdot 10^{-3}$.

MEASUREMENTS AND DATA TREATMENT

Selected fission fragments of mass number $80 < A < 160$ were scanned for the kinetic energy distribution function in steps of $1\ MeV$ over energy regions of about $30\ MeV$. The measured distributions were corrected for the energy dispersion of the spectrometer in dividing the measured count-rate by the respective energy. We furthermore corrected the data for target burn-out. For this purpose the target activity was scanned daily in recording the kinetic energy distribution for mass $A = 130$. The target burn-up is seen in fig. 1. In a final step the experimental data points were corrected for energy loss in the target and the cover foil. This was done by the Monte-Carlo code *SRIM* by Ziegler, [2]. Energy loss values range from about $4.5\ MeV$ for mass $A = 160$ to about $7\ MeV$ for the light masses.

CALCULATIONS WITH THE *REX*-MODEL

In the *REX*-code the excitation energy distribution function of the single fragment characterized by (A,Z) is given by the equation

$$\Phi(E^*) = \frac{1}{N} \cdot \rho(E^*) \cdot exp(-\frac{E^*}{\Xi}) \qquad (1)$$

Here $\rho(E^*)$ is the level density of the fragment at excitation energy E^*. For its functional form we take the Fermi-gas model with the inclusion of pairing. The level density parameter as function of fragment mass is taken from Butz-Jørgenson and Knitter [3]. The exponential function contains the excitation mechanism in nuclear

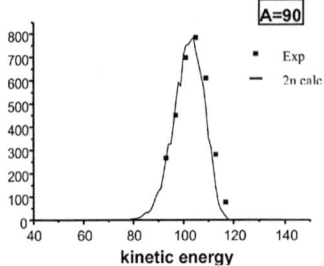 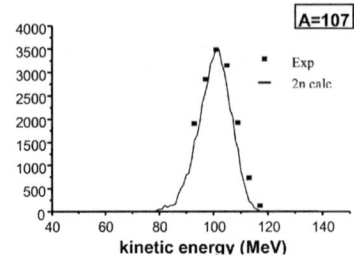

FIGURE 2. Experimental and calculated kinetic energy distributions for masses $A = 90$ and $A = 107$. The energy shift due to the evaporation of 2 neutrons has been accounted for in both masses.

fission, and it was shown that the parameter Ξ depends for spontaneous fission in a linear way on the Q-value of the reaction

$$\Xi_{sf} = \bar{f} \cdot Q \qquad (2)$$

The value of Q is calculated from the ground-state masses of the nuclei involved

$$Q = \Delta_{CN} - (\Delta_{f1} + \Delta_{f2}) \qquad (3)$$

with Δ_{CN} and $\Delta_{f1,f2}$ being the mass excesses for the compound nucleus and the fragments, respectively.
The value of $\bar{f}$ depends on the specific actinide, and we found its value to be $\bar{f} = 0.058$ for the Curium-system.
For neutron induced reactions the excitation energy gained in the neutron capture process has to be included

$$\Xi_r = \sqrt{(\bar{f}Q)^2 + \frac{\varepsilon_c}{a_1 + a_2}} \qquad (4)$$

ε_c is the neutron binding energy for thermal neutron induced fission, and a_1, a_2 are the level density parameter for the two fragments.

The *REX*-code determines from the excitation distribution functions for the fragments, given by eq. 1, by a Monte-Carlo procedure the distribution function for the total excitation energy, the total kinetic energy, and, by applying momentum conservation, the single fragment kinetic energies. In the following we will present calculated kinetic energy distributions with the experimental distributions for selected masses.

RESULTS

Fig. 2 shows the calculated (solid line) and experimental (dots) kinetic energy distributions for masses $A = 90$ and $A = 107$. For both masses we assumed that two neutrons are evaporated, which shifts the kinetic energies by about 2 *MeV* to lower values. A quantitative agreement between the results of the *REX*-model and the experiment is found. The

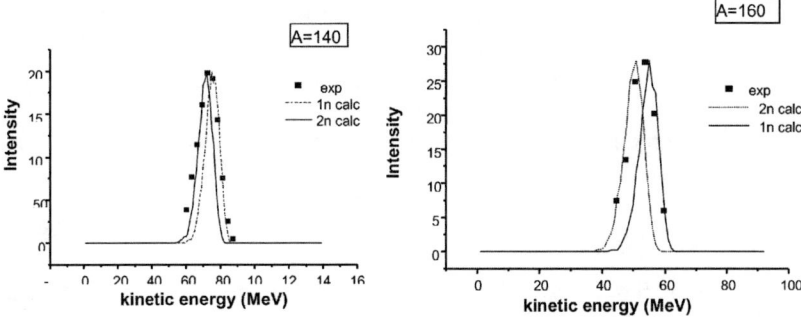

FIGURE 3. Experimental and calculated kinetic energy distributions for masses $A = 140$ and $A = 160$. For both masses calculated energy distributions are shown for 1 neutron and 2 neutrons emitted.

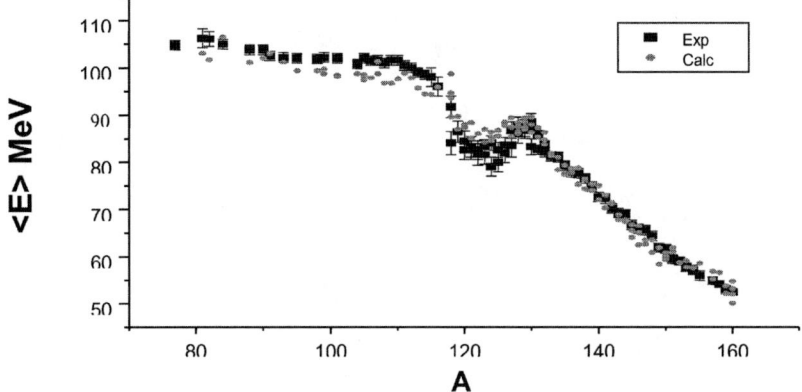

FIGURE 4. Measured and calculated mean kinetic energies for fragments in thermal neutron induced fission of ^{245}Cm.

masses $A = 140$ and $A = 160$ have experimental widths which are somewhat larger than the calculated ones. If we assume that these masses evaporate 1 and 2 neutrons to about the same probability, again very good agreement is obtained. Fig. 3 shows the complete results for the mean single fragment kinetic energies from masses $A = 80$ to $A = 160$. Over the whole region mean kinetic energies, which are computed as the first moments of the calculated and the experimental distributions, agree well. The characteristic feature of near constancy for the light fission wing, the characteristic dip for symmetric masses, and the small peak for the douply magic region around ^{132}Sn is clearly reproduced, as well as the strong decrease for the mean kinetic energies for the heavy masses. These structures observed comes about from the interplay of Q-value, level density parameter, and momentum conservation law.

CONCLUSIONS

Single fragment kinetic energy distribution functions were measured for $^{245}Cm(n,f)$ over mass ranges from $A = 80$ to $A = 160$. Calculations with the *REX*-code show quantitatative agreement with the experimental distributions. It is seen, that in order to calculate the functions for the excitation distributions, the constant which correlates the distribution parameter to the Q-value of the reaction has a value of $\bar{f} = 0.0058$, which is considerably larger than the value which was needed to calculate distributions in the Uranium region.

Unless the reproduction by the *REX*-model of mean values, variances and line shapes for about 80 masses is purely accidental we have to conclude that the underlying assumptions of the model are correct, and the energetics in fission can be calculated with a single constant relating the parameter for the excitation distribution function to the Q-value of the respective mass and charge split.

REFERENCES

1. H. R. Faust and Z. Bao, *Nucl Phys A*, **736**, 55–76 (2004)
2. J. F. Ziegler, J. P. Biersack and U. Littmark, The stopping and range of ions in solids, *Pergamon Press*, 1985
3. C. Butz-Jørgensen and K. H. Knitter, *Nucl. Phys. A*, **490**, 307 (1988)
4. J. K. Tuli, *Nuclear Wallet Cards*, Brookhaven National Laboratory (2000)

MASS AND ENERGY DISTRIBUTIONS

Statistical Approaches to the Even-odd Effect in Fission

K.-H. Schmidt[a], A. V. Ignatyuk[b], F. Rejmund[c], A. Kelić[a], M. V. Ricciardi[a]

[a]*GSI, Planckstr. 1, 64291 Darmstadt, Germany*
[b]*IPPE, Bondarenko Sq. 1, 249020 Obninsk, Kaluga Region, Russia*
[c]*GANIL, BP 5027, 14076 Caen cedex 5, France*

Abstract. Statistical approaches have been widely used for describing the nuclear-fission process. They were quite successful in explaining several prominent features of the global nuclide distributions and many other aspects, while there are controversial conclusions on the origin of the even-odd effect in the nuclear-charge yields. We analyze the ingredients of the main statistical approaches to the even-odd effect in fission and show up that many of their deficiencies rely on unrealistic assumptions. Finally, we demonstrate that the large body of experimental results, obtained in the recent years to a great part at GSI, Darmstadt, is very successfully reproduced by a new stringent statistical approach.

Keywords: Nuclear fission; Pair breaking; Even-odd structure in fission-fragment yields; Statistical model; Superfluid nuclear model
PACS: 25.85.-w; 24.10.Pa; 21.10.Ma

INTRODUCTION

The discovery of the even-odd structure in fission and the evolution of its theoretical interpretation is a fascinating story on the progress in a specific sub-field of research. In many aspects, it is typical for research in general, for the decisive role of experimental data in triggering new ideas and for the gradual progress in the theoretical understanding. On the one hand, it shows the growth in empirical knowledge over the time due to the progress in experimental technique. Major steps have been made, when innovative experimental approaches were introduced. On the other hand, the theoretical understanding did not proceed in a straightforward way. Theories based on inappropriate concepts or inadequate approximations were proposed and survived over long time, until their shortcomings were eventually recognized.

EXPERIMENTS

Early data on fission-fragment yields, obtained with radiochemical techniques, revealed that thermal-neutron-induced fission of ^{235}U produces even-Z fragments more abundantly than odd-Z-fragments[1]. The global even-odd effect, quantified by the following expression

$$\delta = \frac{\sum_i Y_{ei} - \sum_i Y_{oi}}{\sum_i Y_{ei} + \sum_i Y_{oi}}$$

was found to be about 25 %.

Major progress in the yield determination of individual fission fragments, fully identified in atomic number Z and mass number A, was achieved by the experimental program performed with the Lohengrin spectrometer[2] at ILL Grenoble. The combination of the ion-optical deflection by the parabola spectrograph with a precise measurement of energy loss and residual energy provided a full overview on the nuclide production as a function of kinetic energy in the light-fragment group. The first comprehensive experimental study with this new technique[3] was devoted to ^{235}U(n$_{th}$,f), yielding a strong dependence of the proton even-odd effect on the kinetic energy of the light fragment. The experiments at Lohengrin were and still are the most successful measurements for the in-flight identification of fission fragments at the energy provided by the fission process itself.

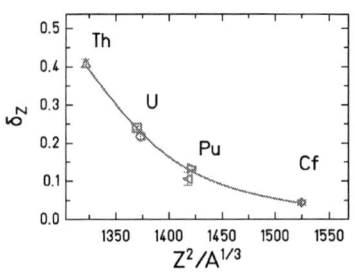

FIGURE 1. Global even-odd effect for thermal-neutron-induced and spontaneous fission.

An overview on the data, measured by this and similar techniques[4] shows a systematic variation as a function of $Z^2/A^{1/3}$ (figure 1) and has also revealed an even-odd effect in the mean kinetic energies of the fragments as a function of Z.

Another major step in the experimental investigation of the even-odd effect in fission has been done by introducing a novel experimental approach. Relativistic beams of fissile nuclei were excited by the electromagnetic interaction with a target nucleus, and the fission fragments were identified in flight[5]. This was the first time that all elements over the whole distribution could be resolved. Among other results, these experiments brought clear evidence for the appearance of a proton even-odd effect also in the fission of odd-Z fissioning nuclei and on systematic variations of the strength of the even-odd effect as a function of the asymmetry of the mass split [6], figure 2.

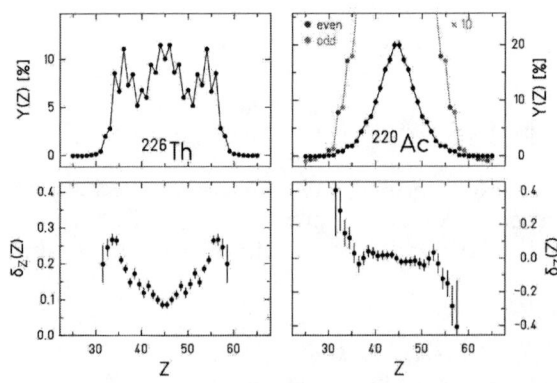

FIGURE 2. Upper part; Elemental yields of fission fragments produced in electromagnetic fission of an even-Z (^{226}Th) and an odd-Z (^{220}Ac) nucleus. Lower part: Local even-odd effect. The data are taken from ref..

THEORY

First, we would like to make some clarifying remarks on the theoretical interpretation of even-odd structure in fission-fragment yields. It is helpful to remind that also in the case of nuclear binding energies the specific structure introduced by pairing correlations in terms of even-odd mass differences has been studied by appropriate filters. In this way, one could show that the gap parameter generally decreases with increasing mass like $1/\sqrt{A}$ and that there is a clear increase of both neutron and proton gap with increasing Z/N ratio. These conclusions could be drawn from investigations just filtering the global trends of the even-odd structure from the complex features inherent in the nuclear binding energies, and they demand for some specific explanations. In this sense, it is very helpful to analyze general trends in global and local even-odd effects in fission-fragment distributions with appropriate filters and to try to find theoretical interpretations which concentrate on this problem.

Our second remark concerns the relevance of the statistical model of nuclear reactions. Like in any problem related to nuclear reactions, the statistical model forms a basic step in the interpretation. We mean this in the sense that any more specific conclusion, e.g. indications for dynamic effects, can only be drawn if this "uninteresting" interpretation does not explain all features of the data. Appropriate efforts to interpret the data with the statistical model of nuclear reactions also help to work out those specific features of the data which contain more specific information. Thus, we stress as a general, very important statement that the statistical model has a key role in the interpretation of nuclear-reaction data.

The statistical model of Fong and its modifications

Fong[7] formulated a statistical model with the aim to calculate the fission-fragment yields on the basis of the number of available states in the scission-point configuration for different splits of the fissioning system in neutron and proton number. He calculated the level density of the fragments with the Fermi-gas model taking into account even-odd effects as a shift of the effective excitation energies. Since such a shift exactly compensates for the even-odd staggering in the Q value as a function of proton or neutron number, this model did not predict any even-odd effect.

We would like to mention here that the shifted Fermi-gas model is not adequate to model the level density of the superfluid nucleus at low excitation energies. This is a flaw which still survives in recent publications[8] and which is erroneously taken as a proof that the statistical model is unable to explain any even-odd effect in fission-fragment yields. In particular, the two-component character of the nuclear system is not properly accounted for. E.g. the assumption that there are no levels below Δ in an even-odd or an odd-even nucleus is not realistic. The Fermi-gas level density should be replaced by a more appropriate formulation of the super-fluid nuclear model e.g. the one developed by Ignatyuk et al.[9].

Wilkins and Steinberg[10] refined Fong's model. Two formulae represent the key equations of the model. The potential energy at scission is formulated as a sum of the liquid-drop ground-state energies of the fragments (V_{LD}), their shell corrections in the neutron and proton subsystem (S), their pairing energy (P), their interactive Coulomb and nuclear potential (V_C and V_n). It is given as a function of neutron number (N), proton number (Z), deformation (β), intrinsic temperature (τ), and neck distance (d). Pairing was considered with the BCS formalism. The energy dependence of the gap parameter Δ was parameterized according to numerical results of Moretto[11]. In a thermodynamical approach, the temperatures of the intrinsic and the collective degrees of freedom are considered as key parameters of the model. The probability for the formation of a specific fragment with neutron number N and proton number Z is given by:

$$P(N,Z,\tau,d) = \int_{\beta_1=0}^{\beta_{max}} \int_{\beta_2=0}^{\beta_{max}} \exp[-V(N,Z,\beta,\tau,d)/T_{coll}] d\beta_1 d\beta_2$$

T_{coll} is the collective temperature, which might be different from the intrinsic temperature τ. That means that the nuclide distribution is determined by a Boltzmann factor with the collective temperature T_{coll}. In contrast to the Fong model, this model predicts an even-odd effect in the fission-fragment yields, both in neutron and proton number, due to the even-odd effect in the binding energies of the fragments.

This treatment is far too simple, in particular in the implicitly used nuclear level densities by means of the thermodynamic Boltzmann factor, to yield realistic values for the even-odd effects. In particular, this formulation severely fails to properly model the characteristic influence of pairing correlations on the level density in even-even, even-odd, odd-even, and odd-odd nuclei in the superfluid phase. In this aspect, it is even less realistic than the shifted Fermi-gas formula used by Fong.

Pommé et al.[12] proposed the following formula for the excitation-energy dependence of the even-odd effect in fission:

$$\delta(E) = \begin{cases} \delta_0 & \text{for } E < B_f + 2\Delta \\ \delta_0 \cdot e^{-\frac{E-B_f-2\Delta}{T}} & \text{otherwise} \end{cases}$$

E = excitation energy of the initial fissioning nucleus above its ground state
B_f = fission barrier
T = temperature parameter
δ_0 = even-odd effect for $E = B_f$

This formulation reminds a modified version of the treatment of pairing in the model of Wilkins and Steinberg. Also this formula lacks good theoretical justification.

The combinatorial model of Nifenecker

Nifenecker et al.[13] introduced a mathematical model, based on combinatorial methods, to explain the even-odd structure in the fission-fragment element yields. Instead of going into any detail, we just mention that Nifenecker's model is based on statistical considerations on the basis of the number of broken pairs. This is a very peculiar kind of statistical consideration, which is not conform with the basic principles of the statistical model of nuclear reaction that is based on the *number of available states* in the final configuration considered. Nevertheless, this model has been used[4,14] to deduce the thermal excitation energy at scission from the magnitude of the even-odd effect in fission-fragment charge distributions.

The statistical model of Mantzouranis and Nix

Mantzouranis and Nix[15] developed a model for the interpretation of the even-odd effect in fission. They based their model on a characteristic value given by the number of quasi-particle excitations, normalized to the number of particle-hole excitations in an equivalent nucleus without pairing correlations. Again, this kind of statistical consideration is not based on the number of available states, and thus it is also not conform to the basic principles of the statistical model of nuclear reaction.

The question of the energy scale

Hambsch et al.[16] and later Bouzid et al.[17] raised the question, what is the appropriate energy scale for analyzing the even-odd structure in the fission-fragment yield as a function of the kinetic energy of the fragments. This discussion has even lead to the provocative title of ref.: "The positive odd-even effects observed in cold fragmentation - are they real?" Since the Q value is modulated by an even-odd structure in the binding energies of the fragments, there is a systematic shift in splits with even or odd neutron or proton numbers if either the sum of the excitation energies of the two nascent fragments or the energy scale of the fissioning system is used as a reference. The relation of the energy reference to the physics of the problem will be discussed below.

The dynamical model of Bouzid et al.

Bouzid et al.[17] were convinced that the statistical model fails to interpret the predominant production of even-Z nuclei in fission. They developed a dynamical model to explain the even-odd effect in fission.

Without commenting the details of this model, we state at this moment that the statistical descriptions mentioned in the previous sections suffer from severe shortcomings. Therefore their failure cannot be taken as a proof for the inadequacy of a statistical description.

The statistical model of Rejmund et al.

Recently, a new statistical approach has been formulated by Rejmund et al.[18]. It is based on a rigorous formulation of the level density in the super-fluid-nucleus model of the fissioning system just before scission.

There is an apparent difference in the formulation of the model of Rejmund et al. compared to the previous formulations. In their paper, the number of available states of the strongly deformed system just before scission is considered. These states are classified according to the number of quasi-particle excitations in the proton and in the neutron subsystem. If a subsystem stays fully paired during scission, only fragments with even numbers of that kind of nucleons are produced. Unpaired nucleons are assumed to be attached to one or the other fragment according to the number of available single-particle states in that fragment. This model predicts that the even-odd effect in proton number is much stronger than that in neutron number, see figure 3. As explicitly discussed in ref., it also explains the appearance of the strong even-odd effect in asymmetric splits for even-Z as well as for odd-Z systems as observed in figure 2. In some sort, this is a kind of dynamical model, since it relates the number of available states before scission to the way the nucleons are attributed to the nascent fragments at scission. In contrast, the previously proposed statistical models assumed that the production of a specific fragment pair is proportional to the phase space given by the number of states available in the two fragments right after scission. In these models, even-odd fluctuations in the yields are related to the fluctuations in the number of available states due to pairing effects in the Q value and in the level densities.

It can be shown that these two approaches give identical results. The approach of Rejmund et al. is based on an energy scale related to the ground state of the fissioning system. The alternative approach starts from the ground-state masses of the fission fragments, but by introducing the Q value, it also shifts the energy scale in each split to the ground-state energy of the fissioning nucleus.

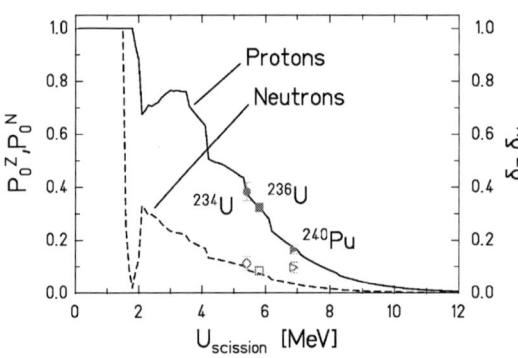

FIGURE 3. Calculated dependence of survival probabilities P_0^Z (full line) and P_0^N (dashed line) of the completely paired proton and neutron configurations on the excitation energy at the effective scission point. The experimental data on the proton and neutron even-odd effects δ_Z and δ_N at fixed kinetic energies of the light fission fragments are shown for the fissioning nuclei ^{234}U (E_{kin} = 111 MeV), ^{236}U (E_{kin} = 108 MeV), and ^{240}Pu (E_{kin} = 111 MeV) by closed and open symbols, respectively.

DISCUSSION

We have seen that the experimental knowledge on the even-odd structure appearing in the nuclide distribution produced in low-energy fission has grown enormously over time. Major steps in the experimental approaches introduced were followed by major steps in the quality and the quantity of the data.

On the theoretical side, many attempts to interpret the observations with the statistical model failed due to inappropriately simplified level-density descriptions used. Several attempts to explain the even-odd structure with statistical arguments, although they were apparently rather successful, suffered from basic errors in the fundamental assumptions. In this situation, it was concluded that even-odd effects in fission cannot be explained by the statistical model of nuclear reactions. Recently, a careful investigation of Rejmund et al. showed up the deficiencies of previous attempts based on the statistical model and succeeded to explain great part of the rich complex signatures of even-odd structure accumulated until now. This does not mean that dynamical effects are absent in creating the even-odd structure of fission-fragment yields, but it needs further efforts to clearly extract specific signatures which go beyond the statistical model of nuclear reactions, before it can be claimed that these data evidence dynamical effects in fission.

SUMMARY

As a summary, we would like to give a classification of the different models, which have been proposed to treat the even-odd effect in fission.

There is general agreement that pairing leads to a staggering of the Q value in fission. In many cases, the Q value can even be deduced from experimental masses.

On this basis there is a first school which treats the problem on the basis of the Q value with the thermodynamical formalism of an ideal gas in the grand-canonical ensemble. In this approach, the nucleus is characterized by its temperature. Energy and particle number are subject to fluctuations. It is assumed that the entropy of the nucleus is well represented by an (ideal) Boltzmann gas, disregarding the specific properties of a Fermionic system and, in particular, disregarding the even more specific properties of a super-fluid Fermi system. This school expects an even-odd effect in the fission-fragment yields which reduces exponentially with increasing excitation energy. This approach is applied by Wilkins and Steinberg, Pommé et al. and remains to be used in many recent models, e.g. the cluster-decay model of Gupta et al.[19].

A second school replaces the Boltzmann statistics by a calculation of the level density of a super-fluid Fermi system with the BCS formalism. This approach is much more realistic. It leads to a compensation of an increased Q value of even-even splits by a reduced level density. Already Fong introduced this idea in one of his early papers. Later, different authors, e.g. Medkour et al., formulated this problem more quantitatively. In this approach, the level density is calculated with a shift of $E_{shift} = n \cdot \Delta$ (n = 0, 1 or 2) for nuclei of different classes, odd-odd, even-odd and odd-even, and finally even-even nuclei. In addition, the super-fluid model also yields a gain in binding energy by the condensation energy, which decreases with increasing excitation energy, until it vanishes at the critical temperature, corresponding to the transition from the super-fluid to the normal-fluid phase. Since this condensation energy has little influence on the even-odd structure in the yields, it is often neglected, and instead the Fermi-gas level density is applied with the above-mentioned energy shifts. In this approach, the even-odd differences in the Q values are exactly compensated by the energy shift of the level density. Thus, this model leaves no room for obtaining an even-odd structure in the fission-fragment yields. This result has been adopted by great part of the scientific community engaged in this problem as a proof that statistical approaches cannot explain the even-odd structure in the fission-fragment yields.

Rejmund et al. recently carried out a more careful investigation of the problem. They developed a more realistic formulation of the number of the first excited levels. Some of the essential ingredients of these considerations are the exact conservation of excitation energy in the isolated nuclear system and the explicit treatment of the quasiparticle excitations of the two subsystems (protons and neutrons) of the nucleus. As an essential characteristic of this formulation, they recognized the importance of configurations, which stay completely paired in one of the subsystems, protons or neutrons, at energies, which exceed the pairing gap. With this stringent formulation of the number of the first excited levels in the two-component super-fluid nuclear system, one predicts a sizeable even-odd structure in the fission-fragment yields, which turns out to be in rather good agreement with the observations.

The work of Nifenecker et al. and the one of Mantzouranis et al. play a special role. Although they took a basis for their statistical considerations, which is not consistent with the statistical model of nuclear reactions, they helped to get some systematic insight into the experimental results.

We conclude that there exists a hierarchy of models according to the degree of approximations. The treatment of even-odd effects in fission proved to be very sensitive to any approximations in the calculation of the number of excited levels. The most stringent calculation of the number of quasi-particle excitations is finally able to explain numerous features of the even-odd effect in the fission-fragment yields in the frame of the statistical model. Dynamical effects are certainly to be expected in addition, but regarding the signatures discussed here any proof for their importance is not evident.

REFERENCES

1. Amiel, S., Feldstein, H., *Proc. Int. Symp. On Physics and Chemistry of Fission* (Rochester 1973) (Vienna: IAEA, 1974) p. 65.
2. Moll, E., et al., *Nucl. Instrum. Methods* **123**, 615 (1975).
3. Lang, W., et al., *Nucl. Phys. A* **345**, 34 (1980).
4. Gönnenwein, F., "Mass, Charge and Kinetic Energy of Fission Fragments" in *The Nuclear Fission Process*, CRC Press, London, 1991, C. Wagemans, ed., pp 409.
5. Schmidt, K.-H., Steinhäuser, S., Böckstiegel, C., Grewe, A., Heinz, A., Junghans, A. R., Benlliure, J., Clerc, H.-G., de Jong, M., Müller, J., Pfützner, M., Voss, B., *Nucl. Phys. A* **665**, 221 (2000).
6. Steinhäuser, S., Benlliure, J., Böckstiegel, C., Clerc, H.-G., Heinz, A., Grewe, A., de Jong, M., Junghans, A. R., Müller, J., Pfützner, M., Schmidt, K.-H., *Nucl. Phys. A* **634**, 89 (1998).
7. Fong, P., *Phys. Rev.* **102**, 434 (1956).
8. Medkour, G., Asghar, M., Djebara, M., Bouzid, B., *J. Phys. G: Nucl. Part. Phys.* **23**, 103 (1997).
9. Ignatyuk, A. V., Sokolov, Yu. V., *Sov. J. Nucl. Phys.* **17**, 376 (1973).
10. Wilkins, B. D., Steinberg, E. P., *Phys. Lett. B* **42**, 141 (1972).
11. Moretto, L. G., *Phys. Lett.* **40B**, 1 (1972.
12. Pommé, S., Jacobs, E., Persyn, K., De Frenne, D., Govaert, K., Yoneama, M.-L., *Nucl. Phys. A* **560**, 689 (1993).
13. Nifenecker, H., Mariolopoulos, G., Bocquet, J. P., Brissot, R., Mme Hamelin, Ch., Crancon, J., Ristori, Ch., *Z. Phys A* **308**, 39 (1982).
14. Rochman, D., Faust, H., Tsekhanovich, I., Gönnenwein, F., Storrer, F., Oberstedt, S., Sokolov, V., *Nucl. Phys. A* **710**, 3 (2002).
15. Mantzouranis, G., Nix, J. R., *Phys. Rev. C* **25**, 918 (1982).
16. Hambsch, F.-J., Knitter, H.-H., Budtz-Joergensen, C., *Nucl. Phys. A* **554**, 209 (1993).
17. Bouzid, B., Asghar, M., Djebara, M., Medkour, M., *J. Phys. G: Nucl. Part. Phys.* **24** 1029 (1998).
18. Rejmund, F., Ignatyuk, A. V., Junghans, A. R., Schmidt, K.-H., *Nucl. Phys. A* **678** 215 (2000).
19. Gupta, R. K., Balasubramaniam, M., Kumar, R., Singh, D., Beck, C., Greiner, W., *Phys. Rev. C.* **71**, 014601 (2005).

Prompt fission neutron multiplicity and spectrum evaluation

Anabella Tudora[a], G.Vladuca[a], B.Morillon[b], F.-J.Hambsch[c] and S.Oberstedt[c]

[a]*Bucharest University, Faculty of Physics, Bucharest-Magurele, POB MG-11, R-76900,Romania*
[b]*Commissariat à l'Energie Atomique, DAM Ile-de-France, BP 12, 91680 Bruyères-le-Châtel, France*
[c]*EC-JRC-Institute of reference Materials and Measurements, Retieseweg 111, B-2440, Geel, Belgium*

Abstract. The prompt fission neutron multiplicity and spectrum for spontaneous fission and neutron induced reactions in the energy range of the first chance fission threshold are evaluated using the "point by point" approach that takes into account all possible fragmentations of the fissioning nucleus, the multi-modal concept of fission and the most probable fragmentation approach. At higher incident energy when only the most probable fragmentation approach can be used, for the first time the model was extended to take into account the fission of the secondary compound nucleus chains formed by charged particle emission. The model parameters and their dependencies on the incident energy are determined by the study of the reactions where the respective nuclei are the main compounds. The linear dependence of the prompt gamma-ray energy on prompt multiplicity is parameterized as function of the mass and charge numbers of the fissioning nucleus. The above models were successfully applied for many neutron induced reactions on actinides, giving also good results in integral benchmarks.

Keywords: Prompt fission neutron multiplicity, prompt fission neutron spectrum
PACS: 24.75.+i; 25.85.-w;25.85.Ca

INTRODUCTION

The Los Alamos (LA) model in its original form [1] and improved versions [2-4] was and is successfully used by the nuclear physics community for the prompt fission neutron multiplicity (PFNM) and spectrum (PFNS) evaluations and is actually the agreed model for data evaluation purposes. That is due to the better physical ingredients compared to other models usually used up to now and to the fact that only few input model parameters are required: the energy release in fission (E_r), the total kinetic energy (TKE) of the fission fragments (FF), the average neutron separation energy (S_n) for the FF, the prompt gamma-ray energy (E_γ) and the level density parameter. In the case of spontaneous fission (SF) and neutron induced reactions in the incident energy (E_n) range of the first fission chance the model features can be used in three ways: a) taking into account only *"the most probable fragmentation"* [1, 4], the model parameters being determined as average values depending on E_n; b) *the multi-modal (MM) concept of fission* when the total PFNM and PFNS are calculated as superposition of the multiplicity and spectra of each fission mode weighted with the modal branching ratios and the model parameters are determined for each mode based on experimental and modal data [5-7]; c) *"the point by point (PbP) approach"* which is the most accurate because the entire FF range is taken into account, the FF mass distribution playing the most important role. The total PFNM and PFNS are calculated as superposition of the multiplicity and spectra of each FF pairs weighted with the charge and mass distribution [8]. At higher E_n where more fission chances are involved neither the PbP nor the MM approaches can be used because it is very difficult to distinguish the FF mass and charge distributions corresponding to each compound nucleus (CN) undergoing fission. The only remaining way is to take into account the most probable fragmentation for each CN involved and the use of the average values of the model parameters. For the first time this last model was extended to provide PFNM, PFNS and other related quantities at high E_n where *fission of the CN from secondary chains occurs*. Presently the fission of the secondary CN formed by: proton emission, neutron evaporation from the nuclei formed by proton emission, deuteron emission, alpha emission and neutron evaporation from the nuclei formed by alpha emission are taken into account [4]. The average model parameters and their appropriate dependence on the excitation energy of the fissioning nucleus (or on E_n) were determined by the study of the neutron induced reaction where the respective nucleus is the main compound. The entire FF range with PbP treatment was used in cases where experimental Y(A) and TKE(A) distributions exist. The "7 points" approach [1] with simulation of Y(A) distribution was used in case of non-existing or scarce experimental data [4, 8]. The linear E_γ dependence on the PFNM is parameterized only as a function of the mass and charge numbers of the fissioning nucleus [3-6]. The above models were successfully applied for the SF of ^{252}Cf and Pu isotopes and for the neutron induced reactions on many actinides e.g. $^{231-233}$Pa, ^{237}Np, ^{233,235,238}U, ^{239,240}Pu, giving also good results in integral benchmarks [8].

PFNM AND PFNS IN THE INCIDENT ENERGY RANGE OF THE 1-ST FISSION CHANCE

The PbP treatment idea of Ref.[9] was applied only to ^{252}Cf(SF) spectrum calculation for a limited number of FF (with a step size of 6 mass units). In Ref.[9] the same value of the level density parameter for all FF pairs was used. Here we developed the idea further and refined it as follow. The entire FF range is taken into account by the use of the whole FF range of available experimental mass distribution Y(A) with a step of 1 mass unit. For each FF mass pair, two isobars per FF mass are taken with values of the nuclear charge chosen as the nearest integer values above and below the most probable charge [5-8]. For instance 106

FF pairs are used for ^{235}U(n,f) [8] and respectively 100 FF pairs for ^{252}Cf(SF). The level density parameter is calculated for each FF of each pair using the super-fluid model of Ignatiuk. Also the PFNM is calculated explicitly with an original determination of the average S_n of each FF pair [8]. Details about the PbP approach are given in Ref.[8]. Fig.1 shows as example the PFNM calculation for ^{235}U(n,f) in comparison with experimental data and other evaluations. The PbP treatment plays the most important role in the improved PFNM evaluation that leads to very good result (Keff=1.00045 [8]) of the integral benchmark GODIVA.

The MM approach was applied to neutron induced reactions taking into account the prominent three fission modes: two asymmetric Standard I and II (S1, S2) and one symmetric super-long (SL). Details are given in Refs.[5, 6]). In the case of ^{252}Cf(SF) five fission modes (four asymmetric S1, S2, S3, SX and one symmetric SL) were necessary to describe the FF experimental data (Y(A) and TKE(A)). The PFNS of ^{252}Cf(SF) was calculated using the MM approach (with refinements concerning the anisotropy effect and a more accurate expression of the FF residual nuclear temperature distribution as it was described in Ref.[7]) and also using the PbP approach (with 100 FF pairs taken into account).

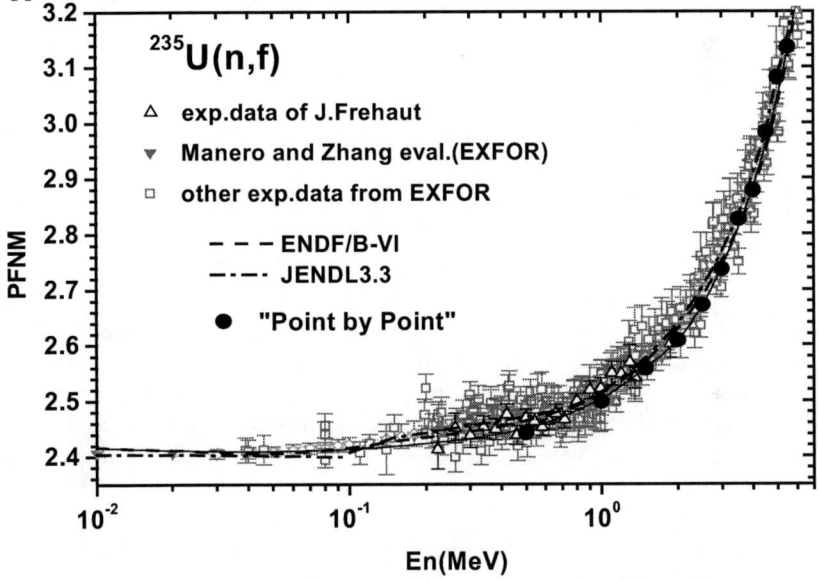

FIGURE 1. PFNM of ^{235}U(n,f) calculated with the PbP approach (full circles) in comparison with the existing experimental data and other evaluations (ENDF/B-VI with dashed line and JENDL3.3 with dash-dotted line)

The new spectrum calculations of ^{252}Cf(SF) are given in Fig.2, where the improved agreement (in comparison with the calculation of Ref.[9]) with the standard point-wise evaluation of Mannhart and recent experimental data is visible.

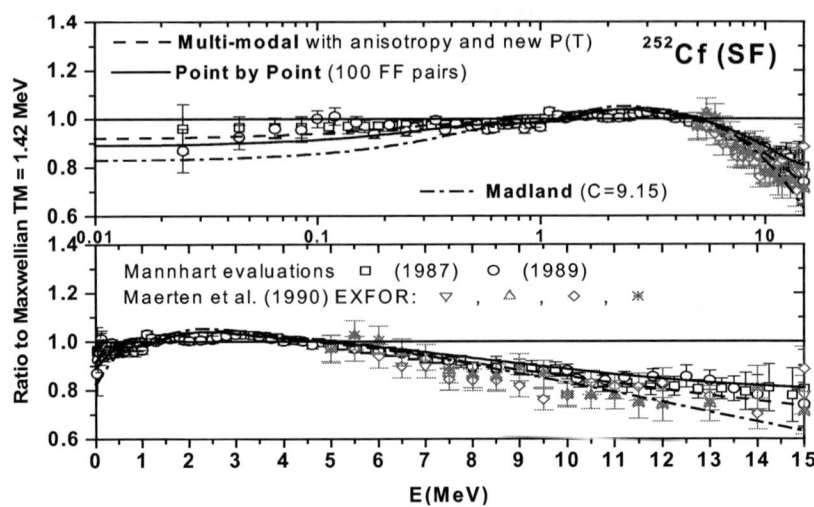

FIGURE 2. PFNS of ^{252}Cf(SF) calculated with PbP (solid line) and MM (dashed line) approaches in comparison with the point-wise Mannhart evaluation and recent experimental data as well as with the Madland calculation [9]

PFNM AND PFNS AT HIGH INCIDENT NEUTRON ENERGY

At E_n above about 25-30 MeV the most probable fragmentation approach [1, 2] was extended to take into account also fission of nuclei of the secondary chains formed by charged particle emission. The recursive equations describing the energetics of the nuclei formed following the 6 ways, neutron evaporation from the main chain nuclei, proton, deuteron and alpha emission from the main chain nuclei as well as the secondary neutron evaporations from the nuclei formed by proton and by alpha emission respectively, are given in detail in Ref.[4]. The PFNM and PFNS of each nucleus of each way depend on these energetics. The total PFNM and PFNS of the reaction are calculated as function of the multiplicities and spectra of the fissioning nuclei involved and as function of their fission probabilities expressed as fission cross-section ratios (RF). To obtain the partial RF, the production cross-sections of the secondary nuclei formed by proton, neutron via proton, deuteron and respectively by alpha and neutron via alpha emission (provided by a statistical-preequilibrium code [8, 10]) are used. The average model parameters and their appropriate dependence on the excitation energy of the fissioning nuclei of each way are determined by the study of the neutron induced reactions where the respective nucleus is the main compound using the PbP treatment when FF experimental data are available. An example of model parameters used in the cases of n+^{238}U reaction up to E_n = 80 MeV [4] and n+^{235}U reaction up to E_n = 30 MeV [8] are given in Fig.3 for the main U chain.

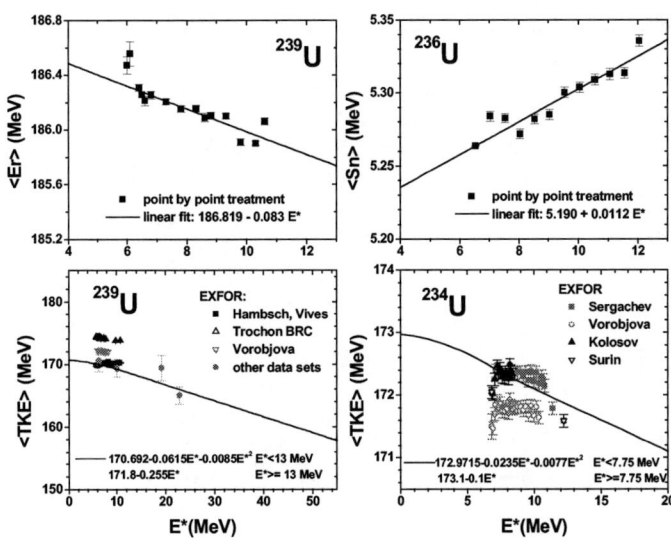

FIGURE 3. Model parameters of the U main chain versus the fissioning nucleus excitation energy

The total PFNM calculated with the new extended model for the neutron induced reactions on ^{238}U and ^{235}U are obtained in good agreement with the EXFOR experimental data and the new measurements [11] performed up to 200 MeV as it can be seen in Figs. 4 and 5.

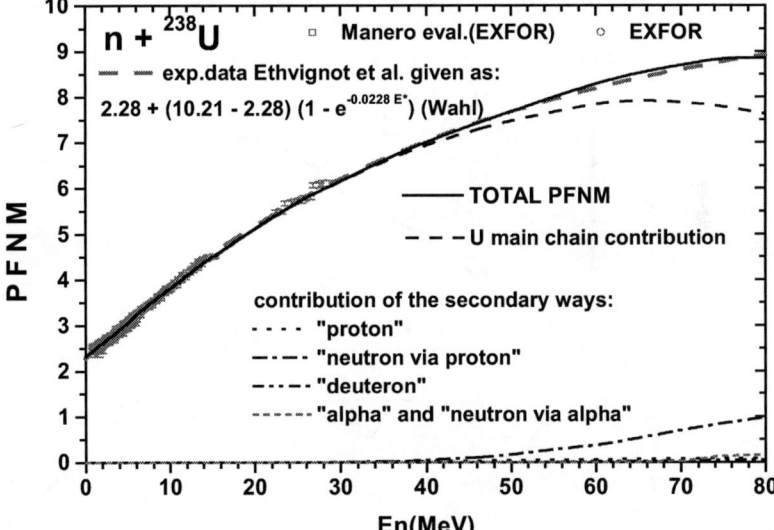

FIGURE 4. PFNM of the n+^{238}U reaction up to 80 MeV incident energy in comparison with the experimental data. The contributions of the U main chain and of the secondary ways (of the Pa and Th nucleus chains) are also given with different symbol lines.

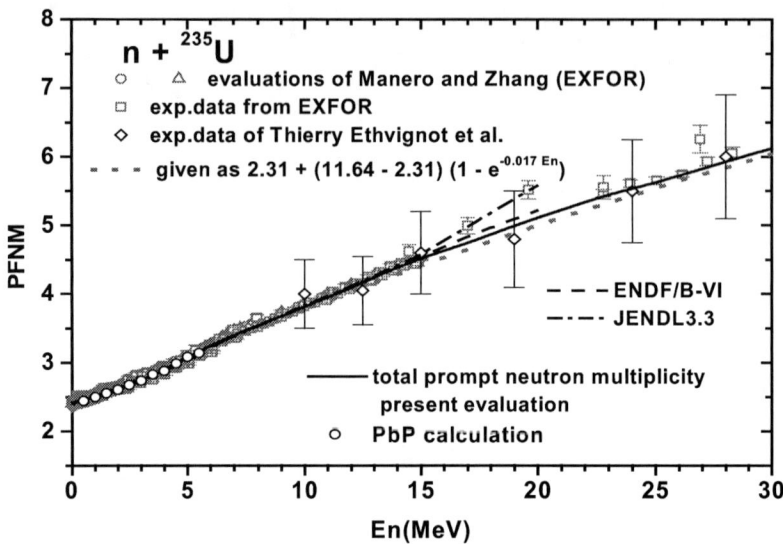

FIGURE 5. PFNM of n+^{235}U in comparison with the experimental data and other evaluations

The extended model also provides other quantities that can be compared with experimental data when they exist. As example the total average E_γ of n+^{235}U reaction is given in Fig.6. For each nucleus involved in the reaction the average E_γ is taken linearly dependent on the prompt multiplicity with the slope and intercept parameterized as it was described in Refs.[3, 5, 6].

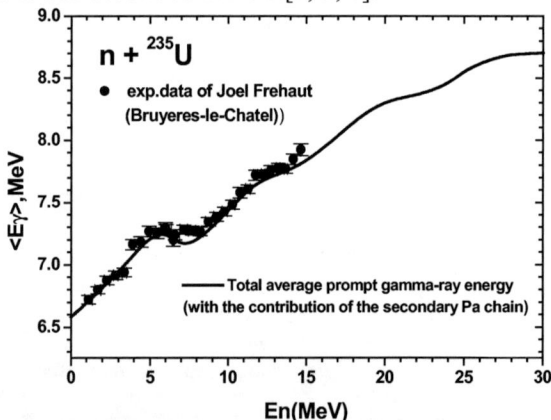

FIGURE 6. Total $<E_\gamma>$ of n+^{235}U reaction in comparison with the existing experimental data

An other quantity provided by the new model is the average TKE of the reaction. This quantity is obtained almost constant with E_n, being in very good agreement with the constant value provided by the intra-nuclear cascade plus evaporation model code BRIC developed at Bruyères-le-Châtel for high energies (see Fig.7).

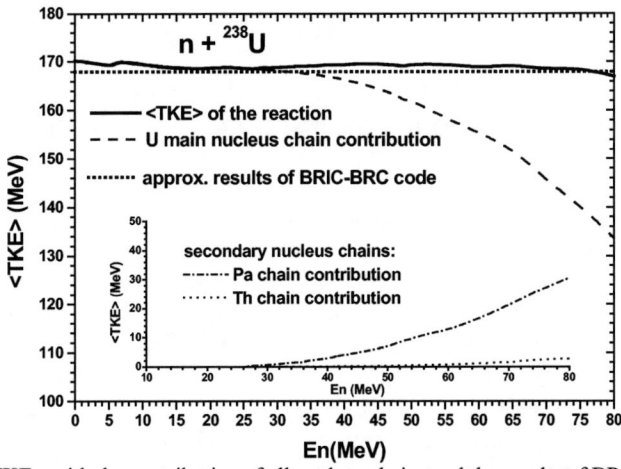

FIGURE 7. <TKE> with the contribution of all nucleus chains and the results of BRIC code

EXAMPLE OF VALIDATION TESTS

The PFNM and PFNS of the ^{235}U(n,f) were introduced in the new ^{235}U BRC evaluated data file [8, 12]. To validate this evaluation two types of criticality benchmarks were studied: the integral critical experiment GODIVA having a fast neutron spectrum and a series of thermal spectrum experiments with enriched ^{235}U fuel [8]. The continuous energy cross-sections of the ^{235}U evaluation were prepared by the NJOY processing system and the Monte-Carlo code TRIPOLI was used to calculate the K_{eff} values. Very good result of the GODIVA benchmark was obtained: K_{eff} = 1.00045. The thermal spectrum benchmarks results were also improved by the use of the new PFNM and PFNS calculations as it can be seen in Fig.8.

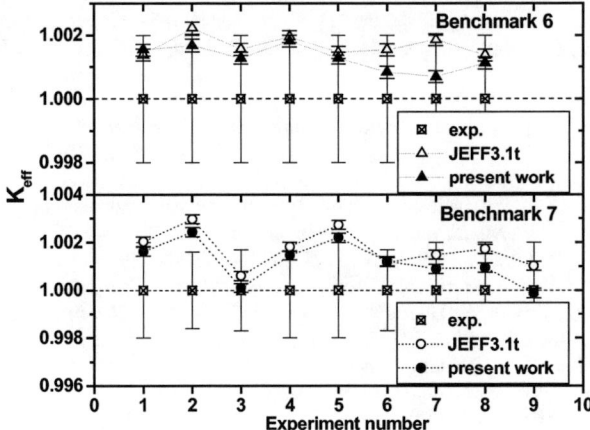

FIGURE 8. K_{eff} results of the thermal spectrum benchmarks (type LEU-COMP-THERM)

CONCLUSIONS

The improvements brought to the PFNM and PFNS model and to the model parameters treatment can be synthesized as follows:

1) *In the E_n range of the 1-st chance: i)* the "point by point" approach takes into account the entire FF range covered by the experimental FF mass distributions, the PFNM expression is given, and a new treatment of the level density parameter is included, *ii)* the "multi-modal fission" approach with multi-modal parameters determined on the basis of modal and experimental data *iii)* the dependence on E_n of the average model parameters entering the most probable fragmentation approach is taken into account and the E_γ linear dependence on prompt multiplicity is parameterized *iv)* refinements consisting in the inclusion of the anisotropy and a new expression of the FF residual nuclear temperature distribution were brought. A computer code describing the above models was written.

2) *In the E_n range of multiple fission chances*: *i)* a new model (and the respective computer code) was developed for higher E_n where the fission of the nuclei of secondary chains and ways formed by charged particle emission occurs, *ii)* the parameters of this model were determined by the study of the neutron induced reactions where the respective fissioning nuclei are the main compounds.

3) The above models were successfully applied for the SF of ^{252}Cf(SF) and Pu isotopes and in neutron induced reactions on many actinides (e.g. $^{231\text{-}233}$Pa, ^{237}Np, ^{239,240}Pu, $^{233\text{-}238}$U) giving good results in integral benchmarks.

REFERENCES

1. D. G. Madland, J. R. Nix, *Nucl. Sci. Eng.* **81**, 213-271 (1982)
2. G. Vladuca, Anabella Tudora, *Comp. Phys. Communic.* **125 (1-3)**, 221-238 (2000)
3. G. Vladuca, Anabella Tudora, *Annals of Nucl. Energ.* **28**, 419-435, 689-700, 1653-1665 (2001)
4. Anabella Tudora, G. Vladuca, B. Morillon, *Nucl. Phys. A* **740**, 33-58 (2004)
5. F.-J. Hambsch, S. Oberstedt, G. Vladuca, Anabella Tudora, *Nucl. Phys. A* **709**, 85-102 (2002)
6. F.-J. Hambsch, S. Oberstedt, Anabella Tudora, G. Vladuca, I. Ruskov, *Nucl. Phys. A* **726**, 248-264 (2003)
7. F.-J. Hambsch, Anabella Tudora, G. Vladuca, S. Oberstedt "Prompt fission neutron spectrum evaluation for ^{252}Cf(SF) in the frame of the multi-modal fission model", *Annals of Nucl. Energ.* (2005) in press
8. Anabella Tudora, B. Morillon, F.-J. Hambsch, G. Vladuca, S. Oberstedt "A refined model for ^{235}U(n,f) prompt fission neutron multiplicity and spectrum calculations with validation in integral benchmarks", *Nucl. Phys. A* (2005) in press
9. D. G. Madland, R. J. LaBauve, J. R. Nix, *IAEA-INDC(NDS)-220*, 1989, pp.259-281
10. M. J. Lopez-Jimenez, B. Morillon, P. Romain, *Annals of Nucl. Energ.* **32**, 195-213 (2005)
11. T. Ethvignot, M. Devlin, H. Duarte, T. Granier, R.C. Haight, B. Morillon, R. O. Nelson, J. M. O'Donnell, D. Rochman, *Phys. Rev. Lett.* 94 052701 (2005)
12. M. J. Lopez-Jimenez, P. Romain, B. Morillon, "Overview of recent Bruyères-le-Châtel actinide evaluations", *International Conference on Nuclear Data for Science and Technology ND2004*, Santa Fe, USA, 2004

Prompt Neutron Emission from Fragments in Spontaneous Fission of $^{244,\,248}$Cm and ^{252}Cf

A.S. Vorobyev[1,*], V.N. Dushin[2], F.-J. Hambsch[3], V.A. Jakovlev[2], V.A. Kalinin[2], A.B. Laptev[1,4], B.F. Petrov[2], O.A. Shcherbakov[1]

[1] *Petersburg Nuclear Physics Institute, Gatchina, Leningrad district, 188300, Russia*
[2] *V.G. Khlopin Radium Institute, St. Petersburg, 194021, Russia*
[3] *EC-JRC-Institute for Reference Materials and Measurements Retieseweg 111, B-2440, Geel, Belgium*
[4] *Japan Nuclear Cycle Development Institute, Tokai-mura, Naka-gun, Ibaraki, 319-1194, Japan*

Abstract. Neutrons emitted in fission were measured separately for each complementary fragment in correlation with fission fragment energies. Two high efficient Gd-loaded liquid scintillator tanks were used for neutron registration. Fission fragment energies were measured using a twin Frisch gridded ionization chamber with a pin-hole collimator. The neutron multiplicity distributions were obtained for each value of the fission fragment mass and energy and corrected for neutron registration efficiency, background and pile-up. The dependencies of these distributions on fragment mass and energy for different energy and mass bins, as well as the mass and energy distribution of the fission fragments are presented and discussed.

Keywords: Spontaneous fission; Fission Fragments; Neutron Multiplicity; $^{244,\,248}$Cm and ^{252}Cf.
PACS: 25.85.Ca, 27.90+b, 21.10.Gv, 29.40.Mc

INTRODUCTION

Until today one of the most interesting problems in the field of low energy fission is the study of the reaction dynamics. Thereby the emphasis is on the careful investigation of the energy balance in every individual fission event for a fixed mass split. The energy released in the fission process, Q, is divided among total kinetic, TKE^*, and total excitation, TXE, energies:

$$\overline{Q}(m_L^*/m_H^*) = \overbrace{E_{pre} + E_{coul}(m_L^*/m_H^*)}^{TKE^*} + \overbrace{E_{def}^L + E_{def}^H + E_{int}}^{TXE}, \quad (1)$$

$$TXE(m_L^*/m_H^*, TKE^*) = \nu_L \cdot (B_{nL} + \eta_{nL}) + \nu_H \cdot (B_{nH} + \eta_{nH}) + E^{\gamma}_{tot}. \quad (2)$$

Here E_{pre} is the "pre-scission" kinetic energy, E_{coul} the Coulomb potential energy; $E_{def}^{L,H}$ the deformation energy of light and heavy fragments, respectively, E_{int} the intrinsic excitation energy. $\nu_{L,H}$, $\eta_{nL,H}$ and E^{γ}_{tot} are the number of neutrons, their mean kinetic energy and total energy carried away by γ-rays from fragments with mass

---

[*] Corresponding author: Tel. +7 813-714-6444, Fax +7 813-713-6041, E-mail: *alexander.vorobyev@pnpi.spb.ru*

$m_{L,H}*$ and total kinetic energy $TKE*$, respectively. $B_{nL,H}$ is the binding energy of a neutron in a nucleus with mass $m_{L,H}*$ averaged over the charge distribution.

Whereas the value of $TKE*$ is mainly determined by Coulomb repulsion of the nascent fragments, the number of emitted neutrons characterizes the fission fragment excitation energy and has a well known "saw-tooth" dependence on fragment mass [1]. Therefore, the number of neutrons emitted by each of the complementary fragments provides direct information on the redistribution of the excitation energy between fission fragments. Because of that, a very efficient way of investigation of the fission fragment mass and energy formation is the study of the energy and mass distribution of fission fragments for fixed pairs of neutron numbers emitted by light and heavy fragments. In other words, to gain a better insight into fission dynamics, the distribution of prompt neutron emission probabilities for specific fission fragments should be obtained.

Recently, measurements of this type have been carried out for ^{252}Cf and $^{244, 248}$Cm(SF) by the KRI-PNPI-IRMM collaboration within the framework of the ISTC-554 Project [2-5]. As a result, the initial prompt neutron multiplicity distributions have been obtained for each isotope investigated. Some characteristics of these neutron multiplicity distributions are presented.

EXPERIMENTAL PROCEDURE AND DATA PROCESSING

The prompt neutron multiplicity distributions have been measured using two large Gd-loaded liquid scintillator counters in conjunction with an ionization chamber for simultaneous registration of the fission fragment kinetic energies and the prompt number of neutrons emitted by complementary fragments in each fission event. In these so-called 2×2π-geometry measurements, the efficiency of the neutron registration was about 55% for each detector. The same experimental set-up was used for the 4π-geometry calibration measurements (total neutron multiplicity distribution).

The fission events were defined by registration of coincident fission fragment pulses from both halves of the ionization chamber. The fission event was considered as "useful" when only one coincident fragment pair was registered in the 20μs neutron measuring time interval opened ~0.3μs after the fission event. In total about $3 \cdot 10^6$ "useful" events for each spontaneous fissioning nuclide investigated have been accumulated. The list-mode data included the pulse-height, the number of emitted neutrons for both fragments and the neutron background multiplicity.

The data processing was performed in several steps for every fission event. First, on the basis of reference values of the mean "post-neutron" mass and energy of fission fragments, the conversion of obtained pulse-height to "provisional" mass, μ, and "post-neutron" total kinetic energies, TKE, was performed. Second, using the measured neutron distributions corrected for pile-up effects, background and neutron detector efficiency, the "pre-neutron" mass, $m*$, and total kinetic energy, $TKE*$, were calculated. Third, the initial prompt neutron multiplicity distributions were unfolded for fixed values of $m*$ and $TKE*$. The applied unfolding procedure was verified using the ^{252}Cf data. It was ascertained that the experimental method as well as the data processing is adequate for an accurate determination of the neutron multiplicity distributions for fixed "pre-neutron" fragment masses and kinetic energies (for more details see [6]).

RESULTS AND DISCUSSION

Correlation between Number of Neutrons Emitted by Complementary Fission Fragments

The main characteristics of the prompt neutron multiplicity distribution unfolded from measured data are presented in Table. 1.

TABLE 1. Prompt Neutron Emission Probability, $P(v)$.

Number of Emitted Neutrons, v	Emission Probability for Light Fragments, $P(v_L)$ (2×2π-Geometry)			Emission Probability for Heavy Fragments, $P(v_H)$ (2×2π-Geometry)			Total Neutron Emission Probability, $P(v_{TOT})$ (4π-Geometry)		
	^{252}Cf	^{248}Cm	^{244}Cm	^{252}Cf	^{248}Cm	^{244}Cm	^{252}Cf	^{248}Cm	^{244}Cm
0	0.0584	0.0769	0.1063	0.1052	0.2008	0.2393	0.0023	0.0061	0.0175
1	0.2520	0.3037	0.3808	0.3436	0.4212	0.4459	0.0290	0.0608	0.1121
2	0.3644	0.3669	0.3597	0.3495	0.2782	0.2438	0.1230	0.2272	0.2996
3	0.2449	0.2008	0.1241	0.1592	0.0853	0.0605	0.2719	0.3460	0.3387
4	0.0688	0.0452	0.0241	0.0363	0.0131	0.0092	0.3052	0.2476	0.1768
5	0.0095	0.0054	0.0041	0.0052	0.0012	0.0011	0.1867	0.0906	0.0473
6	0.0015	0.0009	0.0007	0.0008	0.0001	0.0001	0.0654	0.0190	0.0072
7	0.0003	0.0002	0.0001	0.0001	<0.0001	<0.0001	0.0139	0.0024	0.0007
8	0.0001	<0.0001	<0.0001	<0.0001			0.0021	0.0002	0.0001
9	<0.0001						0.0005	<0.0001	<0.0001
Dispersion, $\sigma^2(v)$	1.125	1.041	0.916	1.042	0.876	0.805	1.623	1.336	1.272
Average, $<v>$	2.051	1.854	1.596	1.698	1.293	1.158	3.756 ±0.031	3.130 ±0.023	2.720 ±0.061

It can be seen that for all three nuclides the sum of variances in the number of neutrons emitted from light, $\sigma^2(v_L)$, and heavy, $\sigma^2(v_H)$, fragments, which were obtained by 2×2π-geometry measurements, is larger than the variance in the total number of neutrons per fission, v_{TOT}, obtained by 4π-geometry measurements. This gives a value of the covariance, cov, defined by:

$$2 \cdot cov(v_L, v_H) = \sigma^2(v_{TOT}) - \sigma^2(v_L) - \sigma^2(v_H), \qquad (3)$$

equal to -0.27 ± 0.07, -0.29 ± 0.05 and -0.22 ± 0.10 for ^{252}Cf, ^{248}Cm and ^{244}Cm, respectively. An appreciable anticorrelation is evident. It should be noted that the $cov(v_L, v_H)$ obtained by the direct method using only the 2×2π-geometry data gives the same results within the indicated errors. This confirms the quality of unfolded data and points to the absence of any essential systematic errors.

As it was pointed out by Nix and Swiatecki [7], the domination of a distortion-asymmetry mode results in an anti-correlation of the fragment excitation energies (or number of neutrons emitted by these fragments) within the framework of the liquid-drop model. At the same time, the value of the coefficient of correlation between the deformation energies of the fission fragments is proportional to the value of the intrinsic excitation energy []. Indeed, it is no surprise that the general trend in *cov* as a function of *TKE** is like the odd-even effect in the charge yield [5], which is directly related to the intrinsic excitation (dissipative) energy.

Cold Compact Fission (neutronless case)

The fission fragment mass distributions for the neutronless case (when no neutrons are emitted by the fission fragments) are shown in Fig. 1.

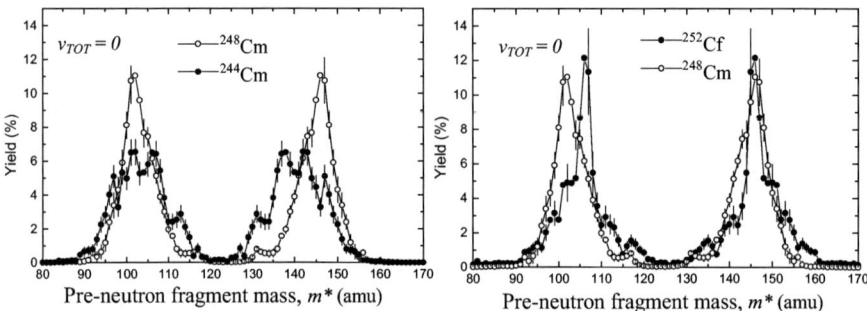

FIGURE 1. The fragment mass distributions obtained by selection of the fission events with zero number of emitted neutrons for spontaneous fission ^{252}Cf and $^{244,\,248}$Cm. The errors shown here are due to the unfolding procedure and the asymmetry of the fission fragment detector as well as that of the neutron counters.

It can be seen that for all nuclides investigated, the yields are higher in the mass regions corresponding to the neutron shell $N = 64 - 68$, $N = 82 - 84$ and $N = 86 - 90$ [8]. Here, the mean total excitation energy, $<TXE(v_{tot} = 0)>$, determined by Eq. 4 is about 10 MeV for all investigated nuclei.

$$<TXE(v_{TOT} = 0)> = <Q_{tab}^{fit}> - <TKE^*(v_{TOT} = 0)>, \qquad (4)$$

where $<Q_{tab}^{fit}> = $ smoothed $(<Q_{tab}>)$,

$$<Q_{tab}(m^*)> = \sum_{Z=\text{int}(Z_p - 4\sigma_Z)}^{\text{int}(Z_p + 4\sigma_Z)} [P_{Gauss}(m^*, Z) Q^{tab}(m^*, Z)],$$

$Q^{tab}(m^*, Z)$ – total available energy calculated from the mass tables [9] at the most probable charge Z_p and the charge dispersion σ_Z of fission fragments taken from Ref. [10];

Then the mean intrinsic excitation energy of fission fragments, estimated as:

$$<E_{int}> = <TXE(v_{TOT} = 0)> - <E_{def}^L> - <E_{def}^H>, \qquad (5)$$

is not smaller than $5 \div 7$ MeV since the deformation energy of the fission fragment ground state, $<E_{def}^L> + <E_{def}^H>$, determined from a systematic of cold fission data is about $3 \div 5$ MeV [11].

Strongly Deformed Fission

Through the systematic study of mass distributions for a fixed number of neutrons emitted from complementary fragments, it was revealed that in the case of $v_{TOT} \geq 5$ structures with a period of about 5 amu appear. The more asymmetry in fragment deformation is available, the more these structures are pronounced (see Fig. 2). A similar phenomenon was observed previously only for ^{252}Cf [12]. It was interpreted as an existence of strongly deformed cold configurations at the scission point. Therefore, it can be concluded that the intrinsic excitation energy depends on the total excitation energy as well as on the fission fragment deformation asymmetry.

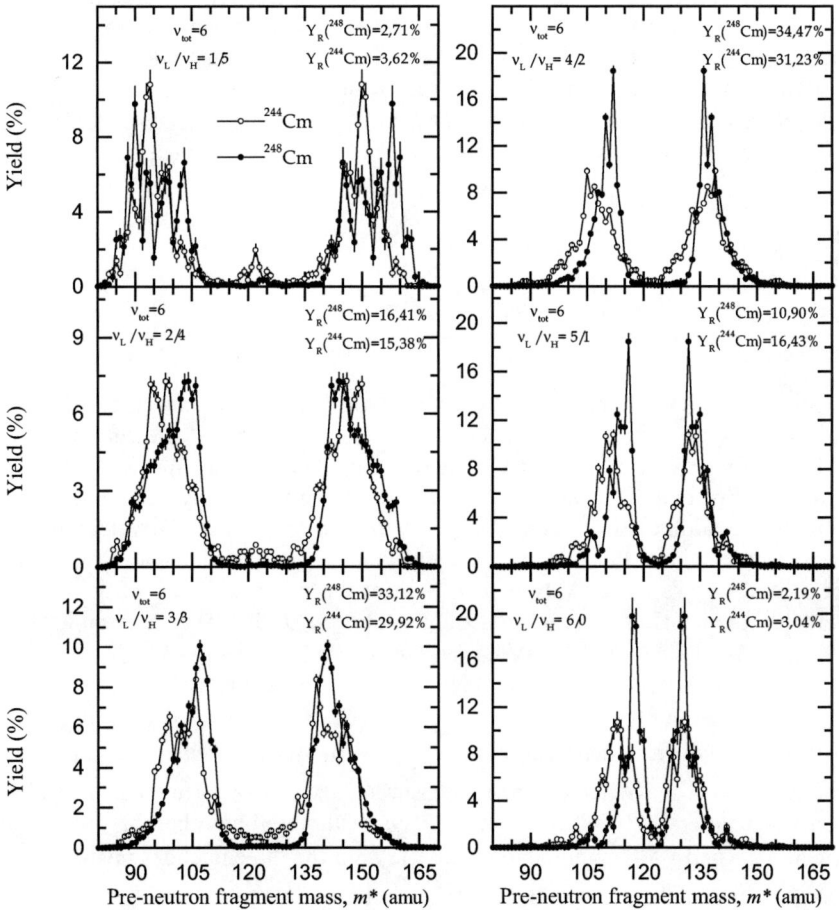

FIGURE 2. Comparison of ^{244}Cm and ^{248}Cm partial pre-neutron mass distributions for a fixed total number of prompt neutrons. The yield in % is normalized to the total number of events corresponding to the indicated numbers of neutrons emitted by the light and heavy fragments, respectively. Y_R is the relative total yield of fission events which are characterized by a given asymmetry in the number of prompt neutrons emitted from complementary fragments. The total number of prompt neutrons is equal to 6 in all cases.

Energy Balance

In order to determine the mean energy required for the emission of one additional neutron, $<S_n>$, the mean TKE^* was calculated for each fission fragment mass at a fixed total number of neutrons, $<TKE^*(m^*: \nu_{TOT})>$. Then the slope $\Delta \nu_{TOT} / \Delta TKE^*(m^*)$ was found using the obtained mean values:

$$[- \Delta \nu_{TOT} / \Delta TKE^*(m^*)]^{-1} = <S_n(m^*)> = <B_n(m^*) + \eta_n(m^*) + E_{n\gamma}(m^*)>, \quad (6)$$

Here $E_{n\gamma}(m^*)$ is the part of $S_n(m^*)$ which is responsible for the presence of n-γ competition. In assumption that the $TKE^*(m^*: \nu_{TOT})$ - distribution is similar to a Gaussian distribution, a statement can be made about the independence of the $<S_n>$ on the fission fragment detector resolution unlike to traditional estimations based on the data on an average total number of neutrons as a function of total kinetic energy of fragments for fixed fragment mass, $<\nu_{TOT}(TKE^*: m^*)>$ (for example, see Ref. [13]). The obtained values of $<S_n(m^*)>$ and the mean neutron binding energy, $<B_n(m^*)>$, calculated by means of Eq. 7 are presented in Fig. 3.

$$<B_n(m^*)> = \sum_{\nu_1,\nu_2} P_{init}(\nu_1, \nu_2, m^*) \left\{ \sum_{S=1}^{\nu_1>0} \left[\sum_{Z=int(Z_p-4\sigma_Z)}^{int(Z_p+4\sigma_Z)} \left[P_{Gauss}(m^*, Z) B_n^{tab}(m^* - (s-1), Z) \right] \right] \right\} \quad (7)$$

Here $P_{init}(\nu_1, \nu_2, m_1^*)$ is the initial neutron multiplicity distribution averaged over the TKE^* distribution; $B_n^{tab}(m^*, Z)$ is the neutron binding energy calculated using the mass tables [9] for a nucleus with mass m^* and charge Z taken from Wahl [10].

It is noteworthy that whereas the calculated $<B_n>$ increases by ~ 0.07 MeV on average with one additional neutron emitted by fission fragments, the values of the $<S_n>$ are constant within the experimental error and not dependent on ν_{TOT}. This behavior of the $<S_n>$ is most likely related both to the weak variation of the charge density of the fragments, Z/m^*, and to a minor decrease in the mean neutron kinetic energy, $<\eta_n>$, (about 0.07 ± 0.03 MeV per neutron) with an increase of TXE.

As it can be seen from the right-hand side of Eq. 6 under the assumption that the n-γ competition does not exist, the estimation of $<S_n(m^*)>$ averaged over fragment mass distribution, which is denoted below as $<\overline{S_n}>$, may be obtained independently by summing the data on $<\eta_n>$ taken from measurements analogous to those of Budtz-Jorgensen and Knitter [14] and the calculated $<B_n>$. It was found that the difference $E_{n\gamma}$ between $<S_n>$ and $<\overline{S_n}>$ is equal to ~ -0.3 MeV for the three nuclides investigated. This may be due to the existence of n-γ competition in the fission fragment de-excitation process as well as possible experimental and systematic errors. To clear this situation, the dependence of the total energy of prompt γ-rays on the total number of prompt neutrons $<E^\gamma_{tot}(\nu_{TOT})>$, found as the difference between energy of reaction $<Q>$ and sum of $<TKE(\nu_{TOT})>$ and $<S_n> \cdot \nu_{TOT}$, was obtained. It was established that when ν_{TOT} increases by one neutron, $<E^\gamma_{tot}>$ decreases by ~ 0.3 ± 0.1 MeV on an average.

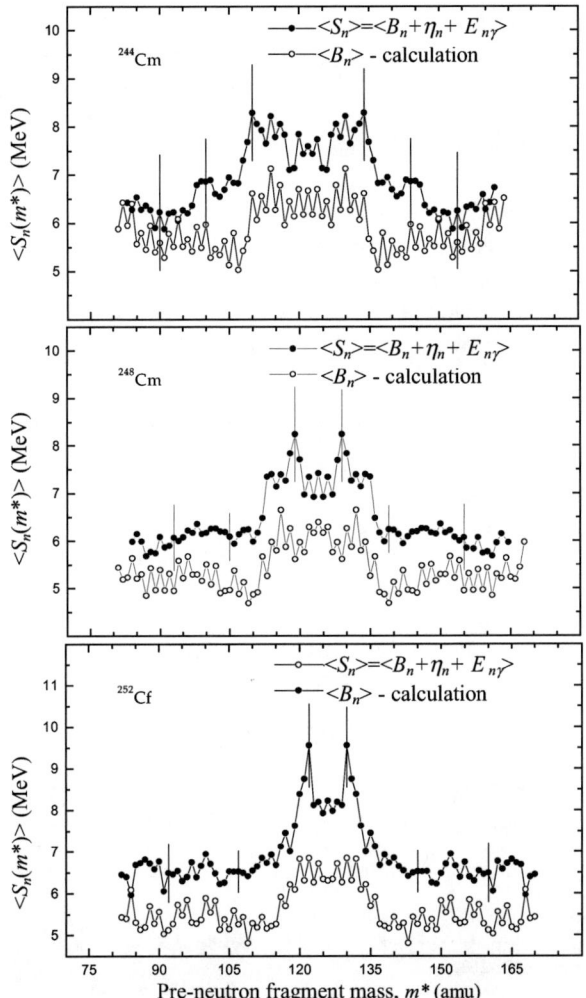

FIGURE 3. Comparison of the mean energy needed to emit an additional neutron, $\langle S_n \rangle$, obtained in this work and the mean neutron binding energy, $\langle B_n \rangle$, calculated from the mass tables [9] on the basis of the $m^* - TKE^*$ distributions for a fixed total number of neutrons in ^{252}Cf(SF) and ^{244,248}Cm(SF).

SUMMARY

The neutron multiplicity distributions have been measured for each value of the fission fragment mass and energy for ^{252}Cf(SF) and ^{244,248}Cm(SF) using the same experimental set-up and data processing.

The anticorrelation between the numbers of neutrons emitted from complementary fragments was confirmed for all three nuclides investigated.

By the investigation of the cold compact fission (without neutron emission) it was revealed that the neutron shells N=64÷66, N=82 and N=88÷90 play a dominant role in the fission fragment formation. At that, the mean intrinsic excitation energy is at least 5-7 MeV.

For the spontaneous fission of $^{244,\ 248}$Cm and ^{252}Cf, a fine structure with a period of 5 amu has been observed in the mass distributions of the strongly deformed fission fragments ($v_{TOT} \geq 5$). This fine structure becomes more pronounced with an increase of the asymmetry of fragment deformation ($v_H/v_L \geq 3$).

An anticorrelation between the total number of emitted neutrons, v_{TOT}, and the mean total energy of fission γ-quanta, E^{γ}_{tot}, was observed. As v_{TOT} increases by one neutron, $< E^{\gamma}_{tot} >$ decreases by ~0.3 MeV on an average.

ACKNOWLEDGMENTS

This work was carried out with the support of the ISTC Project #554. The authors would like to thank many co-workers at KRI and PNPI for their fruitful participation in this Project. The contributions of D.V. Nikolaev, V.I. Petrova, T.A. Zavarukhina, V.E. Sokolov, N.V. Skovorodkin, T.E. Kuzmina and S.V. Davydova are particularly acknowledged. The authors thank F. Gönnenwein, A.A. Goverdovski and G.A. Petrov for useful discussions.

REFERENCES

1. Terrell J., *Phys. Rev.*, **127**, 880-904 (1962).
2. Kalinin, V., Dushin, V., Hambsch, F.-J., *et al.*, *J. Nucl. Sci. and Tech.*, Suppl. 2, v. 1, 250-253 (2002).
3. Kalinin, V.A., Dushin, V.N., Jakovlev, V.A., *et al.*, *In Proc. of the "Seminar on Fission Pont D'Oye V"*, Castle of Pont d'Oye, Habay-la-Neuve, Belgium, 16-19 Sept., 2003, ed. Wagemans C., publ. by World Sci. Publ. Co. Pte. Ltd, 2004, pp.73-82.
4. Kalinin, V.A., Dushin, V.N., Petrov, B.F., *et al.*, *In Proc. of the 5th Int. Conf. on "Dynamical Aspects of Nuclear Fission"*, Častá-Papiernička, Slovak Republic, 23-27 October, 2001, ed. Kliman J., publ. by World Sci. Publ. Co. Pte. Ltd, 2002, pp.252-263.
5. Vorobyev, A.S., Dushin, V.N., Hambsch, F.-J., *et al.*, *In book "IX Int. Seminar on Interaction of Neutrons with Nuclei "Neutron Spectroscopy, Nuclear Structure, Related Topics", ISINN-9*, Dubna, May 23-26, 2001, ed. Sukhovoj, A.M., Dubna, JINR, E3-2001-192, 2001, pp.276-295.
6. Dushin, V.N., Hambsch, F.-J., Jakovlev, V.A., *et al.*, *Nucl. Inst. and Meth.*, **A 516**, 539-553 (2004).
7. Nix, J.R., Swiatecki, W. J., *Nucl. Phys.*, **71**, 1-94 (1965).
8. Wilkins, B.D., Steinberg, E.P., Chasman, R.R., *Phys. Rev*, **C14**, 1832-1863 (1976).
9. Audi, G., Wapstra, A. H., *Nucl. Phys.* **A595**, 409-480 (1995).
10. Wahl, A.C., *Atomic Data and Nuclear Data Tables*, Vol. 39, No. 1, 3-137 (1988).
11. Floresky, A., Sandulescu, A., Coiaca, C., *et al.*, *Journal of Physics*, **G19**, 669-683 (1993).
12. Alkhazov I.D., Kuznetsov A.V., Kovalenko S.S., *et al.*, *In Proc. of Int. Conf. "Nuclear Data for Science and Technology"*, Mito, Japan, May 30 – June 3, 1988, ed. Igarasi S., Saikon Publ. Co. Ltd, 1988, pp.991-994.
13. Nifenecker H., Signarbieux C., Babinet R., Poitou J., "Neutron and Gamma Emission in Fission", *In Proc. of 3rd IAEA Symposium on Physics and Chemistry of Fission*, Rochester, New York, 13-17 August, 1973, printed by the IAEA, Vol. II, Vienna, 1974, pp.117-178.
14. Budtz-Jørgensen C., Knitter H.-H., *Nucl. Phys*, **A490**, 307-328 (1988).

Isotopic Yields of Fission Fragments from Transfer-Induced Fission

F. Rejmund[1,♦], G. Barreau[2], B. Jurado[2], K.-H. Schmidt[3], A. Kelic[3],
M.V. Ricciardi[3], E. Casarejos[4], C.-O. Bacri[5], L. Tassan-Got[5], C. Schmitt[6],
J. Benlliure[4], T. Enqvist[7], N. Alahari[1], M. Rejmund[1], J. Frankland[1],
M. Morjean[1], H. Savajols[1], W. Mittig[1]

[1] GANIL, bd H. Becquerel, 14076 Caen, France
[2] CENBG, Gradignan, France
[3] GSI, Planckstrasse 1, 64291 Darmstadt, Germany
[4] USC, Santiago de Compostella, Spain
[5] IPNO, rue G. Clémenceau, 91406 Orsay, France
[6] IPNL, Lyon, France
[7] CUPP, University of Oulu, Finland

Abstract. It is proposed to use transfer-induced fission in inverse kinematics coupled to the large acceptance spectrometer VAMOS to identify in atomic and mass number the complete distribution of the fission fragments. The measure of the kinetic properties of the transfer partner allows for determining precisely the excitation of the fissioning system. For the first time, the isotopic yields of the heavy and light fragments may be measured as a function of the excitation energy in neutron-rich actinides.

Keywords: Isotopic yields, inverse kinematics, transfer-induced fission.
PACS: 24.75.+i, 25.70.Hi

INTRODUCTION

Nuclear fission is best suited to study the dynamic properties of cold and moderately excited nuclear matter. At low excitation energies, the onset of dissipation in a super-fluid Fermionic system and the influence of quantum-mechanical structure on a large-amplitude collective motion can be studied. Nuclear fission is also a unique tool to explore shell effects at extreme deformation. Despite its general importance, the understanding of the nuclear-fission process both at low and at high excitation energies is still rather limited. This is partly due to the complexity of nuclear dynamics, especially at low excitation energies, where the influence of nuclear structure is strong, but also due to severe experimental constraints. In this contribution, we present a project for a new experimental method to investigate fission fragment distributions in transfer-induced reactions.

♦ frejmund@ganil.fr

Shell Effects in Fission Fragment Yields

The mass distribution of low-energy fission shows an asymmetric split that is understood as the effect of shell gaps on the potential energy landscape of the fissioning nucleus. The liquid drop model of the fission process predicts a symmetric split of equally deformed fragments[1]. The shell closures around N=82 for a spherical shape and N=86 for a deformed nucleus distort this potential energy landscape. Because the slope of the liquid-drop energy surface is steep for a spherical shape, the strong shell effect at N=82 results in a relative minimum of the potential energy. The absolute minimum of the potential energy is then around the deformed shell closure[2] around N=86. However, the exact nucleon number of the deformed shell closure is not precisely determined. Depending on the literature, a number between 86 and 90 is given. Indeed, a correlation exists between the deformation and the nucleon number corresponding to the shell closure. The more the nucleus is deformed, the heavier is the nucleon number of the deformed shell closure. Therefore, depending on the scission elongation and on the mass of the fissioning system, different numbers may be predicted. The confrontation of theoretical predictions to experimental data is difficult, because the data are scarce. First of all, the measure of the heavy fragments isotopic yields is impossible in direct kinematics before their beta-decay. Therefore, the identification is done by radiochemical methods[3], where one has to account for the different lifetimes and branching ratios of the populated levels. The complete isotopic chains cannot be identified, and fluctuations in the yields are large. Since the advent of spectrometers such as LOHENGRIN in ILL, complete isotopic yields can be identified, but are limited to the light fragments[4]. The isotopic distribution of the heavy fragments is then reconstructed from the average neutron number evaporated. However the evaporated-neutron number is known only as a function of the fission fragments mass[5], and many assumptions have to be made to determine the isotopic yields in the heavy fragments distribution. In addition, the systematic for fissioning nuclei is limited to actinides with sufficiently long lifetimes to be separated and produced on a target. Concerning the experiments of ILL, only thermal energy fission could be studied. Only recently could a large systematic be covered, using the secondary beams of fragmentation of a relativistic U beam at GSI, Darmstadt. The nuclear-charge distribution could be determined for the first time for the electromagnetic-induced fission of hundreds of short-lived actinides and pre-actinides[6]. Among other important results, a striking feature is that the average nuclear charge of the heavy fission fragments for the different fissioning systems is astonishingly stable[7] and equals 54. This new result puts into question the influence of neutron shell closure and the possible existence of a shell effect around Z=54.

Even-Odd Staggering In Fission Yields

Another important property of fission at low energy is the even-odd staggering in the element yields. The fission of a completely paired even nucleus excited below the pairing gap into two odd fragments is understood as evidence for the onset of quasi-particle excitations by the coupling between the collective degrees of freedom (deformation) and the intrinsic degrees of freedom. The recent results at GSI[6] show the existence of even-odd staggering in the element yields of even- and odd-fissioning nuclei. This finding sheds new light on variations of the even-odd effect as a function of mass asymmetry. The systematic increase of the even-odd effect at large asymmetry, previously attributed to particularly low dissipation[4], is now at least partly understood in terms of the phase space available in the fissioning system[8]. This result restarted the discussion on the origin of the even-odd staggering[9]. A new model[10] based on a rigorous formulation of the level density in the super-fluid-nucleus model of the fissioning system just before scission is able for the first time to describe the difference in the even-odd staggering in proton and neutron number. The excitation energy gained by the nucleus when deforming from saddle to scission is deduced from the even-odd staggering amplitude observed in the experimental data. The most comprehensive data on the evolution of the even-odd effect as a function of the excitation energy have been produced by Brehmstrahlung induced fission[11], for which the excitation energy distribution is flat. Therefore, no conclusion may be drawn on the reproduction of the data by the model, neither on the evolution of nuclear viscosity with increasing excitation energy.

INVERSE KINEMATICS TRANSFER-INDUCED FISSION

The recent results of GSI on the measure of the element yield of a large systematic of fissioning system[6] and on the complete isotopic cross sections of spallation reactions[12] showed the tremendous advantage of inverse kinematics coupled to spectrometer identification. We propose here to use inverse kinematics at GANIL energy to induce fission after multi-nucleon transfer on a ^{238}U beam, measured with the VAMOS spectrometer[13]. Transfer reactions have been used for a long time in direct kinematics to measure fission probabilities and fission barriers[14-16] with high precision. Multinucleon transfer has proven to be a significant contribution to the total fission probability in heavy ion reactions around the barrier and leads to heavy fissioning systems[14]. Table 1 shows an estimation of the cross section for multinucleon transfer-induced fission, from ref.[17-18]. By measuring the energy and angle of the transfer partner, the fissioning system is completely characterized in atomic and mass number, and in excitation energy. The excitation energy resolution may be better than 1MeV depending on the angular resolution, and the excitation spectrum spread over 20 to 30 MeV, depending on the particle transferred. In inverse kinematics, the transfer-induced fission fragments are focused in forward direction, and their kinetic energy may reach 8 A MeV. The high velocity of the fission fragments allows for their identification in atomic number in an ionisation chamber. The mass and charge state are resolved using the high precision VAMOS spectrometer (dBr/Br=10^{-4}), time of flight (dT/T=10^{-3}) and total energy measurement (dE/E=0.5%). Table 1 shows an example of fissioning systems that may be studied by this technique. For the first time, it will be possible to access the isotopic distribution of fission fragments as a function of the excitation energy of the fission system, for a range of neutron-rich actinides.

TABLE 1. Estimated cross sections of channels for transfer induced fission of ^{238}U on ^{19}F

Transfer Partner	Fissioning system	Fission Cross Section (mb)
^{17}O	^{240}Np	1
^{18}O	^{239}Np	6
^{19}O	^{238}Np	2
^{19}N	^{238}Pu	0.6
^{18}N	^{238}Pu	5
^{17}N	^{238}Pu	30
^{16}N	^{238}Pu	7.5
^{15}N	^{238}Pu	4
^{16}C	^{241}Am	4
^{15}C	^{242}Am	1
^{14}C	^{243}Am	1
^{13}Be	^{244}Cm	1

SUMMARY

It is proposed to use inverse kinematics at GANIL to induce multi-nucleon transfer fission on a U beam. The detection of the transfer partner allows for a complete reconstruction of the fissioning system, in mass, charge and excitation energy, which vary over a neutron-rich actinide region. The high resolution of the spectrometer allows for the identification in flight of the fission fragments before their beta-decay. High precision data will be obtained on the evolution of the pairing correlation with the excitation energy, and shell effects in the heavy fragments as a function of the fissioning system on a wide range of neutron-rich actinides. This project is complementary to the experimental results of GSI that gave the element yields for neutron deficient actinides at an excitation energy defined by the Coulomb interaction.

REFERENCES

1. Cohen, S. and Swiatecki, W. J., Ann. Phys. **22**,406 (1963)
2. Wilkins, B.D, Steinberg, E.P., and Chasman, R.R, *Phys. Rev. C*, **14**,1832(1976)
3. Iyer, R.H. et al., *J. Nucl. Science and Eng.*,**135**,227,(2000).
4. Sida, J.L. et al, *Nucl. Phys. A* **502**, 233c, (1989).
5. Terrell, J., *Phys. Rev.*, **127**, 880, (1962).
6. Schmit, K.-H. et al., *Nucl. Phys. A*, **665**, 221, (2000).
7. Benlliure, J., Junghans, A. R., and Schmidt, K.-H., *Eur. Jour. Phys. A*, **13**, 93, (2002).
8. Steinhaueser, S., et al., *Nucl. Phys. A*, **634**, 89, (1998)
9. Schmidt., K.-H., Ignatyuk, A. V., Rejmund , F., Kelic, A.,,. Ricciardi, M. V, this proceeding
10. Rejmund, F., Ignatyuk, A.V., Junghans, A.R., Schmidt, K.-H., *Nucl. Phys. A*, **678**, 215 (2000).
11. Pommé., et al., *Nucl. Phys. A*, **560**, 689 (1993).
12. Enqvist, T., et al., Nucl. Phys. A
13. Savajols, H. et al.,Nucl. Instr. Meth. A
14. Wing, J., Ramler, W.J., Harkness, A.L., and Huizenga, J.R., *Phys. Rev*, **114**, 163,(1959).
15. Britt,H.C, and Cramer, J.D., *Phys. Rev. C*, **2**, 1758, (1970).
16. M. Petit et al., Nucl. Phys. A **735**, 345,(2004).
17. Videbaek, F., Steadman, S.G., Batrouni, G.G., and Karp, J, *Phys. Rev. C*, **35**, 2333, (1987).
18. Biswas, D.C. et al. , *Phys. Rev. C*, **56**, 1926, (1997).

NEEDS FOR NUCLEAR DATA
AND NEW FACILITIES—I

Status of High Intensity Laser Experiments for Nuclear Fission Investigations

J. Galy, J. Magill

European Commission - Joint Research Centre- Institute for Transuranium Elements,
Postfach 2340, D-76125 Karlsruhe, Germany

Abstract. The constant evolution of the multi-terawatt and petawatt laser performances is opening new paths of research investigations [1,2], particularly in the study of nuclear reactions and, in particular, fission. These experiments may provide an innovative approach for nuclear investigations without access to large facilities such as nuclear reactors or particle accelerators. The Institute of Transuranium Elements, through various collaborations, plays an essentials role in this new field of research. In 2000, laser-induced photo-fission of ^{238}U was for the first time experimentally demonstrated in UK with the giant pulse laser VULCAN [3] in USA, followed by the USA using the NOVA laser [4]. A couple of years later, the first photofission of ^{232}Th induced by Bremsstrahlung generated with a "tabletop" laser [5] was performed at the Jena University in Germany. Using this tabletop laser, photofissions of ^{238}U and ^{232}Th have been observed. New experimental campaigns are foreseen to investigate further the possibilities of the laser induced photofission and laser accelerated proton-fission. Status of the laser-induced fission experiments and perspectives of the laser-induced nuclear reactions are discussed in this paper.

Keywords: Laser – Fission – Actinides – Spectroscopy
PACS: 25.85.Jg, 29.30.Kv, 29.30.Hs, 46.62.Cf

BACKGROUND AND MOTIVATIONS

In 1988, Boyer [6] investigated the possibility that high intensity laser beams could be focused onto solid surface and would trigger nuclear reactions, and in particular the electro and photo-fission of uranium. In 2000, we demonstrated experimentally for the first time laser-induced photo-fission of ^{238}U with the VULCAN [2] laser in UK, followed by the USA with the NOVA laser [3]. Those experiments opened a wide new field of research for the nuclear reactions induced by intense laser sources. In the last few years, the ability to induce a variety of nuclear reactions with very high intensity lasers has then been demonstrated in several laboratories [2]. Laser activation, fission [1-6], fusion [1], long-lived fission product transmutation [7,8] and neutron production have all been demonstrated, through European collaborations with the Institute for Transuranium Elements (ITU) in Karlsruhe, Germany.

In the recent years, a renewed interest on the photo-reactions has developed in various laboratories in the world. Several applications are foreseen: radioactive beam production, sources of neutrons, medical isotope production, transmutation and nuclear material production. Photo-fission is expected to play there a significant role.

EXPERIMENTAL EXPLORATION OF THE LASER INDUCED-FISSION

Experimental evidence of laser induced fission

The photo-fission reactions by means of high intensity laser follow a three cascade scheme. The laser generates ultra short pulses. In the focal spot, matter is instantaneously turned into a hot dense plasma. In this intense light field, electrons are accelerated to relativistic energies. The plasma electrons impinge on a Bremsstrahlung converter from which photons up to the maximum electron energy can be produced. The Bremsstrahlung photons are afterward directed to an actinide target to induce photo-fission. We demonstrated the first laser-induced fission, on uranium 238, in 2000 [3], followed by [4]. The fission evidence was carried through post irradiation detection of the gamma activity of the decay of fission products such as ^{134}I, ^{92}Sr or ^{138}Cs. One shot on the uranium target (of a few grams) gave rise to 10^6 and 10^8 fissions for VULCAN [3] and NOVA [4] respectively. The large developments in laser technology during the last decade led to the creation of powerful terawatt table-top (T^3), capable of high intensity of 10^{19-20} W.cm^{-2}. The first T^3 induced fission of ^{238}U was reported by Malka et al. [9] using the Ti-sapphire laser system at the Laboratoire d'Optique Appliquée (LOA) in France. They produced 2.10^4 fissions per shot at 10 Hz repetition rate. The demonstration of T^3 induced fission of ^{238}U was confirmed using the laser located at the University of Jena, Germany. In the same experimental campaign ^{232}Th was fissioned [4] for the first time by means of high intensity laser. Uranium and thorium experiments were re-iterated the same year at the VULCAN facility [1]. Fig. 1 shows a typical γ-spectrum measured post irradiation of ^{238}U by a laser-induced Bremsstrahlung. Measured fission rates for the different laser-induced fission are summarized in Table 1. In all experiments the amount of uranium was of a few grams.

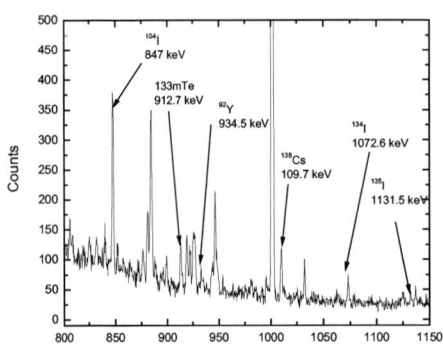

FIGURE 1. Spectrum showing characteristic γ-rays emitted by fission products following the ^{238}U(γ,f) reaction induced with the VULCAN laser facility (ITU results).

TABLE 1. Fission rate comparison of the different laser systems

Laser	Wavelenght (mm)	Pulse (fs)	Energy per pulse (J)	Intensity (W.cm^{-2})	Fission rate of uranium for 1J
LOA	0.8	30	0.5	$2 \cdot 10^{19}$	$4.6 \cdot 10^4$
Jena	0.8	80	0.26	$4 \cdot 10^{19}$	$4 \cdot 10^2$
RAL	1.0	1000	17.5	10^{19}	$6 \cdot 10^4$
NOVA	1.0	450	250	10^{20}	$7 \cdot 10^4$

Major foreseen applications of laser-induced fission

Nuclear data

The first and obvious application which is foreseen is the production of nuclear data. The IAEA has compiled photo-nuclear cross section of about 160 nuclides on which only 9 fissioning systems are available and in most of them experimental data are missing. With the development of the laser systems and their flexibility of utilization in comparison to accelerator systems, cross-section studies can be designed. As an example, we reported the integral (γ,n) cross-section on the long-life fission product ^{129}I [7,8] using both T^3 and giant pulse lasers. In addition to cross-section, fission yields distribution can be investigated. There is a considerable lack of experimental fission yield data for the photo-fission reactions. Those data are of use to understand the photo-fission process, for practical applications and to compare it with the neutron induced fission. Previous experiments have used the fission product activity as evidence to the fission process. Accurate fission yield assay can be extended by means of spectroscopy on-line and off-line.

Detection of Nuclear Material

Different techniques are being developed to detect presence of fissile material by the means of photo-nuclear fission interrogation. The delayed neutron measurement is of particular interest for the fissile material characterization in the frame of non-proliferation. Neutron measurement is, however, quite challenging – delayed neutron production rate is of only about 3 per 100 fissions distributed in 4π. Nevertheless, experiment is under consideration using a pulsed beam in combination with a neutron detection system with a very high efficiency. The feasibility of the detection of delayed neutrons is dependent on the intensity of the photon beam and the amount of material. Measurement of delayed neutrons from ^{238}U photo-fission induced by accelerator bremsstrahlung has been reported by Nikotin [10] and experimental studies are under progress [11] in France. Lasers can be used for the investigations on delayed neutron characterization of different fissile material. Experiments are already planned. An alternative to the delayed neutron characterization is the investigation of the fission product distribution of the fissile materials under interrogation. The center of the light peak distribution of the fission yield is expected to move to higher mass when increasing the atomic number of the fissile material, such as for the neutron

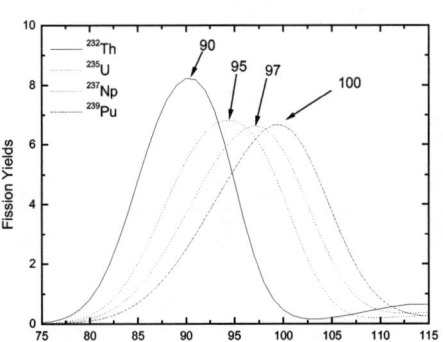

FIGURE 2. Mass yield distribution of compound nuclei ^{232}Th, ^{235}U, ^{237}Np, ^{239}Pu with an excitation energy of 7 MeV calculated with the Z_p model [14].

induced fission process. Studies to show the link between neutron and photo-induced fission are thus of great interest. A first approximate would be to only consider the compound nucleus to determine theoretically the yield distribution. However due to the fact that even if neutron reaction forms the same compound nucleus at the same energy, the spin distributions different, and that may lead to different yield distribution. Preliminary studies have been reported in [11-13]: a 25 MeV Bremsstrahlung fission (roughly equivalent to a 15 MeV mono-energetic photon-induced fission) can be regarded as a 6-8 MeV neutron-induced fission in term of fission yield distribution. Fig. 2. shows the calculated yield distribution for ^{232}Th, ^{235}U, ^{237}Np and ^{239}Pu compound nucleus created by absorption of 7 MeV neutron, using the Z_p model [14]. Laser investigations can play a major role in studying the distribution for different fissile material in view of characterizing those.

CONCLUSION AND PERPECTIVES

High intensity giant pulse and tabletop lasers have in the recent years been used to demonstrate nuclear fission. ^{238}U and ^{232}Th laser induced photo-fission has been investigated. Forthcoming experimental campaigns will extend the investigation to ^{233,235}U, ^{237}Np and ^{242}Pu photofission and attempt to fission ^{232}Th and ^{238}U by laser induced proton-fission Future development in the field of laser nuclear fission investigation will benefit from the constantly fast evolution of the laser technology. These laser experiments offer a new approach of studying the nuclear reactions behavior without resource to large facility such as an accelerator, and the reduced size of T^3 laser make them appropriate, among the large potential field of applications, for detection of nuclear material by means of photo-fission interrogation.

REFERENCES

1. Galy J. et al., "Nuclear Physics and Potential Transmutation with the VULCAN Laser", Central Laser Facility Annual Report 2001/2002, 29, 2002. RAL-TR-2002-013 (2002).
2. Magill, J.,Galy J., "Radioactivity, Radionuclides, Radiation", Springer Eds.,ISBN3-540-21116-0, 92-103,2004.
3. Ledingham K.W.D. et al., Phys. Rev. Lett.. 84, 899 (2000).
4. Cowan T.E. et al., Phys. Rev. Lett.. 84, 903 (2000).
5. Schwoerer H. et al. Europhys. Lett. 61 (1), 47-52 (2003).
6. Boyer K. et al., Phys. Rev. Lett. 60,557, (1988).
7. Magill, J., et al. . Appl. Phys. B. 77(4), 387-390 (2003).
8. Ledingham, K.W.D., et al., J. Phys. D : Appl. Phys. 36,. L79-L82 (2003).
9. Malka G. et al., Phys. Rev. E 66, 66402 (2002).
10. Nikotin O.P. et al., Atom En. 20 268 (1965)
11. David J.-C., et al., "Fission Fragments Distributions and Delayed Neutrons Yields", Proc. of the Int.Conf. on Nucl. Data for Science and Technology, September 27 - October 1 , Santa Fe NM, USA (2004).
12. Giacri M.L. et al., "Development of the Photonuclear Activation Data Library for CINDER' 90", Proc. Of the Int. Workshop on Nuclear Data for the Transmutation of the Nuclear Waste", GSI-Darmstadt, Germany, Sept. 1-5 (2003) ISBN2-00-0122761
13. David J.-C. et al., "Mass and Charge Distribution from Photon-induced Fission: Comparison with experimental Data and Yields from Neutron Induced Fission", Proc. of the XVIth Int. Workshop on Phys. of Nucl. Fission, Oct. 7-10 (2003), Obninsk, Russia.
14. Wahl A., "Systematic of Fission-Product yields", Los Alamos report LA-13928 (2002).

The new pulsed mono-energetic neutron source at the IRMM and the shape isomer search in ^{239}U

S. Oberstedt, G. Lövestam, C. Chaves de Jesus, T. Gamboni, W. Geerts and R. Jaime Tornin

EC - JRC IRMM, B-2440 Geel, Belgium

Abstract. In the frame of the exploratory research initiative of the JRC a new pulsing device has been installed at the Van-de-Graaff of the IRMM to produce pulsed quasi mono-energetic neutron beams in the MeV range. The pulse width may be tuned from 10 μs up to several hundreds of μs with a repetition rate between 1 Hz and 5 kHz. The aim of the device is to study the decay of short-lived activation products between pulses in an essentially neutron-free environment. In a first application the shape isomer in ^{239}U was searched.

Keywords: pulsed neutron beams, tunable neutron pulse, shape isomer, lifetime, (isomeric) γ-decay, neutron-induced reactions, high-resolution γ-spectroscopy.
PACS: 21.10.-k, 21.10.Tg, 27.90.+b, 28.20.-v, 29.25.Dz

INTRODUCTION

Studies of fission isomerism in the actinide mass region flourished in the 1960's and 1970's after its first observation by Polikhanov et al. in 1962 [1]. The existence of fission isomers is a consequence of a second minimum in the potential energy surface first described by Strutinsky [2]. The determination of the parameters of this double humped fission barrier, i.e. the height and penetrability of the inner and outer barrier as well as the ground state of the super-deformed shape-isomeric minimum remains a challenge particularly in ^{239}U. Although the population of the shape isomer in this nucleus has been observed [3,4] (see Fig. 1), its decay remains still unobserved. Predictions of the shape isomer decay half-life cover five orders of magnitude ranging from several μs [5] up to more than 100 ms [6] depending on the decay mode assumed.

Stimulating the investigation is the recent experimental confirmation of the existence of a third and relatively deep minimum at even larger deformation, see e. g. ref. [7], which may imply drastic changes in the traditional picture of the relative barrier heights probably favoring γ-decay back to the normal ground state over delayed fission. Hence, the knowledge of the decay properties of ^{239}U^f will contribute to better understand the structure of the fission barrier as well as of extremely deformed nuclei. However, the measurement of decay processes, with even a small probability, poses a delicate problem to the experimentalist. Neither a traditional VdG-driven pulsed neutron source nor a neutron time-of-flight facility allow the measurement of decay times above a few μs due to secondary neutron-induced reactions, and also the activation technique is limited to half-lives longer than a few seconds.

For this purpose a NEw Pulsed and TUnable NEutron source (NEPTUNE) has been built at the Van de Graaff (VdG) accelerator of the IRMM.

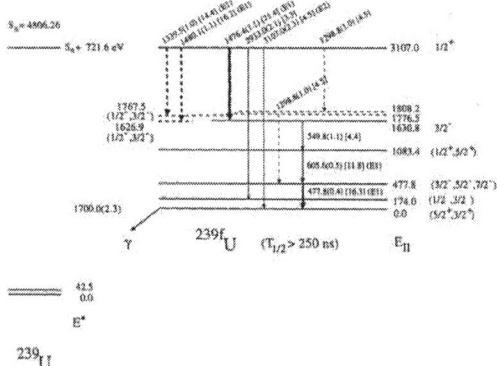

FIGURE 1. Level scheme above the shape isomeric ground state of ^{239}U (based on experimental results from refs. [3,4], reprinted from ref. [7], "Spectroscopy in the Second and Third Minimum of Actinide Nuclei", with permission from Elsevier)

NEW PULSED AND TUNABLE NEUTRON SOURCE

NEPTUNE is based on an electric deflection system integrated into the 0° beam line of the VdG accelerator (see Fig. 2, left). This facility provides pulsed quasi mono-energetic neutron beams at repetition rates between 1 Hz and 5 kHz with a pulse width tunable between 10 μs and the maximum length between two successive neutron pulses. The device enables the measurement of decay processes in the desired time frame in an essentially *background reaction free* environment.

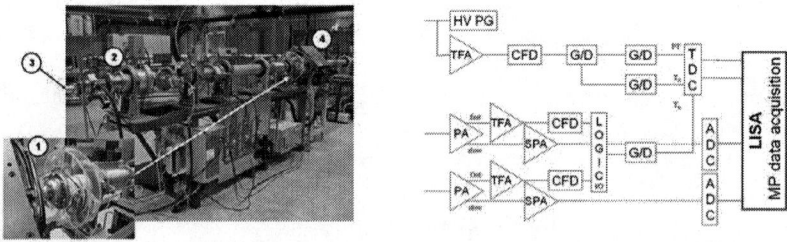

FIGURE 2. Left: View on the 0° beam line of the VdG accelerator: (1) deflection system, (2) position of the tantalum beam-dump, (3) neutron production target, (4) focusing quadrupole magnet **Right:** Electronic set-up which is extendable to maximum 5 detector modules.

The principle is based on the deflection of the charged-particle beam before hitting the neutron production target. The deflection system consists of a parallel-plate steerer to which a high voltage is applied. The high voltage, powered by a high voltage pulse generator [8], is adjusted according to the particle energy to deflect the beam onto a tantalum beam dump at about 3 m distance from the deflection device and is driven by a precision pulse generator with tunable pulse repetition frequency and pulse length. Furthermore, it provides a trigger signal to control the timing between a pulse started and a registered detector event. A schematic of the measurement electronics is shown in Fig. 2 for the use of two detectors. The present set-up may be extended up to maximum 5 detectors.

Up to now NEPTUNE has been tested with ionization chambers, scintillation counters and HPGe detectors, using appropriate shielding. The actually confirmed neutron suppression between pulses is about a factor of 100.

FIRST RESULTS ON THE SHAPE-ISOMER DECAY IN ^{239}U

The experiment is designed to investigate with high energy resolution the possible γ-decay of the shape isomer back to the normal ground state. For this purpose an HPGe-detector with 48 % efficiency relative to NaI was used in conjunction with a heavy Cu-shielding placed between the detector and the neutron source. The measurement was performed at $E_n \approx 1$ MeV, which corresponds to the vibrational resonance at the fission threshold of ^{239}U. The target consisted of 157 g depleted uranium. The energy resolution of the HPGe-detector during *beam OFF* was about 3.6 keV$_{FWHM}$ for γ-rays with an energy of about 1.2 MeV.

Measurements were performed at 1 and 5 kHz with neutron pulse lengths of 300 μs and 60 μs, respectively. γ-ray spectra were acquired for beam *ON, OFF* and from the sample activity.

For the time being the 5 kHz run has been analyzed. The analysis of the OFF-spectra was done in 3 steps:

1. Correction for the sample activity,
2. Search for γ-decay lines, which exclusively appear in the *OFF*-spectra, and
3. Comparison of the line intensity for two different time bins for half-life determination.

The *ON*-spectra suffered from bad energy resolution due to dead-time effects and were only used to estimate the prompt neutron suppression in the beam *OFF* mode.

At least two γ-rays could be identified, where the energy difference to the shape isomeric ground state, i.e. $E_{II} = 1.7$ MeV [4,7], fits a level above the normal ground state in ^{239}U as shown in the left part of Fig. 3.

An analysis of the corresponding time spectra turned out to be extremely difficult. Therefore, the deduced shape isomer half-life $T_{1/2} \sim 100$ μs is very tentative and will have to be tested when analyzing the 1 kHz run.

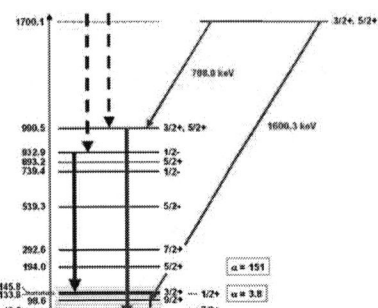
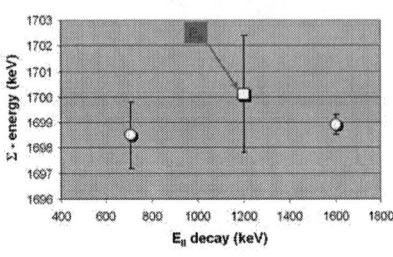

FIGURE 3. Left: Presumable γ-ray transitions connected to shape-isomeric decay back to the normal ground state (red), transitions above the normal ground state populated via prompt neutron capture (blue). **Right:** Sum energy obtained from the transition and the populated level above the normal ground state. Circles correspond to the red transition cascades in the decay scheme. $E_{II} \pm \sigma_E$ is given as a square, too.

PROSPECTS FOR THE FUTURE

Before NEPTUNE may become a standard facility for the investigation of decay processes with half lives below 1 s, the following improvements are foreseen to be implemented:

1. Optimized beam transport and diagnostics to guarantee a beam ON/OFF suppression factor of at least 1000.
2. Fast ADCs to reduce the system dead time in order to achieve a reasonable detector resolution when the beam is in ON-mode.
3. Use of several γ-ray detectors to record γ-γ correlations allowing for further improved background suppression, which is essential for a reliable decay half-life determination.

ACKNOWLEDGMENTS

This work has been funded within the EC-JRC exploratory research initiative 2004. We are indebted to F.-J. Hambsch, V. Fritsch and A. Plompen for fruitful discussions.

REFERENCES

1. S. M. Polikhanov et al. *Sov. Phys. JETP 15* (1962) 1016
2. V. M. Strutinsky, *Nucl. Phys. A95* (1967) 420
3. S. Oberstedt, F. Gunsing, *Nucl. Phys. A589* (1995) 435-444
4. S. Oberstedt, F. Gunsing, *Nucl. Phys. A636* (1995) 129-138
5. S. Oberstedt et al., *Nucl. Phys. A573* (1994) 467-485
6. H. Weigmann and J. P. Theobald, *Nucl. Phys. A187* (1971) 305-313
7. P. Thirolf and D. Habs, *Progress in Particle and Nuclear Physics 49* (2002), pp. 325-402.
8. GBS-Elektronik GmbH, http://www.gbs-elektronik.de

Delayed Neutrons from High Energy Fission-Spallation Reactions

D. Ridikas[a], P. Bokov[a], J.-C. David[a], D. Doré[a], M.-L. Giacri[a], X. Ledoux[b],
A. Plukis[c], R. Plukiene[c], A. Van Lauwe[a]

Corresponding author: ridikas@cea.fr
[a]*CEA Saclay, DSM/DAPNIA/SPhN, 91191 Gif/Yvette, France*
[b]*CEA Bruyères, DIF/DPTA/SPN, 91680 Bruyères-le-Châtel, France*
[c]*Institute of Physics, Savanoriu pr. 231, 02300 Vilnius, Lithuania*

Abstract. The next generation spallation neutron sources, neutrino factories or RIB production facilities currently being designed and constructed world wide will increase the average proton beam power on target by a few orders of magnitude. Increased proton beam power results in target thermal hydraulic issues leading to new target designs, very often based on liquid metal technologies such as Hg, Pb, or PbBi. Radioactive nuclides produced in liquid metal targets are transported into hot cells, into pumps or close to electronics with radiation sensitive components. Besides the considerable amount of photon activity in the irradiated liquid metal, a significant amount of the Delayed Neutron (DN) precursor activity can be accumulated in the target fluid. The transit time from the front of a liquid metal target into areas, where DNs may be important, can be as short as a few seconds, i.e. well within one half-life of many DN precursors. Therefore, it seems very important to evaluate the DN flux as a function of position and determine if DNs may contribute significantly to the activation and dose rates. The multi-particle transport code MCNPX combined with the material evolution program CINDER'90 is used to predict the DN precursors and construct the DN tables. These DN tables are employed within the generalized geometrical model of the MegaPie spallation target at PSI (Switzerland). We show that the contribution of DNs and prompt spallation neutrons to the total neutron flux is comparable at the very top of the liquid PbBi loop. We also demonstrate that these estimates of DNs within MCNPX are very much model-dependent. No experimental data are available for DN yields and time spectra from high energy fission-spallation reactions. An experiment to perform these measurements is proposed.

Keywords: high energy fission, spallation reactions, delayed neutrons.
PACS: 25.85.-w; 25.40.Sc.

INTRODUCTION

The realization of ideas on the next generation spallation neutron sources with nominal beam power of the order of 1 MW or higher were started with the SINQ facility (PSI, Switzerland). While SINQ began operations with a solid target, it was well known that this facility was actively pushing towards a liquid metal technology. In this context, the MEGAPIE irradiation experiment is planned at PSI in 2006. MegaPie (**Mega**watt **Pi**lot **e**xperiment) is a joint initiative by a number of research institutions world wide to design, build and explore a liquid lead-bismuth spallation target for 1 MW of beam power [1]. In addition to MegaPie, the 1 MW beam power or greater spallation source designs for SNS (US), JSNS (Japan), ESS and EURISOL (both Europe), all focus on liquid metal targets.

Liquid metal targets can solve some difficult problems for high-power spallation sources. On the other hand, they also introduce some new issues that must be addressed (e.g., corrosion, resistance of the target window or interface accelerator – windowless liquid metal, etc.). In addition, radioactive nuclides produced in liquid metal targets are transported into hot cells, into pumps or close to electronics with radiation sensitive components. Besides the considerable amount of decay gamma activity in the irradiated liquid metal, a significant amount of the Delayed Neutron (DN) precursor activity can be accumulated in the target fluid. The transit time from the front of a liquid metal target into areas, where DNs may be important, can be as short as a few seconds, i.e. well within one half-life of many DN precursors. Therefore, it seems very important to evaluate the DN flux as a function of position and determine if DNs may contribute significantly to the activation and dose rates. The main goal of this work is to provide quantitative estimates of the neutron fluxes due to DNs as a function of time and position in the MegaPie target fluid loop and compare it with the contribution due to prompt neutrons (PNs) at the same location.

CALCULATION PROCEDURE

Liquid PbBi metal loop in the case of the MegaPie spallation target, as in most of the high power spallation targets based on liquid metal technologies (e.g., SNS [2]), extends much further compared to the primary proton – heavy metal interaction zone (~27 cm long and defined by the 575 MeV proton stopping range).

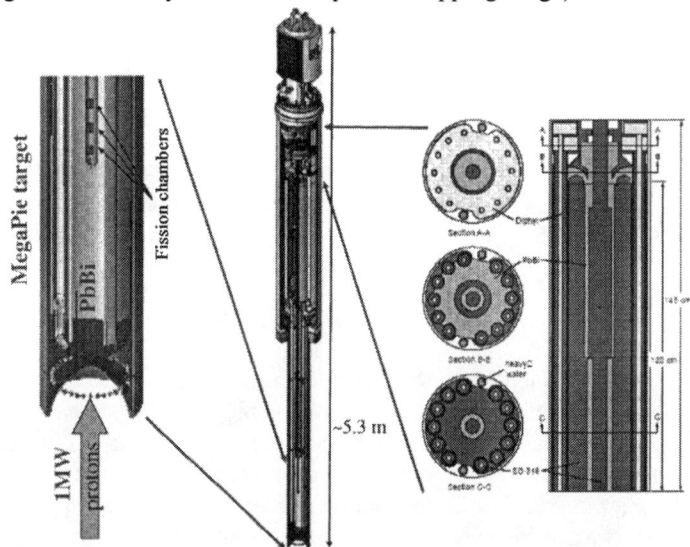

FIGURE 1. In the middle: a schematic view of the entire MegaPie target (5.3 m long). On the left: zoom of the lowest part of the target – proton-PbBi interaction zone. On the right: zoom of the target at the level of the heat exchanger (~400 cm above the target widow).

As it is presented in Fig. 1, the activated PbBi reaches as high as 400 cm arriving in the heat exchanger, from where it returns to the initial position. It takes ~20 s for the entire ~82 liters of PbBi to make a "round trip" at a flow rate of ~4 liters/s. It is clear that a big part of the DN precursors, created in the interaction region via high energy fission-spallation, will not have enough time to decay completely even at the very top location of the circulating liquid metal. The main question is how much DNs contribute to the total neutron flux at the very top position of the heat exchanger (see Fig. 1)?

For this purpose similarly as in [2] we employ the multi-particle transport code MCNPX [3] combined with the material evolution program CINDER'90 [4]. The DN data (emission probabilities and decay constants) were based on the ENDF/B-VI evaluations [5]. For the MegaPie target characteristics we used 575 MeV (1.75 mA; 1 MW) proton beam interacting with the liquid PbBi target. We modeled the 3-D geometry of the target in detail by taking into account all materials used in the design as shown in Fig. 1 [6, 7].

The following procedure was applied to estimate the DN parameters for MegaPie:
- calculation of independent fission fragment and spallation product distributions with MCNPX;
- calculation of cumulative fission fragment and spallation product yields with CINDER'90;
- identification of all known DN precursors and construction of the 6-group DN table.

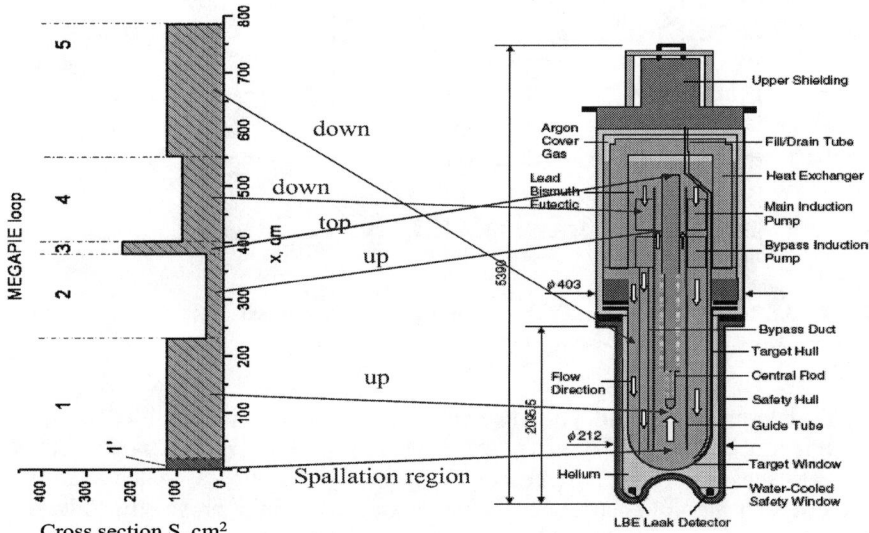

FIGURE 2. On the right: a schematic view of the MegaPie target with dimensions given in (mm). PbBi flow directions are indicated by arrows. On the left: cross section of the liquid PbBi loop as a function of the PbBi geometrical position – trajectory x *(cm)*. Also see text for details.

After having built the DN table we developed a generalized geometrical model to estimate the DN activity densities at any position x of the MegaPie target loop as presented in Fig 2. Within this model the DN activity at position x can be expressed as

$$a(x) = \sum_{i=1}^{n} a_i(x) = \sum_{i=1}^{n} a_i^a \frac{1-\exp(-\lambda_i \tau_a)}{1-\exp(-\lambda_i T)} \exp(-\lambda_i \tau_d(x))$$

with τ_a, s - activation time (PbBi under irradiation); T, s - total circulation period of PbBi (duration of the "round trip"); τ_d, s - transit (decay) time to reach the point x; λ_i, s^{-1} - decay constant of the DN precursor i; a_i, $n/(s\ cm^3)$ - density of DNs due to the precursor i.

RESULTS

By the use of the above equation and Table 1 we found that at the very top position of the MegaPie PbBi loop (400 cm level above the target window) the DN activity is of the order of 2×10^5 $n/(s\ cm^3)$ if the INCL4 model is used (see Fig 3 and Table 1).

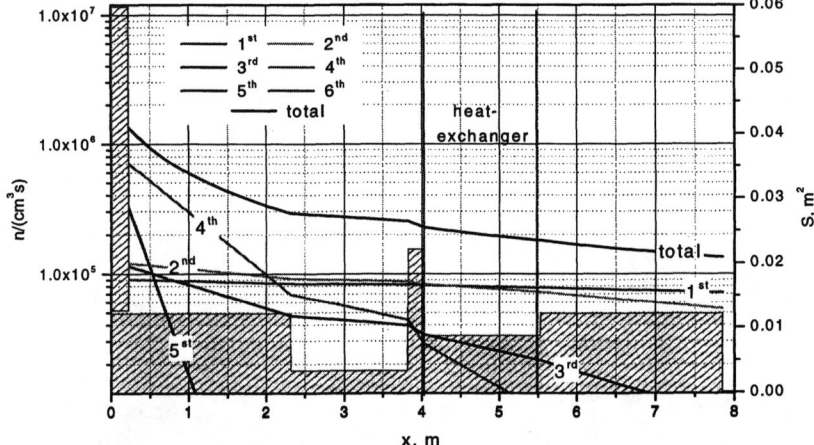

FIGURE 3. DN activity as a function of the PbBi geometrical position x (see Figs. 2). Contributions due to the individual DN groups (from 1 to 6) are also presented.

This intermediate result permitted us to estimate the neutron flux level in the MegaPie heat exchanger. For this purpose, again using MCNPX, we modeled in detail the heat exchanger geometry (see Fig. 1, on the right) and recalculated neutron flux now inserting the volumetric DN source as a function of x provided in Fig. 3. We found that the neutron flux at this position due to DNs and prompt spallation neutrons is of the same order of magnitude, both equal to a few 10^6 $n/(s\ cm^2)$. This result shows that activation and dose rates due to DNs should not be neglected. In addition, prompt neutron energy spectrum at this position is very close to thermal (we remind that the MegaPie spallation target is surrounded by a heavy water moderator-reflector). On the other hand, the DN energy spectrum at this level is not "perturbed" yet, i.e. with the average energy of the order of 400-600 keV. These fast neutrons will have considerably higher penetration power compared to the thermal ones.

TABLE 1. DN parameters (6-group representation) in the case of the MegaPie target. Note that DN yields are normalized for 10^6 incident protons.

	INCL4 model		CEM2k model	
Group	$T_{1/2}$, s	a_i, n/p times 10^6	$T_{1/2}$, s	a_i, n/p times 10^6
1	55.49	0.87	55.60	6.78
2	16.29	0.89	16.35	15.25
3	4.99	0.44	4.66	23.58
4	1.90	1.19	1.63	174.24
5	0.52	0.21	0.45	129.95
6	0.20	0.00	0.11	233.52
Total (averaged)	(18.703)	3.59	(1.903)	583.35

MODEL DEPENDENCE OF DN ESTIMATES

The only measurement, to our knowledge, on DNs from high energy fission-spallation was done by Carpenter [8], where the author obtained the fraction of delayed neutrons from p(300MeV)+^{238}U reaction. In other words, the ratio of DNs to the prompt neutrons (PNs) was measured, i.e. β=DN/PN=0.0053 in this case. By analyzing the experimental conditions we note that PNs are produced by neutron and proton induced fission, as well as produced directly and evaporated in the high-energy nucleon cascade and the prompt (n,xn) and (γ,xn) reactions in both target (^{238}U) and beryllium reflector materials. Using MCNPX and CINDER'90 we modeled precisely the above experiment with our predictions summarized in Table 2.

TABLE 2. DN yields from p(300MeV)+^{238}U, normalized per incident proton. Contributions from low energy neutron induced fission and proton or high energy neutron induced fission-spallation are given separately. Note that in all cases the same number of PNs was used, namely PN_{mcnpx} = 5.5 n/p, as predicted by MCNPX. This number is nearly independent on the physics models employed within MCNPX. The "experimental" DN yield is then DN=β· PN_{mcnpx}.

	INCL4	ISABEL	CEM2k	Experiment
p & high energy n	0.0039	0.0046	0.0291	
low energy n	0.0190	0.0186	0.0171	
TOTAL	0.0229	0.0212	0.0462	0.0291

In brief, we found that DN yields from proton induced reactions on uranium are dominated by fission induced by low energy secondary neutrons[*], i.e. this experiment is of little use for proton induced reactions on Pb or Bi. Nevertheless, we can conclude

---

[*] The only exception is the DN yield given by CEM2k model, which, as it will be shown below, is not reliable for isotopic fission yield predictions.

that both ISABEL and INCL4 give reasonable results underestimating the experiment by only 25-30 % (see the last line of Table 2).

Unfortunately there is no experimental data available for DN yields from high energy fission-spallation reactions on "non-fissile" targets as Pb and Bi to test directly our predictions related to the MegaPie target. In addition, the best available data for isotopic fission fragment distributions [9] contain only one DN precursor for each Z, being the last measured isotope on the neutron rich-side (e.g. ^{87}Br and ^{92}Rb in Fig. 4). Furthermore, in the case of a thick spallation target one would need data for all energies to take into account the proton slowing down and corresponding transport phenomena including fission-spallation by secondary neutrons. Therefore, again a number of different intra-nuclear cascade models within MCNPX [2] were tried in order to evaluate possible uncertainties in our results.

Fig. 4 presents fission fragment distributions for the p(1 GeV) + Pb reaction in the case of Br and Rb isotopes: both theoretical and experimental values [9] are shown. After examination of these and other isotopic distributions with the DN precursors we choose only two physics models to construct the delayed neutron tables, namely INCL4 and CEM2k. The INCL4 predictions are the closest to the experimental data, while CEM2k results in systematic overestimation of neutron rich nuclei, consequently of the DN precursors. In other words, the DN estimates based on these two models would give limiting values of possible DN fluxes within MCNPX.

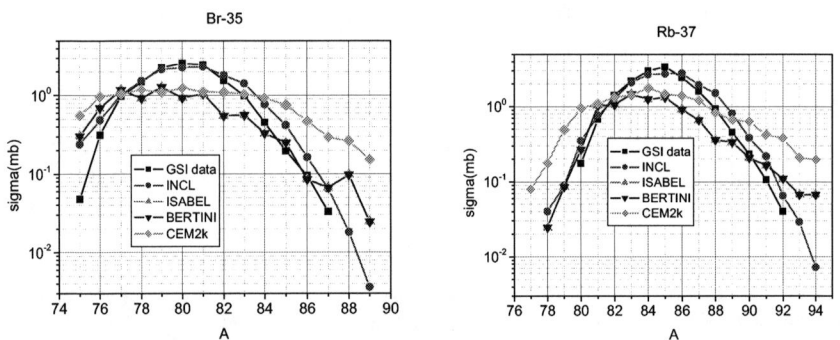

FIGURE 4. Fission fragment distributions form the p(1 GeV) + Pb reaction for Br and Rb isotopes. Both theoretical model calculations and experimental data [9] are presented.

In Table 1 we provide the 6-group DN parameters based on two different physics models, namely INCL4 and CEM2k as discussed above. In brief, both the yields and time spectra of DNs are model-dependent nearly by two orders of magnitude. This analysis shows that DN yields and time spectra from high energy fission-spallation reactions should be measured since no data of this type are available. This data would certainly constrain the physics models within the MCNPX code, and the fission fragment distributions on the neutron rich-side in particular.

DISCUSSION

In the above context we propose to perform DN measurements using 1 GeV protons interacting with massive Pb and Bi targets of variable thicknesses at PNPI Gatchina, St Petersburg, Russia. Targets of different length will allow studying the dependence of DN yields and their time spectra on the excitation energy of nucleus. DNs will be detected with optimized He-3 counters following specific irradiation intervals, namely with long (~300 s), short (~10 s) and very short (~0.2 ms) irradiation periods in order to enhance the contribution of different periods of DN precursors. With available proton beam intensity of ~500 nA we expect to produce up to ~10^6-10^7 n/s of DNs over 4π. Estimated counting rate with a single He-3 detector is of the order of ~100 n/s at t = 0. Not only the total yield of the DNs but also their decay curves will be measured in order to extract the corresponding decay constants and single group weighting factors within the 6-group DN model. We note separately that similar approach was successfully tested and validated by measuring the DN yields and time spectra from photo-fission of ^{238}U [10].

CONCLUSIONS

The DN flux and corresponding time spectra were estimated in the case of the MegaPie spallation target at PSI (Switzerland). For this purpose a generalized geometrical model was built, and could be used for other liquid metal targets. The DN tables within 6-group model were constructed for the first time for high energy fission-spallation reactions.

We showed that the neutron fluxes at the very top position of the MegaPie due to DNs and due to prompt spallation neutrons were of the same order of magnitude, i.e. both equal to a few 10^6 n/(s cm^2). This result warns that DNs can result in operational issues, which must be taken into account in detailed design studies of high power spallation targets based on liquid metal technologies.

We also demonstrated that the final estimates of DNs were very much model-dependent within the MCNPX code. No experimental data are available for DN yields from high energy fission-spallation reactions on Pb and Bi targets. The experiment to perform these measurements is proposed and will be carried out in 2005-2006.

ACKNOWLEDGEMENTS

This work was in part supported by the French Ministry of Foreign Affairs within the ECONET program (Ref. No 08160UK).

REFERENCES

1. G. S. Bauer, M. Salvatores, G. Heusener, "MEGAPIE, a 1 MW Pilot Experiment for a Liquid Metal Spallation Target", Proceedings of ICANS-XV (2000) (Eds. J. Suzuki and S. Itoh, JAERI, Tsukuba, Japan, 2001).
2. P. D. Ferguson, W. B. Wilson, F. X. Gallmeier, E. B. Iverson, "Analysis of Delayed Neutrons in Liquid Metal Spallation Targets", Proceedings of the XVI Meeting of the Int. Collaboration on Advanced Neutron Sources, 12-15 May 2003, Neuss, Germany.
3. MCNPX Team, "MCNPX, Version 2.4.0", LA-UR-02-5253, Los Alamos National Laboratory, August 2001.
4. W.B. Wilson and T.R. England, "A Manual for CINDER'90 Version C00D and Associated Codes and Data", LA-UR-00-Draft, April 2001.
5. W.B. Wilson and T.R. England, "Delayed neutron Study Using ENDF/B-VI Basic Nuclear Data", Progress in Nuclear Energy 41 (2002) pp. 71-107.
6. T.V. Dury, "CFD Design Support at PSI for the International MEGAPIE Liquid-Metal Spallation Target", Journal of Nuclear Science and Technology, Vol. 41, No. 3 (2004) 285-295.
7. M. Fadil, D. Ridikas et al., "Feasibility study of new microscopic fission chambers dedicated for ADS", Proceedings of 7th Information Exchange Meeting on Actinide and Fission Product P&T (NEA/OECD), Jeju, Korea, 14-16 October, 2002.
8. J. M. Carpenter, "The delayed neutron fraction in a pulsed spallation neutron source", Nuclear Instruments and Methods 175 (1980) 287.
9. T. Enquist et al., "Isotopic yields and kinetic energies of primary residues in 1 A GeV ^{208}Pb + p reactions", Nuclear Physics A686 (2001) pp. 481-524.
10. M. L. Giacri, PhD thesis report, CEA Saclay, DSM/DAPNIA/SPhN, France; in preparation (2005).

IAEA Coordinated Research Project On Fission Product Yield Data For Minor Actinides Up To 150 MeV

M. Lammer, A.L. Nichols

Nuclear Data Section, Division of Physical and Chemical Sciences,
Department of Nuclear Sciences and Applications, International Atomic Energy Agency,
Wagramer Strasse 5, A-1400 Vienna, Austria

Abstract. An IAEA Co-ordinated Research Project (CRP) on fission-product yield data was initiated in 1997, with the primary aim of developing systematics and nuclear models to assist in the evaluation of energy-dependent fission yields for incident neutron energies up to 150 MeV. A multinational team with the appropriate expertise participated in this work programme. New concepts for both systematics and theoretical models were developed. Various predictions of the fission-product mass distributions were compared in a benchmark exercise that gave remarkably good results below 30 MeV. Reasons for the discrepancies at higher energies and some failures of the model predictions will be discussed, pointing the way towards future investigations and fission-yield evaluations

Keywords: Fission yields, modeling, systematics, database.
PACS: 24.75.+i; 25.85.Ec

INTRODUCTION

Several concepts have been proposed for the transmutation of nuclear waste, using thermal and fast reactors and accelerator-driven systems as incinerators. While adequate nuclear data exist for thermal and fast reactors, important data (and fission yields in particular) are required at higher energies for accelerator-driven systems. An international working group studied the overall problem, and recommended the assembly of the desired nuclear data at intermediate incident neutron energies [1]. Staff at the Nuclear Data Section of the International Atomic Energy Agency (IAEA) had previously organized a Co-ordinated Research Project (CRP) on "Compilation and evaluation of fission yield nuclear data" [2], and subsequently brought together in 1997 the necessary expertise for a further CRP to consider fission yields up to an incident neutron energy of 150 MeV.

Traditionally, fission-yield evaluations have consisted of individual data sets for thermal, fast and 'high' (14 - 15 MeV) neutron energies. However, new concepts are required in presenting energy-dependent fission yields:

- mathematical functions that describe the energy-dependent yields,
- sets of yields at energy intervals, with interpolation formulae,

- computer programs to calculate fission yields on the basis of systematics and/or nuclear physics models.

Furthermore, standard evaluation procedures are not feasible because experimental data are scarce. Fission-yield measurements are extremely difficult or impossible for certain targets, and therefore nuclear models and systematics need to be developed from the available experimental data so that fission yields can be calculated and utilized in evaluations of their energy dependency. Therefore, an IAEA CRP was launched to study the problems involved in the development of nuclear models and systematics, and to develop a suitable evaluation methodology for such fission yields. While the ultimate goal of this initiative was a new evaluation of energy-dependent neutron-induced fission yields up to 150 MeV, the CRP participants were acutely aware that the necessary models and systematics did not exist at such high energies, and consequently the outcomes of the work were unpredictable. Principal participants in the CRP were as follows:

J.O. Denschlag, Institut für Kernchemie, Johannes Gutenberg-Universität Mainz, Germany;
M.C. Duijvestijn, Nuclear Research and Consultancy Group, The Netherlands;
Th. Ethvignot, CEA, Bruyères-le-Châtel, France;
A.A. Goverdovski, Institute of Physics and Power Engineering, Obninsk, Russian Federation;
F.-J. Hambsch, EC-JRC, Institute for Reference Materials and Measurements, Geel, Belgium;
J. Katakura, Nuclear Data Centre, Japan Atomic Energy Research Institute, Japan;
Yu.V. Kibkalo, Ukrainian Nuclear Data Centre, Ukraine;
M. Lammer, Nuclear Data Section, IAEA, Vienna, Austria;
Liu Tingjin, China Nuclear Data Centre, China Institute of Atomic Energy, PRChina;
V.M. Maslov, Radiation Physics and Chemistry Problems Institute, Belarus;
R.W. Mills, British Nuclear Fuels plc, Sellafield, UK;
F. Storrer, CEA-Saclay, France;
A.C. Wahl, Los Alamos National Laboratory, United States of America;
S.V. Zhdanov, Institute of Nuclear Physics National Nuclear Centre, Kazakhstan.

CO-ORDINATED RESEARCH PROJECT: SUB-PROGRAMMES

Available experimental data for energy-dependent neutron-induced fission yields are insufficient for the development of systematics and the derivation of model parameters. Therefore, the CRP studies were extended to yield data from photon- and light charged-particle-induced fission for comparison with neutron-induced fission reactions, in order to assess the possibility and correctness of their combined use in systematics. Relevant data for all actinides were compiled, and measurements were also recommended to improve systematics and model developments, as described below.

Bibliographic database of experimental fission-yield data from neutron-, photon- and light charged particle-induced fission has been assembled, and will be published in the final report [3] and as a CD-ROM.

Experimental yield data from neutron- and light charged-particle-induced fission have been collected and compiled in different data files. The neutron-induced fission yields are available in the EXFOR database [4], while the non-neutron fission yields will be converted into EXFOR format and added to the database; these data files will also be available on CD-ROM.

Sets of reference fission yields have been assembled as a consequence of extensive evaluation efforts: these yields have been derived with high accuracy through careful evaluation of individual reference fission products, and by making full use of correlations and covariance information.

Measurements

Several experiments were performed to support the investigations of the CRP (measurements initiated by Duijvestijn, Ethvignot, Goverdovski and Zhdanov). These results have been used to develop systematics, derive model parameters, and check the validity of the fission-yield predictions [3].

Differences Between Neutron-Induced And Other Fission Reactions

The possibility of joint use of neutron- and non-neutron-induced fission yields in developing systematics was explored, with the following results:

- same models can be used for the analysis of fission yields with different projectiles;
- trends observed in systematic studies of yield distributions for non-neutron-induced fission are also valid for neutron-induced fission.

However, not all functional dependencies, model parameters and numerical results can be adopted.

Influence Of Multi-Chance Fission

With increasing incident neutron energy, larger numbers of neutrons are emitted prior to scission (known as 'multi-chance' or 'emissive' fission). Hence, the nuclides that undergo fission have lower mass and excitation energy, and other characteristics (e.g., angular momentum) that differ from the same parameters of the original compound nucleus. Consequently, the observed mass distribution is composed of contributions from different fissioning nuclides. The effects that have to be considered (particularly in systematics) can be subdivided into three main categories:

- preferred mode of fission (symmetric or asymmetric) – changes with the reduction of the mass and excitation energy of the fissioning nucleus;

- mass peaks and the point of mass symmetry – shifted towards lower masses with the reduction in the mass of the fissioning nuclide, which results in a broadening of the composite mass distribution;
- prompt neutron emission fragments – observed distribution is composed of contributions from different fissioning nuclides.

A theoretical study of the fission mechanisms and the emissive fission contributions to the total fission cross section was undertaken in order to obtain the contributions of the fissioning nuclides to the total fission-yield distributions.

Systematics And Phenomenological Models

At the beginning of the CRP, existing systematics and models for mass and charge distributions had only been developed and validated for low-energy fission up to 14 MeV. One agreed aim of the new project was to investigate how far such methods of analysis could be adapted and modified for intermediate energy fission, or whether new approaches would have to be explored. The systematic behavior of the energy dependence of the experimental neutron-, photon- and charged-particle-induced fission yields was studied for several fissioning nuclides, and global systematics were derived for the dependencies of the fission-product mass distributions on the fissioning nuclide and excitation energy. Since there are insufficient experimental data that address fission-product charge distributions, this phenomenon was not considered further within the CRP.

Two new phenomenological models were developed for the analysis of experimental yield distributions from neutron-, proton- and alpha-particle-induced fission:

1. Dependence of the formation cross section of the fissioning nucleus and the fission fragment mass and energy distributions on the excitation energy and total angular momentum (Kibkalo *et al.*):

    - transferred angular momentum increases with the mass of the projectile, and the total angular momentum of the compound nucleus depends on the mode of formation;
    - angular momentum of some projectiles possess a critical value, below which compound nucleus formation occurs; however, fission can take place without compound nucleus formation if the projectile transfers sufficient energy;
    - above another critical value of the angular momentum for each type of projectile, the asymmetric fission channel ST-1 is converted into the symmetric channel SL over the whole excitation energy range.

2. Fragment mass and energy distributions (Zhdanov *et al.*): development of a calculational procedure to derive quantitative information of the dependencies of the basic characteristics of distinct fission modes (fragment mass, charge and energy distributions, their average values and variances, etc.) - a new method of multi-component analysis was formulated that is free from any

assumptions concerning the shapes of mass distributions of distinct fission modes:

- derivation of functional dependencies of the modal distribution parameters on the fissioning nuclide and excitation energy that have been used to develop the systematics for modal yields;
- mass distributions can be represented by a single Gaussian symmetric mode and one asymmetric mode described by Charlier's peak function with a low-energy tail.

Prediction Of Fission Yields At Intermediate Incident Particle Energies

Duijvestijn *et al.* developed a theoretical approach for the prediction of fission yields that combines the following two models:

- quantification of the fission cross section for all fissioning nuclides that contribute to a specific fission reaction as a function of their excitation energies, taking into account emissive fission;
- calculation of the fission-fragment and fission-product yields for each set of fissioning nuclides in a form that incorporates the variation of the fission characteristics as a function of excitation energy.

The original multi-modal random-neck rupture model of Brosa *et al.* [5] was extended in order to produce an improved description of nucleon-induced fission up to 200 MeV. Temperature dependence was added to calculate the change in fission mode with increasing excitation energy (i.e., disappearance of the asymmetric fission modes). Additional modifications permit the calculation of the relative contributions of the different fission modes and the rupture probability as a function of location at the neck. Links with the ALICE-91 [6] and TALYS [7] fission cross-section codes allow the calculation of any desired total pre- and post-neutron emission mass yields from 15 to 200 MeV. Satisfactory agreement has been achieved by the adoption of this approach, but specific discrepancies merit further investigation.

DISCUSSION

Benchmark exercises were undertaken to help reveal errors and incorrect assumptions in the models, and to give some kind of quantitative information about the accuracy of the predictions. One of these studies involved an inter-comparison with the experimental data for the neutron-induced fission of U-238, while the other was designed for fissioning systems with no known experimental data. A combination of these benchmark exercises and the earlier individual studies within the CRP provides a number of serious points for extensive debate, as outlined below.

Systematic approaches to fission-yield analyses are only able to reproduce the experimental data on which they are based. Certain assumptions are made about the fitted parameters, the correctness of which can only be studied with the aid of a theoretical model. Also, only a theoretical model can be used to investigate in detail certain features of the mass distributions and all associated contributions (e.g., multi-chance fission and fission modes) and the reasons for the observed discrepancies. Unfortunately, the theoretical model of Duijvestijn is not yet able to reproduce the experimental data (which are also unreliable). These difficulties are attributed to incorrect values for parameters (in particular, post-scission neutron emission) and assumptions about the contributions of dominant fission modes (particularly, emissive fission). Nevertheless, the model is qualitatively able to depict the trends. Overall, the predictions of Wahl and Duijvestijn, and Katakura at higher energies, fail to reproduce the experimental data (Fig. 1). The other models are only in agreement with the physical measurements of fission fragments, but not with the γ-ray measurements of the fission products. Altogether, the present comparison is inconclusive about the model predictions, and many more exact experiments are needed at intermediate energies.

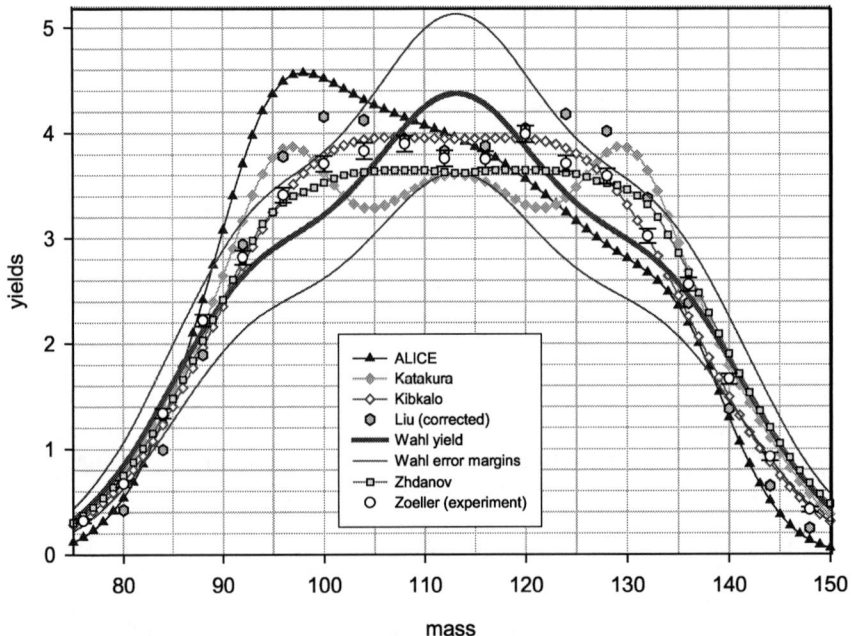

Figure 1. ^{238}U post-neutron emission yields at an incident neutron energy of 160 MeV.

The expected fine structure arising from the two asymmetric fission modes was only visible in the predicted mass distributions of Wahl and Katakura. This observation can be explained on the basis of the narrower Gaussians used in these systematics, whereas the fine structure is 'washed out' in other predictions that use broader peak functions. However, the predicted structures do not agree with the measurements (Fig. 2). All models can account for the disappearance of the asymmetric fission modes with increasing excitation energy, although the domination by symmetric fission begins at different excitation energies. Models with narrower peak functions show three distinct peaks, while models with broader functions exhibit wide and flat plateaus. The latter prediction is supported by measurements, although these experimental data suffer from incomplete mass resolution. Only the predictions of Wahl exhibit a dominating symmetric peak at 160 MeV, although this same behavior is predicted by the others for higher energies beyond the range of the present study.

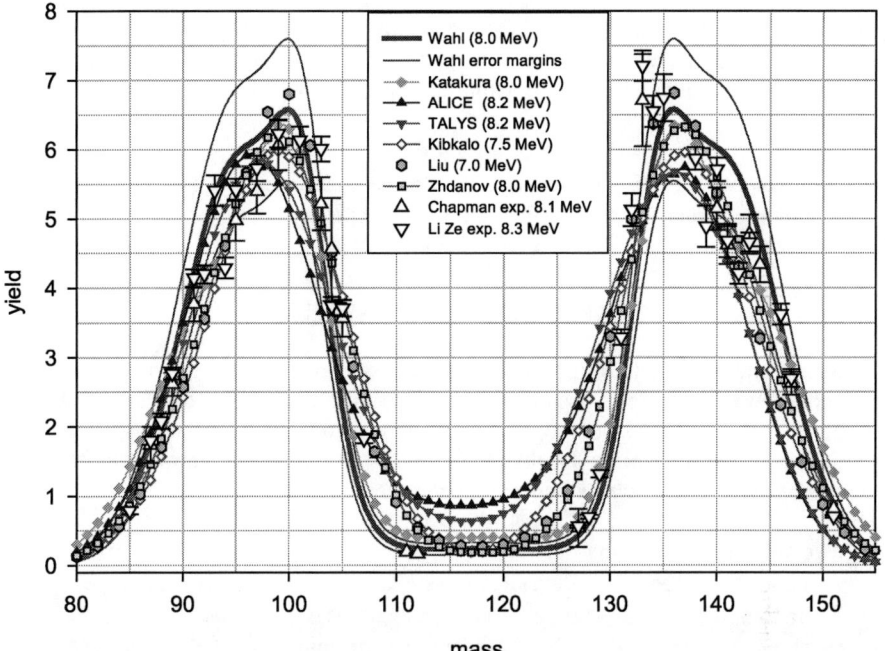

Figure 2. ^{238}U post-neutron emission yields at an incident neutron energy of approx. 8 MeV.

The expected effects of multi-chance fission on the observed mass distribution have only been qualitatively confirmed by calculation, and our present knowledge of these effects is insufficient to predict the real resulting shapes. This unsatisfactory situation is compounded by the serious lack of reliable experimental data, so that only qualitative statements can be made about the incorrect nature of the assumed shape symmetry of the light and heavy mass peaks and the calculated point of symmetry.

At energies less than 30 MeV, the deviations from Gaussian and shape symmetry create uncertainties that are negligible compared with other uncertainties. Thus, discrepancies up to 2.5% absolute yield around the peaks and in the slopes to the valley are mainly due to differences in the positions of peak maxima, the peak shapes and the widths of valleys (Fig. 2), while the peak heights are not as discrepant. At still lower energies below 10 MeV, structure is visible in the peaks due to the two asymmetric fission modes, although these features are "washed out" in several models because the peaks are extremely broad. A detailed analysis should clarify the shape and position of the structure, and determine the energy at which the asymmetric fission modes start to become insignificant. Survey plots of the energy dependencies of the mass distribution parameters reveal deviations from smooth curves for some of the calculations between 13 and 20 MeV. The nature of these 'irregularities' cannot be explained, and this phenomenon needs to be investigated further.

At energies above 30 MeV, the predicted shapes are discrepant. While impossible to define what the correct shapes should be and which model gives the best predictions, the representations of Wahl, Katakura and the theoretical model seem to be far from observed reality (Fig. 1). The following questions need to be addressed in detailed analyses (particularly for the high-energy region):

- how significant is the peak broadening ?
- at what energy does the symmetric peak begin to dominate ?
- do three peaks become visible at a particular energy, or does the peak broaden due to multi-chance fission to give a flat plateau ?

CONCLUDING REMARKS

Models and systematics have been developed that can in principle be used to evaluate energy-dependent fission yields. However, the results of the benchmark exercises are not conclusive enough to recommend one such method at the present time. The same uncertainty is also true for the computer programs. Under these circumstances, the following studies are recommended:

More accurate measurements
- up to about 20 MeV to investigate the exact structure in the peak shape;
- up to about 100 MeV to determine the exact peak positions;
- above about 20 MeV for the accurate determination of the peak shapes to reveal possible broadening;
- different measurement methods above 30 MeV to determine the real shape of the mass distributions;
- systematic measurements for U-238, and also similar measurements for other fissioning nuclides.

Theoretical models
Systematic studies of the important quantities should be performed - one possibility would be to vary parameters and investigate their influence on the calculated mass distribution.

Neutron emissions
- study of the number of pre-scission neutrons, in particular those emitted between saddle and scission, and multiplicity distributions of neutrons emitted from fragments;
- more measurements up to an incident neutron energy of 150 MeV;

- detailed and systematic theoretical analyses to develop more reliable models for predictions (neutron distributions used in models have a significant influence on calculated mass distributions);
- systematic measurements for U-238 and other fissioning nuclides.

Systematics and their models for mass distributions
- what are the reasons for the observed irregularities?
- why such large differences in the predicted shapes of mass distributions?
- can asymmetric peak shapes be introduced, and do they give better results?

Discrepancies among calculations for U-238 fission at low energies are as high as 2.5% in many cases, and nuclides studied in the second benchmark gave even poorer agreement. Since 25% relative accuracy for peak yields of 6-7% amounts roughly to 1.5% absolute fission-yield uncertainty, a target relative accuracy of 25% for the important fission yields has not yet been met. Since a comparison with experiments is also not conclusive, we are not yet in a position to recommend any of the fission models or computer programs for use in applied calculations. On the other hand, models for fission-yield predictions at intermediate energies have been developed for the first time that have the potential to give reliable predictions after further improvements.

ACKNOWLEDGEMENT

The authors would like to thank all participants of the IAEA CRP on "Fission product yield data for minor actinides up to 150 MeV" for their many significant contributions to this work programme, including measurements, developments of models and systematics, and assembly of databases, as discussed briefly in this summary paper.

REFERENCES

1. Koning, A. J., Fukahori, T. and Hasegawa, A., *International Evaluation Co-operation, Volume 13: Intermediate Energy Data*, NEA/WPEC-13, ECN-RX-98-014 (1998).
2. International Atomic Energy Agency, Final report of a co-ordinated research project on *"Compilation and evaluation of fission yield nuclear data"*, IAEA-TECDOC-1168, December 2000, ISSN 1011-4289.
3. International Atomic Energy Agency, Final report of a co-ordinated research project on *"Fission product yield data for minor actinides up to 150 MeV"*, in preparation.
4. McLane, V., *EXFOR basics, a short guide to the nuclear reaction data exchange format*, BNL-NCS-63380, IAEA-NDS-206 (2000); see also: http://www-nds.iaea.org/reports/nds-206.pdf
5. Brosa, U., Grossmann, S. and Müller, A., *Nuclear scission*, Phys. Rep. **197**, 167-262 (1990).
6. Blann, M., *Recent progress and current status of pre-equilibrium reaction theories and computer code ALICE*, pp. 517-586 in Proc. Workshop on Computation and Analysis of Nuclear Data Relevant to Nuclear Energy and Safety, 10 February – 13 March 1992, ICTP, Trieste, Italy, Editors: Mehta, M. K. and Schmidt, J. J., World Scientific Publishing Co. Ltd., Singapore (1993) ISBN 981-02-1224-0.
7. Koning, A. J., Hilaire, S. and Duijvestijn, M. C., *TALYS: Comprehensive nuclear reaction modeling*, Proc. Int. Conf. on Nuclear Data for Science and Technology – ND2004, 26 September – 1 October 2004, Santa Fe, USA, to be published.

ANGULAR MOMENTA AND
FISSION AT HIGHER ENERGIES—II

Scission configurations and the spin of fission fragments

L. Bonneau*, P. Quentin[†,]* and I.N. Mikhailov[**,‡]

*Los Alamos National Laboratory, Theoretical Division, MS B283, Los Alamos, New Mexico 87545 USA
[†]Centre d'Etudes Nucléaires de Bordeaux-Gradignan, CNRS-IN2P3 and Université Bordeaux I, BP 120, 33175 Gradignan, France
**BLTP, JINR, 141980 Dubna, Russia
[‡]CSNSM, Bât. 104, 91495 Orsay-Campus, France

Abstract.
After a brief presentation of the orientation pumping mechanism as a mean to generate finite average angular momenta in oriented systems, some consequences are drawn for the spin of fission fragments. Through a crude model approximation for the scission configurations, the results of microscopic calculations of fragment deformabilities are then used to deduce from the above mechanism, a distribution of fission fragment spins as a function of the total fragment excitation energy. A fair qualitative agreement with available data is demonstrated.

Keywords: scission, fission fragments, angular momentum
PACS: 21.60.Gx, 21.60.Jz, 25.85.Ca, 27.90.+b

1. INTRODUCTION

It has been well known for a long time from various experimental sources that fission fragments (even when resulting from the spontaneous fission of an even-even nucleus) possess quite sizeable angular momenta, typically up to about $10\,\hbar$ [1]. Recent detailed data from GAMMASPHERE experiments for the spontaneous fission of ^{252}Cf have exhibited an increase of this angular momentum with the total fragment excitation energy [2].

Fission fragment spins and related theoretical explanations

The current theoretical explanation takes stock on the thermal excitation of collective angular momentum carrying modes of which the most effective one for this purpose seems to be the bending mode [3]. Refined calculations along these lines have been performed by M. Zielinska-Pfabe and K. Dietrich years ago [4] and are still utilized [5]. This approach meets with, at least, three difficulties. First to account for the average spin values one has to resort to a vastly too high temperature (about 3 MeV) at least for reasonable collective phonon energies as given in [4]. Then it is unable to provide any explanation on the above quoted dependence [2] of the spin with the excitation energy. Finally it is inconsistent with the observed similarities between angular gamma ray distributions of both binary and ternary fissions [6] while the latter should clearly perturb the bending mode excitation with respect to what is expected in the former case.

It is the point of our approach to insist that in most cases the bulk of the fission fragment spins is due to quantal fluctuations rather than to thermal fluctuations. Actually there is more to angular momentum than a mere rotation of a matter spatial distribution, as experienced in so-called magnetic rotations [7] or intrinsic vortical modes [8] for instance. Here, we make use of the Heisenberg uncertainty principle as applied to systems whose orientation is somewhat fixed. Since, in that case, some angular information is known, one gets, as a result, a quantal distribution of the canonically conjugated variable, hence a finite average angular momentum. It is in that sense that one may say that the orientation "pumps" angular momentum [9]. This may be illustrated in the example of a quantal pendulum in the small oscillation angle θ limit, whose Hamiltonian writes [10]:

$$H = -\frac{\hbar^2}{2\mu}\left[\frac{1}{\theta}\frac{\partial}{\partial\theta}\left(\theta\frac{\partial}{\partial\theta}\right) + \frac{1}{\theta^2}\frac{\partial^2}{\partial\phi^2}\right] + \frac{1}{2}C\theta^2. \qquad (1)$$

The corresponding ground state wave-function is given by $\Psi(\theta) = N_0 \exp(-\gamma\theta^2)$, where γ is proportional to the rigidity parameter C of the oscillator. The orbital momentum (l) expansion of the latter

$$\Psi(\theta) = \sum_{l=0}^{\infty} a_l Y_l^0(\theta) \qquad (2)$$

exhibits a weighted gaussian distribution in l for large l values with respect to $\sqrt{\gamma}$

$$a_l \sim \left(l+\frac{1}{2}\right)\exp\left(-\frac{(l+1/2)^2}{4\gamma}\right). \qquad (3)$$

This is an example of this angular moment pumping for such a system constrained to move in a restricted θ angular domain.

Orientation pumping mechanism

Let us come back now to our fission context and assume that the scission configuration is reasonably well described as a product of two separated wave-functions, as BCS wave-functions. Upon projecting on good angular momentum each of these wave-functions, and coupling them to a total zero angular momentum (we assume that we describe here the spontaneous fission of an even-even nucleus and that the relative angular momentum is vanishing so as to minimize any other sources of angular moment generation in the fragments but the orientation pumping) we have obtained in [9] for a pure rotor distribution of projected energies that

$$\langle J^2 \rangle = \left(\frac{1}{\langle J_1^2 \rangle_{\text{intr}}} + \frac{1}{\langle J_2^2 \rangle_{\text{intr}}}\right)^{-1} \qquad (4)$$

where $\langle J_i^2 \rangle_{\text{intr}}$ stands for the intrinsic expectation value of the $\hat{\mathbf{J}}^2$ operator computed for the wave-function of fragment i. It is to be noted that this value is i-independent as it should for angular momentum conservation reasons. Furthermore if one of the fragment is spherical then this formula yields vanishing fragment spins. However we clearly reach one limit of our model assumptions, namely the rotational character of the projected spectra which should not be valid in this case.

2. SEMI-MICROSOPIC DESCRIPTION OF SCISSION CONFIGURATIONS

Let us consider a nucleus undergoing a fission process all along which axially symmetrical shapes are assumed. We decompose the total energy of the fissioning nuclear system in the following way:

$$E_{\text{tot}}(Q_{20}^{(1)},Q_{20}^{(2)},D) = \sum_{i=1}^{2} E_i(Q_{20}^{(i)}) + E_{\text{mut}}^{(Coul)}(Q_{20}^{(1)},Q_{20}^{(2)},D) + E_{\text{mut}}^{(nucl)}(Q_{20}^{(1)},Q_{20}^{(2)},D) \quad (5)$$

where $Q_{20}^{(i)}$ denotes the axial quadrupole moment of the fragment i – chosen to represent the elongation of the fragment – and D is the distance between the fragments centers of mass. With obvious notations, $E_{\text{mut}}^{(Coul)}$ and $E_{\text{mut}}^{(nucl)}$ stand for the mutual Coulombian and nuclear interaction energies, respectively, whereas E_i denotes the (deformation) energy of the fragment i. In the expression (5) of E_{tot}, it is assumed that E_i depends only on $Q_{20}^{(i)}$, which means that the polarization of both fragments would potentially be only considered through this quadrupole mode only. In the present preliminary crude study, however, we go as far as to neglect any polarization altogether by computing E_i, for given $Q_{20}^{(i)}$ values corresponding to fully separated nuclei, in the Skyrme–Hartree–Fock plus BCS approach used in [11] for fission barriers calculations and in another contribution to this workshop [12]. In the calculation of E_i, the pairing force parameters for the corresponding nucleus and the factor $1/A_i$ in the one-body term of the center of mass correction are used. Moreover, we have assumed axial and left-right symmetries. The latter hypothesis sounds reasonable since the ground state shapes of the considered nuclei, namely ^{106}Mo and ^{146}Ba, do not exhibit any energetically significant octupole distortions [13] and that we neglect any such polarization effet which would arise essentially from the Coulomb repulsion of the fission partner. The deformation energy curves $E_i(Q_{20}^{(i)})$ obtained for the ^{106}Mo and ^{146}Ba isotopes are displayed in Fig. 1.

To arrive at a quantitative definition of scission, we start from the idea that, at scission, the nuclear interaction acting between two nucleons whose wave functions are localized in different fragments becomes negligible. Moreover we consider that a scission configuration should be accessible from the initial state, chosen to be the ground state of the fissioning nucleus since we are interested in spontaneous fission. This means that the total energy of such a configuration should be lower than the ground state one. Any

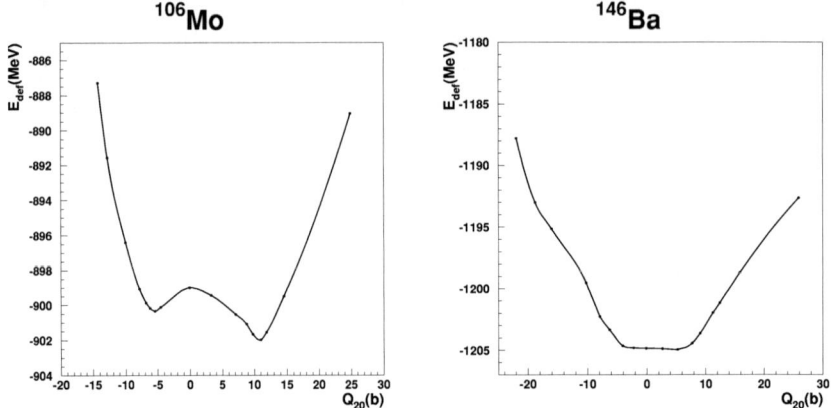

FIGURE 1. Deformation energy curves of the ^{106}Mo (left) and ^{146}Ba (right) nuclei.

scission configuration should then satisfy the following two relations:

$$\left| \frac{E_{\text{mut}}^{(nucl)}}{E_{\text{mut}}^{(Coul)}} \right| = \varepsilon \ll 1 \tag{6}$$

$$E_{\text{tot}}^{(sc)} \leqslant E_{\text{tot}}^{(GS)}. \tag{7}$$

In the above Eq. (6), ε is chosen to be 1%. This quantitative condition for scission is of course somewhat arbitrary. Eqs. (6) and (7) lead to a set of solutions $(Q_{20}^{(1)}, Q_{20}^{(2)}, D)$ where D is a function of $(Q_{20}^{(1)}, Q_{20}^{(2)})$ which both vary in limited ranges because of (7).

As a first approach to the above problem, we approximate $E_{\text{mut}}^{(Coul)}$ by the expression given in Ref. [14] obtained by likening the fragments to homogeneously charged, spheroidal droplets with collinear symmetry axes – namely the fission direction:

$$E_{\text{mut}}^{(Coul)}(Q_{20}^{(1)}, Q_{20}^{(2)}, D) = \frac{Z_1 Z_2 e^2}{D} S(x_1, x_2) \tag{8}$$

where the eccentricitiy-like variable x_i is defined through its square:

$$x_i^2 = \frac{c_i^2 - a_i^2}{D^2} \tag{9}$$

and the (dimensionless) function S, expressing the departure from two spherical fragments, takes the form:

$$S(x,y) = \frac{3}{40} \left[\frac{1 + 11(x^2 + y^2)}{x^2 y^2} + P_x P_y \left(\frac{(1+x+y)^3}{x^3 y^3} \times \ln(1+x+y)\left(1 - 3(x+y) + 12xy - 4(x^2 + y^2)\right) \right) \right] \tag{10}$$

in which $P_x[f(x)]$ means the even part of the function $f(x)$. In Eq. (9) c_i and a_i denote the semi-axes along the axial symmetry – corresponding to the fission direction – and in the perpendicular direction of the fragment i, respectively. As can be easily seen, prolate shapes ($c_i > a_i$) lead to real x_i-values, whereas x_i is purely imaginary for oblate shapes ($c_i < a_i$). As expected, we have $S(0,0) = 1$ for two spherical shapes. As for the attractive nuclear interaction between both fragments we chose the proximity potential of Blocki and Swiatecki [15, 16] expressed as, with their notations:

$$E_{\text{mut}}^{(nucl)} = 4\pi \bar{R} \gamma b \Phi(\zeta). \qquad (11)$$

In this expression, $\bar{R}$ is a kind of mean curvature radius and has been calculated exactly for two spheroids from Eq. (4) of Ref. [15] as:

$$\bar{R} = \left(\frac{a_1}{c_1^2} + \frac{a_2}{c_2^2} \right)^{-1}. \qquad (12)$$

The values of the γ and b parameters of Eq. (11) have been taken from Ref. [15], whereas the parametrisation of the universal proximity fucntion Φ is the one from Ref. [16].

It is worth mentioning that the above modelisation of scission configurations is formarly similar to the one sketched in Ref. [17]. An important difference lies in the way D is treated (or equivalently, the tip distance d). In the present model, the condition (6) leads to a distribution of D-values, thence a distribution of tip distances, whereas in the work of Ref. [17] a particular d-value has been considered.

3. FISSION FRAGMENT ANGULAR MOMENTA

In the above described "orientation pumping" mechanism and in the absence of any other source of angular momentum, both fragments have the same spin J_{frag} defined as:

$$J_{\text{frag}}(J_{\text{frag}} + 1) = \langle J^2 \rangle \qquad (13)$$

where $\langle J^2 \rangle$ is calculated within the crude approximation of Eq. (4). In the same spirit as what has been done for the energy E_i of the fragment i in Sect. 2, we have neglected the polarization effects in the calculation of $\langle J^2 \rangle_{\text{intr}}^{(i)}$ entering Eq. (4) and used the BCS wavefunction of the isolated nucleus i to compute the expectation value of the $\hat{J}^2$ operator. This has been done for a number deformations $Q_{20}^{(i)}$, for both fragments. The resulting J_{frag}-value has been obtained as a function of $(Q_{20}^{(1)}, Q_{20}^{(2)})$ and plotted as a contour map in Fig. 2.

In view of calculating the average J_{frag}-value as a function of the fragment deformation energy $E_{\text{def.frag}}$ defined by:

$$E_{\text{def.frag}} = \sum_{i=1}^{2} \left(E_i - E_i^{(GS)} \right), \qquad (14)$$

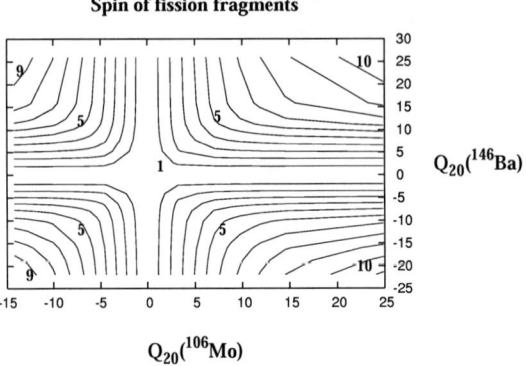

FIGURE 2. Contour map of J_{frag} as a function of $(Q_{20}^{(1)}, Q_{20}^{(2)})$ for the fragmentation ^{106}Mo+^{146}Ba.

we need to assign a weight to each configuration satisfying the two scission conditions (6) and (7), and corresponding to a fixed $E_{\text{def.frag}}$-value. Before further discussing this, it is worth noting that $E_{\text{def.frag}}$ depends only on the fragment deformations $Q_{20}^{(i)}$. Lines of equal total deformation energies in the $(Q_{20}^{(1)}, Q_{20}^{(2)})$ plane are reported in Figure 3 for configurations satisfying the scission conditions (6) and (7). The distribution over

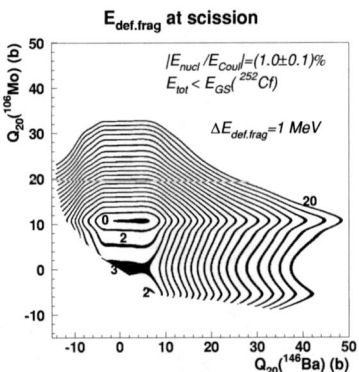

FIGURE 3. Contour lines of equal $E_{\text{def.frag}}$-values from 0 to 20 MeV for ^{106}Mo+^{146}Ba scission configurations.

various scission configurations having the same total deformation energy should result ultimately from an appropriate quantal calculation for the collective motion. To mock it up, in the present preliminary stage, we resort rather to a "Boltzman" weight $e^{-E_{\text{tot}}/T}$ where T should be a priori of the order of 1 MeV (value chosen here). This enables to compute the mean J_{frag}-value for a given $E_{\text{def.frag}}$-value as:

$$\langle J_{\text{frag}} \rangle = \frac{1}{N} \sum_{Q_{20}^{(1)}, Q_{20}^{(2)}} e^{-E_{\text{tot}}/T} J_{\text{frag}}(Q_{20}^{(1)}, Q_{20}^{(2)}) \quad (15)$$

where we have discretized the $(Q_{20}^{(1)}, Q_{20}^{(2)}, D)$ space and used the scission condition (6) to deduce D for each $(Q_{20}^{(1)}, Q_{20}^{(2)})$ pair. The constant N is a normalization coefficient, equivalent to a partition function:

$$N = \sum_{Q_{20}^{(1)}, Q_{20}^{(2)}} e^{-E_{\text{tot}}/T}. \quad (16)$$

Note that even though the above simulation of the spreading among scission configurations seem to refer to the approach of Wilkins and collaborators [18], it is quite different in spirit. We are not resorting here to a thermal distribution but to a quantal fluctuation. The latter yields a Gaussian distribution provided that, at a given $E_{\text{def.frag}}$, one makes a harmonic approximation around the minimum of E_{tot}. Upon varying $E_{\text{def.frag}}$, we have obtained an increasing trend for J_{frag} as shown in Fig. 4 where the reported error bars

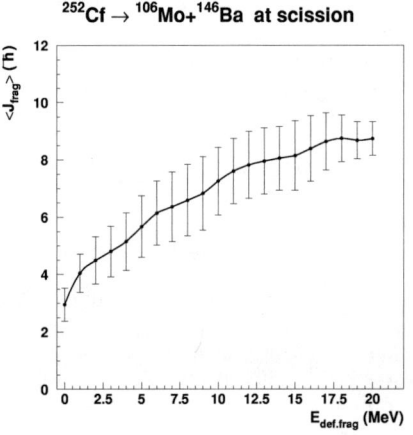

FIGURE 4. Variation of the average angular momentum $\langle J_{\text{frag}} \rangle$ at scission with the total fragment deformation energy $E_{\text{def.frag}}$ for the ^{106}Mo+^{146}Ba fragmentation.

are calculated as the variance of the corresponding spin distribution:

$$\Delta J_{\text{frag}} = \sqrt{\langle (J_{\text{frag}} - \langle J_{\text{frag}} \rangle)^2 \rangle}. \quad (17)$$

Assuming that the total excitation energy of fission fragments is stored in deformation, which sounds reasonable at low TXE-values, we can deduce from Fig. 4 that the mean fission fragment spin increases with TXE, which is compatible with the similar experimental trend reported in Ref. [2]. Interestingly, our average J_{frag}-value obtained for cold fission events (no neutrons emitted) is finite and very close to the experimental one. It seems however that our spin values might rise somewhat too fast as a function of TXE. Nevertheless one should keep in mind the very crude approximations made here with respect to the fragment deformations, the angular momentum projection properties of their intrinsic wave-function descriptions and the over-simplified definition of TXE.

4. CONCLUSION

We have proposed a quantitative criterion for scission configurations (in terms of the nuclear and Coulomb mutual energies) and implemented the orientation pumping mechanism in a semi-microscopic scission point model. Upon identifying the total excitation energy of the fragments with their deformation energy, we have finally shown that the orientation pumping mechanism is able not only to account for the order of magnitude of the fission fragment spins, but also to reproduce the experimental increasing trend of the average fragment spin as a function of TXE.

ACKNOWLEDGMENTS

We gratefully acknowledge enlightening discussions with F. Gönnenwein. One of the authors (Ph. Q.) thanks the Theoretical Division at LANL for the excellent working conditions extended to him during numerous visits. This research has been supported by the CENBG and the U.S. Department of Energy under contract W-7405-ENG-36.

REFERENCES

1. P. Armbruster et al., *Z. Naturforsch. Teil A* 26, 59 (1971)
2. G.M. Ter Akopian et al., *Proc. of the Second Int. Conf. on Fission and Properties of Neutron-Rich Nuclei*, (World Scientific, Singapore, 1998) p. 645
3. M.M. Hoffman, *Phys. Rev.* 133B (1964) 714; J.R. Nix and W.J. Swiatecki, *Nucl. Phys.* 71, 1 (1965)
4. M. Zielinska-Pfabe and K. Dietrich, *Phys. Lett. B* 49, 123 (1974)
5. S. Misicu et al., *Phys. Rev. C* 60, 034613 (1999)
6. Yu. Kopach et al., *Phys. Rev. Lett.* 82, 303 (1999)
7. See, e.g., S. Frauendorf, *Z. Phys. A* 358, 163 (1997)
8. See, e.g., I.N. Mikhailov, P. Quentin and D. Samsoen, *Nucl. Phys. A* 627, 259 (1997)
9. I.N. Mikhailov and P. Quentin, *Phys. Lett. B* 462, 7 (1999)
10. N. Lo Iudice and F. Palumbo, *Phys. Rev. Lett.* 41, 1532 (1978)
11. L. Bonneau, P. Quentin and D. Samsœn, *Eur. Phys. J. A* 21, 391 (2004)
12. See the contribution to this workshop by L. Bonneau and P. Quentin, *Microscopic calculations of potential energy surfaces: fission and fusion properties*
13. P. Möller, J. R. Nix, W. D. Myers and W. J. Swiatecki, *At. Dat. Nucl. Dat. Tab.* 59, 185 (1985)
14. P. Quentin, *J. Physique* 30, 497 (1969)
15. J. Blocki, J. Randrup, W. J. Swiatecki and C. F. Tsang, *Ann. Phys.* 105, 427 (1977)
16. J. Blocki and W. J. Swiatecki, *Ann. Phys.* 132, 53 (1981)
17. Ş. Mişicu and P. Quentin, *Eur. Phys. J. A* 6, 399 (1999)
18. B. D. Wilkins, E. P. Steinberg and R. R. Chasman, *Phys. Rev. C* 14, 1832 (1976)

Capture and Fusion-Fission Processes in Heavy Ion Induced Reactions

M. G. Itkis[1], S. Beghini[2], B.R. Behera[3], A. A. Bogatchev[1], V. Bouchat[4],
L. Corradi[3], O. Dorvaux[5], E. Fioretto[3], A. Gadea[3], F. Hanappe[4],
I. M. Itkis[1], M. Jandel[1], J. Kliman[1], G. N. Knyazheva[1], N. A. Kondratiev[1],
E. M. Kozulin[1], L. Krupa[1], A. Latina[3], V.G. Lyapin[6], T. Materna[4],
G. Montagnoli[2], Yu. Ts. Oganessian[1], I. V. Pokrovsky[1],
E. V. Prokhorova[1], N. Rowley[5], V.A. Rubchenya[7], A. Ya. Rusanov[1],
R.N. Sagaidak[1], F. Scarlassara[2], C. Schmitt[5], A.M. Stefanini[3], L. Stuttge[5],
S. Szilner[3], M. Trotta[8], W.H. Trzaska[6], V. M. Voskresenski[1]

[1]*Flerov Laboratory of Nuclear Reactions, JINR, 141980 Dubna, Russia*
[2]*INFN and Universita di Padova, Padova, Italy*
[3]*INFN, Laboratori Nazionali di Legnaro, Legnaro (Padova) Italy*
[4]*Universite Libre de Bruxelles,1050 Bruxelles, Belgium*
[5]*Institut de Recherches Subatomiques, F-67037 Strasbourg Cedex, France*
[6]*Department of Physics, University of Jyväskylä, Finland*
[7]*Khlopin Radium Institute, St.-Petersburg, Russia*
[8]*Dipartimento di Fisica and INFN, Napoli, Italy*

Abstract. Results of the experiments aimed at the study of fission and quasi-fission processes in the reactions ^{12}C+^{204}Pb, ^{48}Ca+^{144,154}Sm, ^{168}Er, ^{208}Pb, ^{238}U, ^{244}Pu, ^{248}Cm; ^{58}Fe+^{208}Pb, ^{244}Pu, ^{248}Cm, and ^{64}Ni+^{186}W, ^{242}Pu are presented. The choice of the above-mentioned reactions was inspired by the experiments on the production of the isotopes ^{283}112, ^{289}114 and ^{283}116 at Dubna [1,2] using the same reactions. The ^{58}Fe and ^{64}Ni projectiles were chosen since the corresponding projectile-target combinations lead to the synthesis of even heavier elements. The experiments were carried out at the U-400 accelerator of the Flerov Laboratory of Nuclear Reactions (JINR, Russia), the XTU Tandem accelerator of the National Laboratory of Legnaro (LNL, Italy) and the Accelerator of the Laboratory of University of Jyvaskyla (JYFL, Finland) using the time-of-flight spectrometer of fission fragments CORSET [3] and the neutron multi-detector DEMON [4, 5]. The role of shell effects and the influence of the entrance channel asymmetry and the deformations of colliding nucleus on the mechanism of the fusion-fission and the competitive process of quasi-fission are discussed.

Keywords: Fusion-fission, quasi-fission, mass-energy and angular distributions, capture and fusion-fission cross-sections.
PACS: 25.70.Jj; 25.70.Gh; 25.85.-w; 21.10.Gv

THE FUSION-FISSION PROCESS OF SUPERHEAVY NUCLEI WITH Z=102-122

The experiments devoted to the study of the fusion-fission dynamics of superheavy nuclei with Z=102-122 were carried out at the U-400 accelerator of the FLNR JINR,

Dubna. Mass-energy distributions (MED) of the fragments were obtained with use of kinematic coincidence method, the capture and fusion–fission cross-sections were deduced. Figure 1 presents the MED of the fragments of the elements with Z= 102-116, produced in the ^{48}Ca induced reaction on the targets ^{208}Pb, ^{238}U, ^{244}Pu and ^{248}Cm at the excitation energy E*≈ 33 MeV. On the top of the Fig.1 two-dimensional matrixes of Total Kinetic Energy (TKE)/Mass are presented, the mass yields of the reaction products are shown in the bottom. The main feature of the data is the sharp change of the MED triangular shape for the reaction ^{48}Ca+^{208}Pb, where fusion-fission (FF) process dominates, to the quasi-fission (QF) [6,7] shape of MED for the ^{286}112-^{296}116 nuclei.

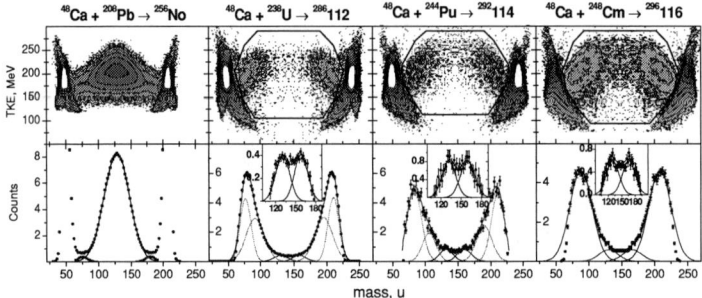

FIGURE 1. Two-dimensional matrixes TKE-Mass (top panels) and Mass yields of the fragments (bottom panels) for the reactions ^{48}Ca+ ^{208}Pb, ^{238}U, ^{244}Pu, ^{248}Cm at excitation energy E*≈33 MeV.

The distinctive feature of the quasi-fission process for these superheavy nuclei is the wide two-humped mass distribution with high peak of heavy fragment near double magic lead (M$_H$≈208). In spite of the dominating role of the quasi-fission process for these reactions we assume that in the symmetric region of the fragment masses (A/2±20) fusion-fission process coexists with QF. In the framing on the bottom of Fig. 1 the mass yield of the fusion-fission process obtained as difference between experimental spectra and quasi-fission peak description are shown. One can see on the framing that the mass distribution of the fusion-fission is asymmetric in shape with nearly constant mass of the light fragment M$_L$≈132-134 amu.

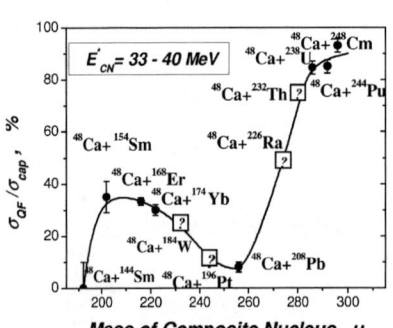

FIGURE 2. The ratio of σ$_{QF}$/σ$_{cap}$ as function of the composite nucleus mass number for the reaction with ^{48}Ca-ions and different targets.

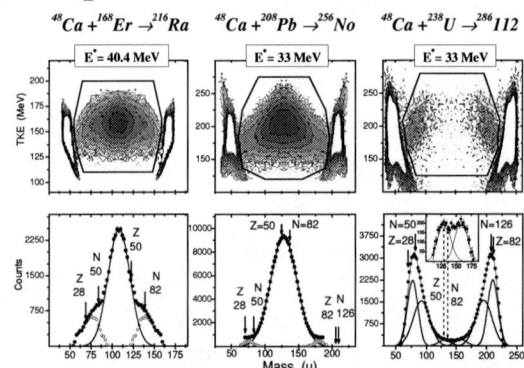

FIGURE 3 Two-dimensional matrices TKE-Mass (top panels) and mass yields (bottom panels) of fission fragments of ^{216}Ra, ^{256}No, ^{286}112 nuclei produced in the reactions with ^{48}Ca ions.

In case of superheavy elements the light spherical fragment with $M_L \approx 132$-134 amu plays a stabilizing role whereas the heavy fragment with $M_H \approx 140$ is playing the same role for actinide nuclei.

In the Fig.2 we show the ratio of the quasi-fission to the capture cross-sections σ_{QF}/σ_{cap} as the function of the mass of composite nucleus for the reaction with ^{48}Ca-projectiles at the excitation energies $E^* = 33$-40 MeV. Solid circles are measured reactions; the question signs are the reactions to be investigated. Our prediction on the behavior of this curve is shown with line. The most possible explanation of such a nontrivial behavior of the ratio σ_{QF}/σ_{cap} is the corresponding probability of formation of the different spherical shells in the nascent fragments. The graphic example of the shell influence on the MED of the fission fragments is shown on Figure 3. One can see that in the case of heaviest targets ^{238}U, ^{244}Pu and ^{248}Cm the ratio σ_{QF}/σ_{cap} changes weakly, and that tendency was observed in the σ_{ER} excitation functions for the superheavy elements [8].

Figure 4 shows the data for the reactions of ^{58}Fe and ^{64}Ni projectiles on ^{232}Th, ^{242,244}Pu and ^{248}Cm targets, leading to the formation of the compound system from ^{290}116 up to ^{302}120 and ^{306}122 (where N = 182-184), i.e. to the formation of the spherical compound nucleus (CN), predicted by theory [9]. As seen from Fig. 3, we observe here even stronger manifestation of the asymmetric mass distributions of ^{306}122 and ^{302}120 fission fragments with the light fragment mass $M_L \approx 132$ amu.

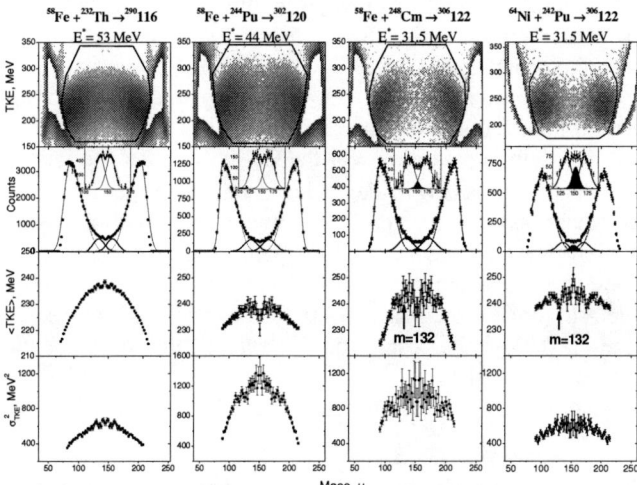

FIGURE 4. Two-dimensional TKE-Mass matrixes, the mass yields, $\langle TKE \rangle (M)$ and $\sigma_{TKE}(M)$ for ^{290}116, ^{302}120 and ^{306}122 nuclei

The corresponding structures are seen well in the $\langle TKE \rangle (M)$ dependence. Only for the reaction ^{58}Fe + ^{232}Th → ^{290}116 (E*= 53 MeV) the valley in the region of M=A/2 disappears – this is seen from the mass distribution as well as from the $\langle TKE \rangle$ (M) and σ^2_{TKE} (M) dependences. This fact is connected with a damping of the shell effects with increase of the excitation energy. On the right-hand side of Fig. 4 the characteristics of ^{306}122 fission fragments, formed in the reactions ^{64}Ni+^{242}Pu and ^{58}Fe+^{248}Cm are demonstrated. Despite the fact that the compound nucleus ^{306}122 undergoes fission at approximately the same excitation energy E*=31.5 MeV and the coefficients of the entrance channel asymmetry differ slightly nevertheless the form of the energy distributions changes to more flat $\langle TKE \rangle (M)$ dependence in case of ^{64}Ni-induced reaction.

Figure 5 shows the results of measurements of the capture σ_{cap} and the fusion-fission σ_{FF} cross sections (defined as $\sigma_{A/2\pm20}$) for a few studied reactions as a function of the initial excitation energy of the compound systems.

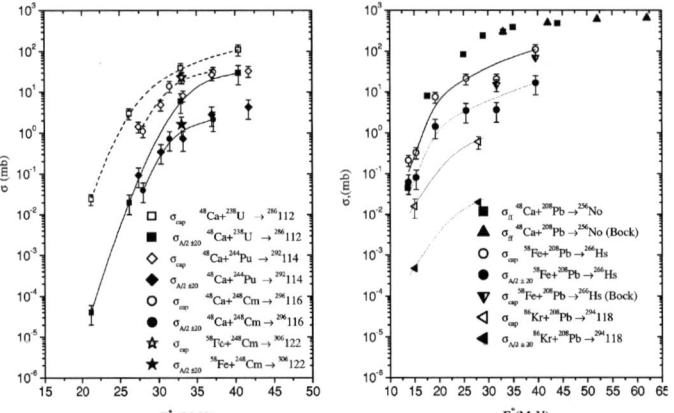

FIGURE 5. The capture cross section σ_{cap} and the fusion-fission cross section σ_{ff} for the reaction of production 102-122 elements as a function of the excitation energy

Comparing the data on the cross sections σ_{FF} for the warm fusion (the reactions from ^{48}Ca + ^{238}U to ^{58}Fe+^{248}Cm at $E^* \approx 33$ MeV) and for the cold fusion (the reactions ^{58}Fe+^{208}Pb and ^{86}Kr+^{208}Pb at $E^* \approx 14\text{-}15$ MeV), one can obtain the following ratios: $\sigma_{FF}(112)/\sigma_{FF}(122) \approx 4\text{-}5$ in the first case and $\sigma_{FF}(108)/\sigma_{FF}(118) \geq 10^2$ for the second one, however the value of Z of the compound nucleus changes by the same 10 units in both cases. That makes the use of asymmetric warm fusion reactions for the synthesis of spherical superheavy nuclei quite promising.

As the quasi-fission is the competitive process with the CN fusion, its properties, distinctive features, and relative contribution into the reaction total cross-section has to be studied for the finding of the optimal conditions for the CN formation. The ratio of these both processes σ_{FF}/σ_{QF} has to be tested in dependence on the properties of the reaction entrance channel (mass asymmetry, colliding nuclei deformations, shell structure, Z/N-ratio), excitation energies and angular momentum. Besides the cross-sections of the superheavies formation are extremely slow, the use of the lighter compound system can be helpful tool for the study of the fusion-fission dynamics and FF/QF competition. We present below the results of the experiments on the study influence entrance channel on the formation ^{266,274}Hs, ^{250}No and ^{216}Ra nuclei; the influence of the target deformations is considered for the ^{48}Ca + ^{144,154}Sm reaction.

THE INFLUENCE OF THE REACTION ENTRANCE CHANNEL ON THE FUSION-FISSION DYNAMICS

The ^{56}Fe+ ^{208}Pb and ^{26}Mg+ ^{248}Cm reactions

Figure 6 shows the formation of the isotopes of ^{266,274}Hs in the "cold fusion" (^{58}Fe+^{208}Pb → ^{266}Hs) and "hot fusion" (^{26}Mg+^{248}Cm→^{274}Hs) reactions. Though in the former reaction compound nucleus ^{266}Hs is produced at lower excitation energies, that should increase the CN survivability in the de-excitation process, nevertheless the quasi-fission process is the main reaction mechanism at the asymmetry of the entrance channel ($\eta=(M_{tar}-M_{ion})/(M_{tar}+M_{ion})=0.56$).

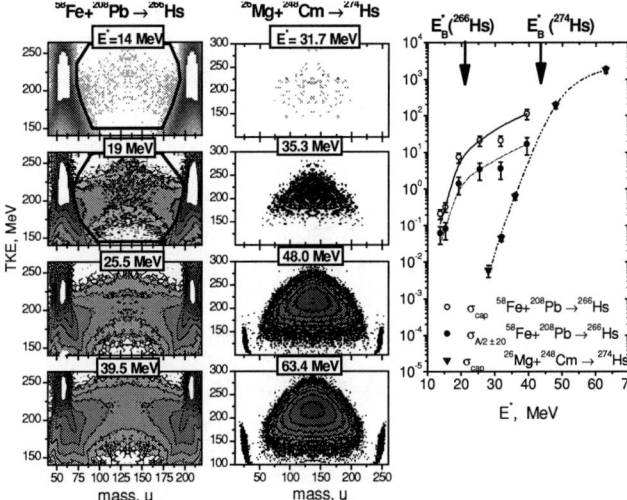

For the ^{26}Mg-induced reaction more n-rich isotope ^{274}Hs is formed with the asymmetry coefficient $\eta=0.81$. As one can see from the central part of the Fig.6 TKE/mass matrixes have the triangular shapes typical for fission of heated nuclei, described by Liquid Drop Model (LDM) [10]. Only at lower excitation energies E* = 31.7 - 35.3 MeV some difference in matrix shape from the LDM predictions appears [11]. The

FIGURE 6. Two-dimensional TKE/Mass matrixes for the reaction ^{58}Fe+^{208}Pb→^{266}Hs at E*=14–39.5 MeV (left panels) and for the reaction ^{26}Mg+^{248}Cm→^{274}Hs at E*=31.7–63.4 MeV (central panels). The excitation functions for both reactions are shown at the right panel.

multimodal fission of ^{274}Hs was revealed, and also the quasi-fission component appears with decrease of excitation energy. The contribution of QF equals $\sigma_{QF}\approx 12\%$ from σ_{cap} at the E*=35.3 MeV. Right panel of Fig.6 shows the excitation functions for both reactions.

The ^{44}Ca + ^{206}Pb and ^{64}Ni + ^{186}W reactions

The reactions ^{44}Ca + ^{206}Pb with mass asymmetry $\eta = 0.648$ and ^{64}Ni + ^{186}W with $\eta = 0.488$ lead to the formation of the same compound nucleus ^{250}No*. Notice that ^{186}W is a strong deformed nucleus, whereas ^{206}Pb is a spherical one. In the study of the spontaneous fission properties of heavy actinide nuclei (Z > 98) it was found that the transition from asymmetric to symmetric fission in the No isotopes takes place somewhere at N = 154 [12]. So, the ^{250}No (N = 148) spontaneous fission should has an asymmetric mass division.

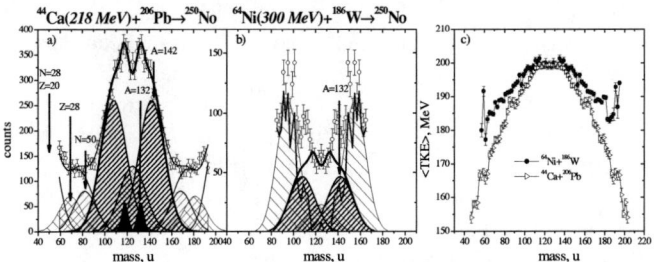

Figure 7 (a) and (b) show the mass distributions, (c) shows ⟨TKE⟩ distributions as a function of the fragment mass for ^{44}Ca + ^{206}Pb and ^{64}Ni + ^{186}W at the

FIGURE 7. Mass distributions for the reactions ^{44}Ca+^{206}Pb (a) and ^{64}Ni+^{186}W (b) and average kinetic energies <TKE> as a function of fragment mass (c) for these reactions at excitation energy about 30 MeV.

energies close to the Bass barrier (the CN excitation energy is about 30 MeV). One can see that mass-energy distributions for these systems are very different. In the case of ^{44}Ca mass distribution has a complicated structure: i) the asymmetric fission connected with the formation of the deformed shell near the heavy fission fragment mass 140; ii) the symmetric fission component determined by the effect of the Z = 50 proton shell; and iii) the quasi-fission component, visible around Z = 20, 28 and N = 28, 50. In contrast to this reaction, the contribution of the quasi-fission component into the total mass distribution in the case of ^{64}Ni + ^{186}W greatly increases. This observation is confirmed by the different behavior of the ⟨TKE⟩ distributions for fission fragments in the systems (see Figure 5c). In the mass region $A_{CN}/2\pm20$ the ⟨TKE⟩ distributions are similar for both reactions, while for the asymmetric mass region ⟨TKE⟩ for ^{64}Ni + ^{186}W is higher than the one for ^{44}Ca + ^{206}Pb. Our analysis shows that only a small part (~25%) of the fission cross section can be associated with complete fusion for the ^{64}Ni + ^{186}W system, the remainder should be attributed to quasi-fission. In the case of ^{44}Ca + ^{206}Pb, the contribution of CN-fission component into the total mass distribution is ~70%.

The ^{48}Ca + ^{168}Er and ^{12}C + ^{204}Pb reactions

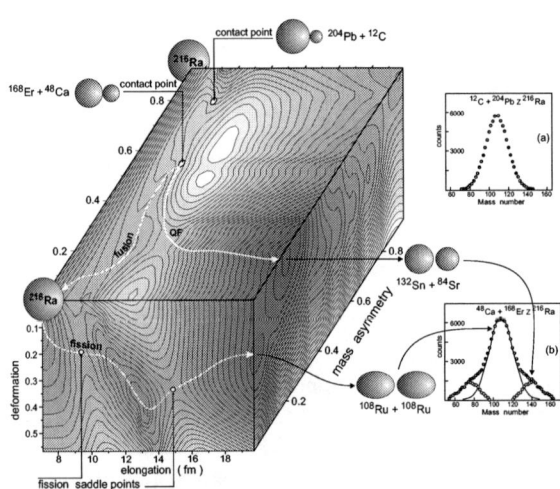

FIGURE 8. Potential energy surface as a function of mass asymmetry of the entrance channels, elongation and deformation for ^{216}Ra. The dark color corresponds to the lower potential energy

The fission cross-section was measured together with the evaporation residue cross-section for the reaction ^{48}Ca + ^{168,170}Er in a wide energy range of ^{48}Ca ions. At the same time, mass-energy distributions of fission fragments, taken alone, are also of great interest. In the recent years our efforts were concentrated on the investigation of the multi-modal structure of the fission fragment MED in the region of transition nuclei 213<A_{CN}<226, which so far has been poorly studied. Thus, MED of the ^{216,218,220}Ra compound nuclei produced in reactions with ^{12}C were studied. It was found that the contribution of the asymmetric fission mode in the case of the ^{12}C+^{204}Pb reaction is 1.5 %[13], and it is 30% in the case of the ^{48}Ca+^{168}Er reaction [14]. We connect such a sharp increase in the yield of asymmetric products in the reaction ^{48}Ca+^{168}Er with the quasi-fission process, MED of which have a clearly expressed shell structure.

The properties of the MED of fission fragment are such that they allow their interpretation by analogy with the low-energy fission of heavy nuclei as a manifestation of an independent mode of nuclear decay which competes with the classical fusion-fission process.

To explain such a difference in the MED a theoretical calculation of the potential energy was made within the three-dimensional version of the "nucleon collectivization" model [15] based on the two-center shell model idea [16]. Figure 8 shows the result of this calculation. One can see that in the case of $^{12}C+^{204}Pb$, mass distribution of which is shown in insert panel (a), the process of fusion-fission is the dominating one. In the case of the $^{48}Ca+^{168}Er$ reaction, there are two optimal paths at the potential energy surface. One of these paths (the white line marked as QF) leads to the $^{132}Sn+^{84}Sr$ exit channel without the compound nucleus formation. This potential energy minimum is determined by shell effects of the doubly magic ^{132}Sn. The second path corresponds to the classical fusion-fission process and leads to the symmetrical exit channel $^{108}Ru+^{108}Ru$, the same as in the reaction with ^{12}C ions. That is why mass distribution of the $^{48}Ca+^{168}Er$ reaction products (insert panel (b)) is the superposition of both processes – the symmetrical classical fission and quasi-fission "shoulders".

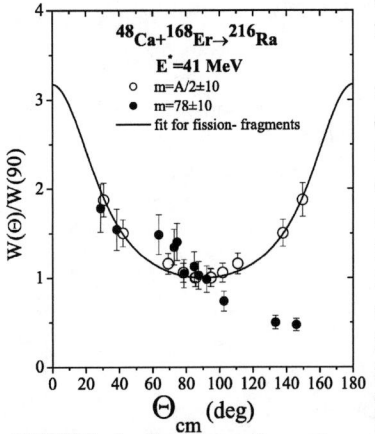

FIGURE 9. Experimental angular distributions of fission fragments from the reaction $^{48}Ca+^{168}Er$ at E_{lab}=195 MeV.

Angular distributions for different fragment masses were measured in order to find experimental evidence of the quasi-fission nature of the "shoulders". Fig. 9 shows the angular distributions for the symmetrical mass range (M=A/2 ± 10, solid circles) and for the "shoulders" (M=78 ± 10, open circles). The solid curve shows the results of calculations made in the framework of the Transition State Model [17]. As one can see from Fig. 9, the solid line fits well into the data for the symmetrical part of mass distribution, whereas the angular distribution for the "shoulders" has a pronounced forward-backward asymmetry, which is one of the distinctive features of the quasi-fission process.

INFLUENCE OF DEFORMATION AND FRAGMENT SHELL CLOSURES ON QUASIFISSION

The effects of the deformations of the colliding nuclei on the reaction dynamic were studied with the reactions $^{48}Ca + ^{144,154}Sm$, going from the closed-shell spherical ^{144}Sm nucleus ($\beta_2 = 0$, $\beta_4 = 0$ [18]) to the well-deformed ^{154}Sm one ($\beta_2 = 0.27$, $\beta_4 = 0.133$). Mass-energy distributions of fission fragments, fission cross sections have been measured in the $^{48}Ca + ^{144,154}Sm$ reactions from well below to well above the Coulomb barrier. For the $^{48}Ca + ^{154}Sm$ reaction we observed an asymmetric component in the fission-fragment mass distributions. A contribution of this component into the total mass distribution increases with respect to the symmetric CN-fission as the projectile energy decreases. At the Coulomb barrier energy, such mass-asymmetric contribution corresponds to about 30%. The symmetric fission components are described by a Gaussian shape with mass variances σ_M^2 =174 amu² at E_{lab} = 202 MeV, 151 amu² at E_{lab} = 195.2 MeV, 123 amu² at E_{lab}=183.2 MeV. These values of the variance are consistent with the results of Ref. [19].

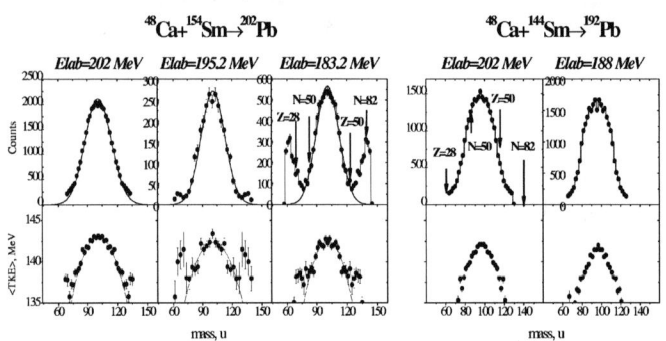

FIGURE 10. Fission fragments mass-energy distributions at different energies of ^{48}Ca for the ^{48}Ca + $^{144, 154}$Sm reactions.

In the case of the ^{48}Ca + ^{144}Sm → ^{192}Pb* reaction no asymmetric "shoulders" are observed. The same behavior one can see in average kinetic energy distribution ⟨TKE⟩ as a function of the fragment mass (Figure 2). The arrows in Figure 2 show the position of the spherical closed shells with Z = 28, 50 and N = 50, 82 derived from the simple assumption of charge/mass equilibration. One can see that in the case of ^{48}Ca + ^{154}Sm the shell structure of the fragments strongly favors the "asymmetric shoulders". We connect this yield of asymmetric products with the quasi-fission process, for which manifestation of shell effects become evident.

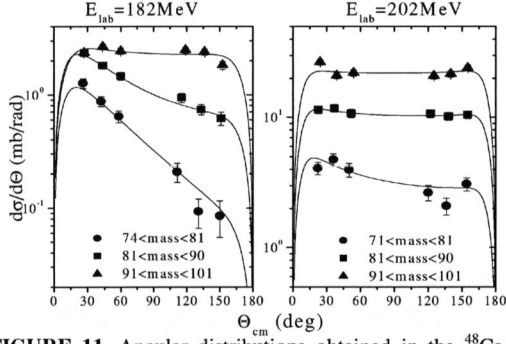 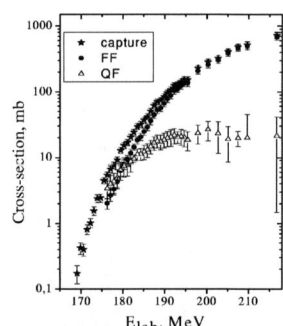

FIGURE 11. Angular distributions obtained in the ^{48}Ca + ^{154}Sm reaction for fission-like fragments corresponding to the selected fragment mass bins. The solid lines are the best fits to the data using Eq (1).

FIGURE 12. Fusion-fission, quasifission and capture cross sections for ^{48}Ca + ^{154}Sm.

Angular distributions for different fragment masses were derived in attempts to find experimental evidences for the quasi-fission nature of the mass-asymmetric "shoulders". In Figure 3 the angular distributions for the selected mass bins of fission-like fragments are shown. The solid curves are fits to the experimental data given by

$$d\sigma/d\theta = 2\pi \sin\theta (a + be^{\beta(\theta-\pi/2)})W(\theta),$$

where $W(\theta)$ is the angular distribution of the CN-fission [20], β is a slope parameter in an exponential decay function reproducing the evident forward-backward asymmetry.

a and b are normalization parameters corresponding to symmetrical and asymmetrical parts of angular distributions.

As we see, for masses around 70–80 at E_{lab} = 182 MeV the angular distribution is very asymmetric. The contributions of quasi-fission extracted from angular distributions are ~32% at E_{lab} = 182 MeV and ~2% at E_{lab} = 202 MeV. Figure 4 presents fusion-fission (circles), quasi-fission (open triangles) and capture (stars) cross sections for the ^{48}Ca + ^{154}Sm reaction as a function of the projectile energy.

Fusion-evaporation cross sections have been measured for the ^{48}Ca + ^{154}Sm reaction in a parallel experiment at LNL. The comparison of fusion-evaporation cross section for this reaction with ^{16}O + ^{186}W, leading to the same ^{202}Pb* CN, shows an inhibition of fusion for the reaction with Ca ions [21].

CONCLUSION AND OUTLOOK

Mass-energy and angular distributions of fragments, fission and quasi-fission cross sections, multiplicities of neutrons and γ-quanta have been studied for a wide range of nuclei with Z = 82-122 produced in reactions with ^{12}C, ^{22}Ne, ^{26}Mg, ^{48}Ca, ^{58}Fe, ^{64}Ni and ^{86}Kr ions at energies close and below the Coulomb barrier.

In the case of the fission process as well as in the case of quasi-fission, the observed peculiarities of mass and energy distributions of the fragments, the ratio between the fission and quasi-fission cross sections, in dependence of the nucleon composition and other factors, are determined by the shell structure of the formed fragments. It is important to note that in the case of the quasi-fission process the influence of the shell effects on the observed characteristics is much stronger than in the case of classical fission of compound nuclei.

For the reaction of the formation of ^{266,274}Hs, ^{250}No and ^{216}Ra nuclei it was found that the decrease of the asymmetry of reaction entrance channel leads to the increase of the QF process. The influence of the nuclei deformation on the FF/QF competition was studied for the ^{48}Ca + ^{144,154}Sm reaction. Fusion suppression and the presence of quasi-fission at energies near and below the Coulomb barrier were observed in the ^{48}Ca + ^{154}Sm reaction on the deformed target nucleus. In the ^{48}Ca + ^{144}Sm reaction no evidence of quasifission was found at the same CN excitation energy and angular momentum as in the case with ^{154}Sm.

A further progress in the field of synthesis of superheavy nuclei can be achieved using hot fusion reactions between actinide nuclei and ^{48}Ca ions as well as actinide nuclei and ^{58}Fe or ^{64}Ni ions.

ACKNOWLEDGMENTS

This work was supported by the Russian Foundation for Basic Research under Grant No 03-02-16779 and by INTAS under Grant N 03-51-6417

REFERENCES

1. Yu.Ts.Oganessian et al., *Eur. Phys. J*, **A5** (1999) 63.
2. Yu.Ts.Oganessian et. al., *Nature* **400** (1999) 242
3. N. A. Kondratiev et al., *Proc. of Int. Conf. DANF'98*, (World Scientific, Singapore, 2000) p. 431.
4. M.Moszinski et al., *Nucl. Instr.Meth*, **A350** (1994) 226.
5. I.Tilquin et al., *Nucl. Instr. Meth*, **A365** (1995) 446.
6. W. Q. Shen et al., *Phys. Rev.* **C 36** (1987) 115.
7. J.Töke et al., *Nucl.Phys.* A440, **327** (1985)
8. V.I. Zagrebaev, *Phys. Rev.* **C64** 034606 (2001)
9. Z. Patyk, A. Sobiszevski, Nucl. Phys. **A 533**, 132 (1991).
10. J. R. Nix and W. J. Swiatecki, Nucl. Phys. **71**, 1 (1965)
11. E.V.Prokhorova et al., *Proc. of Int. Symp. on Exotic Nuclei (EXON-2004)*, Peterhof, July 5-12, 2004, to be published.
12. E.K. Hulet, *Yad. Fiz.* **57**, 1165 (1994).
13. I. V. Pokrovsky et al., *Phys. Rev.* **C60** (1999) 041304.
14. A. Yu. Chizhov et. al., *Phys. Rev.* **C67** (2003) 011603.
15. V. I. Zagrebaev, *Phys. Rev.* **C64** (2001) 034606; *J.Nucl. Radiochem.Sci.*, **3** (2002)13.
16. U. Mosel et al., *Phys. Lett.* **B34** (1971) 587; J.Maruhn and W.Greiner, *Z.Physik*, **251** (1972) 431.
17. B. B. Back et al., *Phys. Rev.* **C32** (1985)195.
18. P. Moller et al., *At. Data Nucl. Data Tabl.* **59**, 185 (1995).
19. M.G. Itkis and A.Ya. Rusanov, *Fiz. Elem. Chastits At. Yadra* **29**, 389 (1998).
20. I. Halpern and V.M. Strutinski, *Proc. of the Second UN Intern. Conf. on the Peaceful Uses of Atomic Energy*, Geneva, 1957, p.408 (UN, Geneva, 1958).
21. M. Trotta et al., *Nucl. Phys.* **A 734**, 245 (2004).

ISOLDE beams of neutron-rich zinc isotopes: yields, release, decay spectroscopy

U. Köster*, T. Behrens†, C. Clausen*,**, P. Delahaye*, V.N. Fedoseyev*,
L.M. Fraile*, R. Gernhäuser†, T.J. Giles*, A. Ionan‡, T. Kröll†, H. Mach§,
B. Marsh*,¶, M. Seliverstov‖,‡, T. Sieber*, E. Siesling*, E. Tengborn*,††,
F. Wenander* and J. Van de Walle‡‡

*CERN, ISOLDE, 1211 Genève 23, Switzerland
†Technische Universität München, Physik-Department, 85748 Garching, Bavaria
**Aarhus Universitet, Institut for Fysik og Astronomi, 8000 Aarhus C, Denmark
‡Petersburg Nuclear Physics Institute, 188300 Gatchina, Russia
§Uppsala University, Department of Radiation Sciences, 75121 Uppsala, Sweden
¶Nuclear Physics Group, University of Manchester, M13 9PL Manchester, UK
‖Johannes-Gutenberg Universität, Institut für Physik, 55128 Mainz, Germany
††Chalmers University of Technology, Dept. of Experimental Physics, 41296 Göteborg, Sweden
‡‡K.U. Leuven, Instituut voor Kern- en Stralingsfysica, 3001 Heverlee, Belgium

Abstract. Intense radioactive ion beams of the neutron-rich zinc isotopes $^{69-81}$Zn have been produced at the isotope separation on-line facility ISOLDE at CERN. The combined use of spallation-neutron induced fission of ^{238}UC$_x$ targets and resonant laser ionization provided sufficient suppression of disturbing isobars (mainly gallium and rubidium) to perform decay spectroscopy up to ^{81}Zn.

Keywords: isotope separation on-line, zinc, resonant laser ionization, neutron-induced fission, 81Zn
PACS: 23.20.Lv, 25.85.Ec, 27.50.+e, 29.25.Rm, 32.80.Fb

INTRODUCTION

Various scientific communities demand radioactive ion beams for a multitude of applications. Depending on the position on the chart of nuclides different nuclear reactions may be used to produce the radioactive isotopes. Neutron-rich isotopes of medium mass are best produced by fission. ISOLDE at CERN provides ISOL (isotope separation on-line) beams of fission products since more than thirty years [1]. With up to 4 μA of 1.4 GeV protons onto a 50 g/cm^2 ^{238}UC$_x$ target slightly over 10^{13} fission products per second are produced. This is orders of magnitude higher than what other operating ISOL facilities can deliver today, and combined with efficient and selective ion sources it makes ISOLDE the world-leading radioactive ion beam facility.

For users the decisive quality parameters of radioactive ion beams are the intensity of the beam, its purity and emittance. It is well known that for ISOL beams the latter is excellent: at several ten keV beam energy the transverse emittance is typically of the order of $\varepsilon_{95\%} \approx 10-20\pi$ mm mrad [2] while the energy spread normally does not exceed few eV. The radioactive ion beam intensity scales obviously with the target thickness, the primary beam intensity and the production cross-section of the isotope in question. However, it is also strongly affected by the release and ionization efficiency, which just like the beam purity depends on the element and isotope in question and on the type of the employed target and ion source unit.

Therefore we discuss the characteristic yields and release parameters of ISOLDE beams in individual papers treating a given element or group of homologue or chemically similar elements. Previous reports discussing in detail the beams produced since 1992 at ISOLDE-PSB (using as driver accelerator a stack of four proton synchrotrons that provide 2 μs long pulses of ca. $3 \cdot 10^{13}$ protons every 1.2 seconds) cover the fission products Mn [3], Ni [4], Ga [5], Kr and Xe [6], Cd [7] and In [8]. In the present paper we will discuss beams of neutron-rich[1] Zn isotopes.

PRODUCTION AND SEPARATION

For the production of beams of neutron-rich Zn isotopes UC_x/graphite ($x \approx 4$) targets have been used in combination with different types of ion sources. Several measures can be used individually or combined to achieve satisfactory beam purity.

Chemically selective transfer line

The chemical elements differ strongly in their volatility. A high volatility translates at a given temperature into a higher vapor pressure and a shorter residence time on the surface. The group 12 elements (Zn, Cd, Hg) are characterized by an atomic subshell closure $d^{10}s^2$. This assures a clearly higher volatility compared to the neighboring elements. Hence, a transfer line (i.e. the tube connecting the target with the ion source) kept at a temperature just sufficient to allow the transport of short-lived Zn isotopes will retain neighboring isobars by condensing them on the surface of the tube. Thus, only the Zn isotopes are reaching the ion source while short-lived radioisotopes of less volatile elements are kept back in the transfer line sufficiently long to decay there. This chemical separation method is known as isothermal vacuum chromatography. In the case of zinc the lower (Cu, Ni, ...) and higher isobars (Ga, Ge, As, Se) are retained completely. Occasionally bromine may be transmitted through the line in form of a volatile molecule, but krypton is always transmitted. This is not very disturbing for decay spectroscopy experiments with very neutron-rich Zn isotopes (e.g. 80 to 82), since the respective krypton isotopes are stable or very long-lived[2]. Thus already at ISOLDE-SC relatively clean Zn beams have been produced using a temperature controlled transfer line and used for half-life measurements by beta-delayed neutron detection up to ^{81}Zn [9]. However, more recent applications of radioactive ion beams include direct precision mass measurements with a Penning trap or experiments with post-accelerated beams like Coulomb excitation. In these cases stable and radioactive isotopes represent equally disturbing background. Thus, another or additional measure has to be used to eliminate also krypton or bromine.

---

[1] ISOLDE beams of neutron-deficient Zn isotopes have been discussed in ref. [3].
[2] The background of krypton isomers can be strongly reduced by beta-gating.

Resonant laser ionization

With the resonance ionization laser ion source (RILIS) [10], atoms of a given chemical element are selectively ionized by stepwise resonant excitation of a valence electron into the continuum. The excitation energy is provided by laser light tuned to strong atomic transitions in the respective element. For other elements the laser light is out of resonance and will therefore cause only negligible excitation and ionization. Thus the RILIS provides intrinsically extremely pure beams.

For the resonant laser ionization of zinc a formerly developed ionization scheme [11] has been used: frequency-tripled dye laser light at 213.86 nm excites from the atomic ground state $3d^{10}4s^2\ ^1S_0$ to the $3d^{10}4s\ 4p\ ^1P^o_1$ state. The second transition is performed with dye laser light at 636.23 nm to the $3d^{10}4s\ 4d\ ^1D_2$ state from where the atoms are ionized non-resonantly to the continuum with the green copper vapor laser light at 511 nm. The off-line measured ion source efficiency was 5%.

Since presently only pulsed lasers can provide a sufficient peak power to saturate the atomic excitations, the atoms effusing from the target have to be "stored" for the time between two laser pulses to have at least one chance to interact with the laser light. Typically a long thin tube ("line" in ISOLDE jargon) is used to constrain the effusion of the atoms and serve as interaction region with the laser light.

To avoid "sticking" of the atoms to the surface of the tube and to ensure sufficient electron emission to provide conditions for ion repelling from the surface, the tube is typically made from a refractory metal and kept at temperatures of the order of 2000 °C. One has to keep in mind that resonant laser ionization enhances the ionization efficiency of the element to which atomic transitions the lasers have been tuned, but it does not reduce the surface ionization of other elements in the hot cavity. Hence, surface ionized background of isobars with low ionization potential will be omnipresent. In the case of zinc this concerns particularly rubidium (4.18 eV ionization potential) which is efficiently surface ionized ($> 65\%$ efficiency) on any hot surface of a refractory metal (Nb, Ta, W, ...) and also gallium (6.00 eV ionization potential) which is ionized with an efficiency of the order of 1 % in typical RILIS cavities at ca. 2000 °C.

While the rubidium ionization efficiency is practically temperature independent, lowering the line temperature can reduce the gallium background. However, for too low line temperatures also the electron emission and, hence, the ion repelling and Zn ionization efficiency are compromised. Figure 1 shows the relative ^{76}Zn and ^{76}Ga intensities in dependence of the line temperature. Below 1950 °C the Zn ionization is compromised, but due to the faster drop of the Ga ionization efficiency the Zn/Ga ratio can be improved by further reduction of the line temperature. For other isotopes the behavior is qualitatively the same, only for the most short-lived ones the release efficiency is significantly affected at very low line temperatures.

Although the RILIS does not suppress all background, it can allow identifying it by comparing "laser on" and "laser off" spectra. For the latter, one or all the ionizing laser beams are blocked or detuned in frequency. The difference between "laser on" and "laser off" spectra shows then the net signal due to the laser-ionized elements.

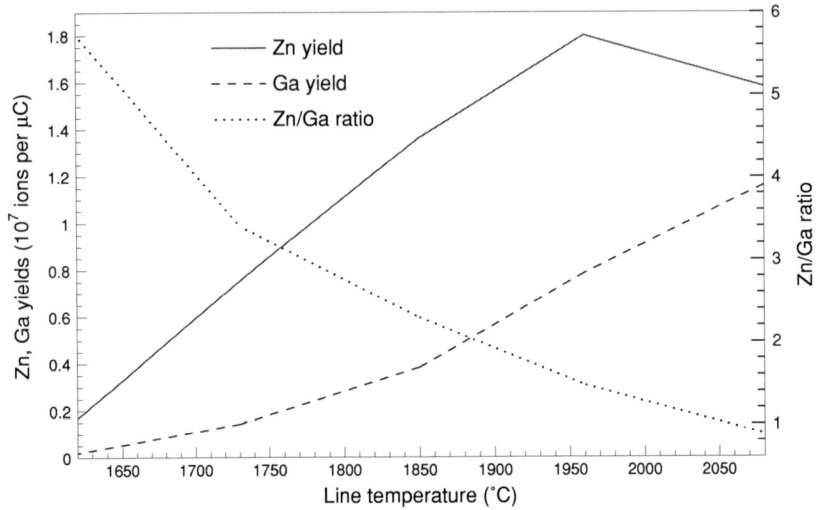

FIGURE 1. Dependence of the zinc (red) and gallium (blue) yields on the "line" temperature

Neutron converter

When inducing fission of ^{238}U by GeV protons, there will be a strong contribution of spallation-fission, i.e. part of the target nuclei will first evaporate a significant number of mainly neutrons and then fission in a later stage. This leads to a shift of the fission yields towards less neutron-rich or even to neutron-deficient species [12]. In the mass region $A \approx 80$ this effect leads to significant background from neutron-deficient rubidium isotopes [5]. To avoid such background it is preferable to use low-energy fission, which only produces neutron-rich species. This can be done by sending the GeV protons not directly onto the UC_x target, but onto a heavy metal rod (e.g. tungsten) mounted close to the target. The spallation neutrons produced in the latter have mainly energies in the MeV region [13]. Part of them are emitted in radial direction and can thus interact with the UC_x target mounted parallel to the metal rod. The latter thus acts as an efficient "converter" of GeV protons into MeV neutrons and is therefore often called "neutron converter". The use of the neutron converter is not sufficient for a unique selection of an isotope, but helps as a supplementary measure to enhance the selection power of the two previously mentioned methods.

Target UC2.280 was equipped with a 125 mm long W converter of 12.7 mm diameter, mounted parallel to the target at a distance of 23.5 mm (center to center). Longitudinally the middle of the converter and of the target coincided.

Macro-Time structure

Due to the different diffusion and effusion times, different elements are released more or less quickly. In the present case zinc shows a clearly quicker release than the unwanted gallium, see table 1 and figure 2. Thus Ga background can be suppressed by an additional factor two to four by sending the radioactive ion beam only during the first 0.2 to 0.6 s after proton impact to the experiment.

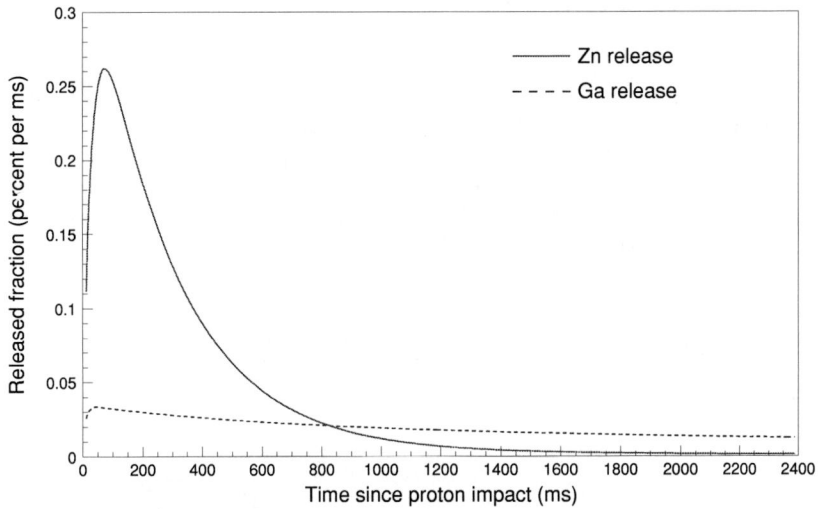

FIGURE 2. Typical release time profile of zinc (red) and gallium (blue).

YIELDS AND RELEASE

The yields (i.e. radioactive ion beam intensity normalized to 1 μA intensity of incident proton beam), have been determined either on- or off-line by Faraday cup measurements or beta- or gamma spectrometry respectively. To determine the release speed only one proton pulse every 14.4 s, 16.8 s or 19.2 s was sent onto the target. The amount of released radioactive ions was then measured as function of time since the last proton pulse. Figure 2 shows representative release time profiles of stable Zn and Ga and table 1 gives the release parameters obtained by fitting the measured curves with the empiric four-parameter formula [14]:

$$P(t) = N \left(1 - e^{-t\ln(2)/t_r}\right) \left[\alpha e^{-t\ln(2)/t_f} + (1-\alpha)e^{-t\ln(2)/t_s}\right]$$

As the release parameters and yields evolve during an on-line run, the time is given for which the target has been exposed to the proton beam. One observes that the release gets slower with an ageing target. The ^{74,76}Zn yields of target UC2.259 dropped by a factor two to three during the 7 days of operation. The target deterioration is mainly due to the thermal spikes and mechanical stress during the impact of the intense proton pulses.

TABLE 1. Target and ion source characteristics and measured release parameters.

Element	Target Number	On-line run time (h)	Target temp. (°C)	Line temp. (°C)	proton beam (1.4 GeV)	t_r (ms)	t_f (ms)	t_s (ms)	α
Zn	UC2.259	1	2000	2030	onto target	26	125	2000	0.991
Zn	UC2.259	120	1900	2120	onto target	23	190	1500	0.989
Ga	UC2.259	120	1900	2120	onto target	6	400	3060	0.40
Zn	UC2.280	1	2060	2040	onto converter	30	200	1600	0.98

In all cases where both isomers were observed for the odd mass zinc isotopes, the yield of the high spin isomer was higher than of the low spin isomer.

With the known release function, one can calculate for each isotope what fraction will decay inside the target before it is released. Correcting for this loss and assuming an ionization efficiency of 5%, one can calculate back to the "in-target production rates" of the Zn isotopes. Comparing the latter with the production rates calculated from cross-sections measured in inverse kinematics with 1 GeV/nucleon at GSI-FRS [12] indicates that the on-line ionization efficiency was probably slightly higher (7-8%). For gallium a good agreement between our in-target production rates and the GSI cross-sections is obtained by assuming 1.5% ionization efficiency.

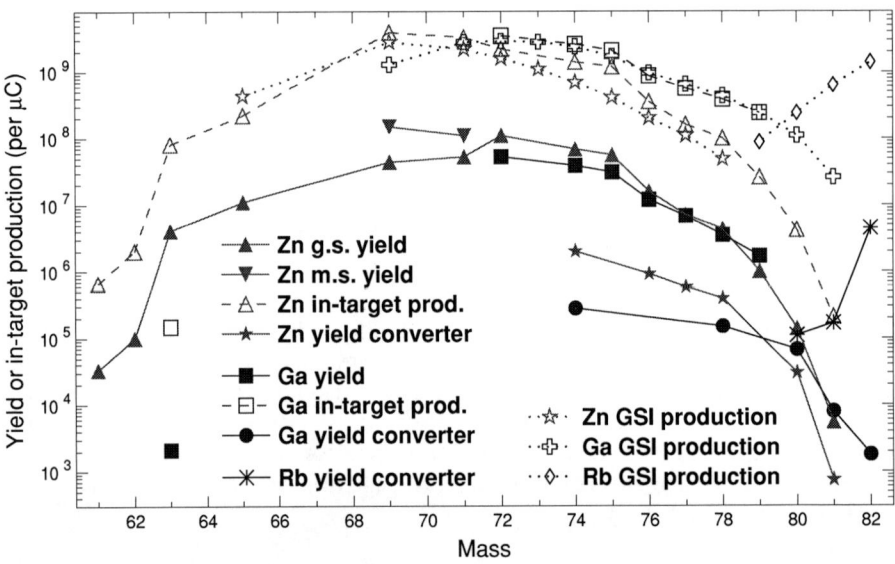

FIGURE 3. Yields (filled symbols) and estimated in-target production rates (open symbols) of Zn, Ga and Rb isotopes. The "converter yields" were measured with the proton beam sent onto the converter of the target UC2.280, all other yields with the proton beam sent directly onto the target UC2.259.

The yields measured with the proton beam sent onto the target UC2.280 were a factor two to three lower compared to those shown from target UC2.259. This might be due to a lower ionization efficiency. Both targets had a thickness of 52 g/cm^2 ^{238}U.

BEAM PURITY

Beams of ^{74,76,78}Zn have been post-accelerated with REX to measure the B(E2) values by Coulomb excitation. At masses 74 and 76 the beam purity was sufficient, even when directing the proton beam directly onto the target. At mass 78 however, strong Rb and Ga background was present. When directing the proton beam onto the neutron converter the Rb background practically vanished. Moreover the Zn to Ga ratio was improved. Fig. 4 shows as $E_{res.}$-ΔE plot the composition of the beam post-accelerated to 2.86 MeV per nucleon and measured with an ionization chamber. The figure shows the beam composition when directing the proton beam onto the target or converter respectively.

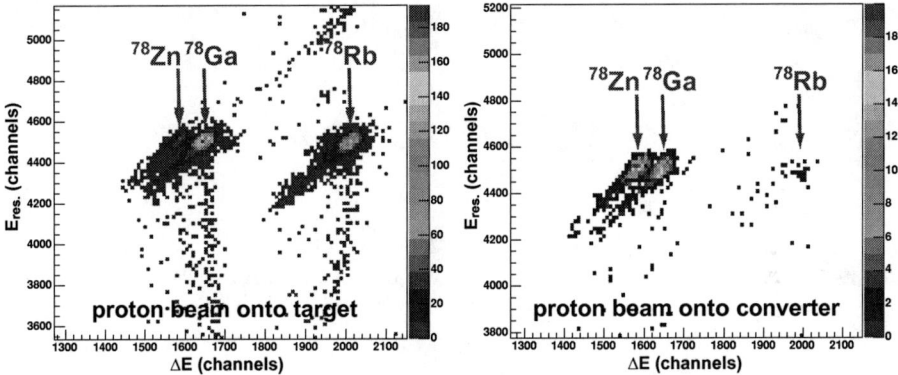

FIGURE 4. Beam composition of post-accelerated mass 78 beams with the proton beam sent onto the target (left) and the converter (right) respectively.

DECAY SPECTROSCOPY

While the yield measurement of longer-lived, more abundantly produced isotopes can be easily performed by the methods mentioned before, the detection of the most exotic isotopes requires a more elaborate set-up. In the case of zinc this becomes necessary for ^{81}Zn. Figure 5 shows gamma spectra taken on mass 81. Comparing the "laser on" (96 pulses of 3.5 μC protons each, 403 s measurement) and "laser off" spectra (61 pulses of 3.5 μC protons each, 255 s measurement) allows to clearly identify the 351 keV and 452 keV lines belonging to the decay of ^{81}Zn. This confirms the assignment from a recent measurement of the ^{81}Zn decay performed at the PARRNe separator at IPN Orsay [15].

The ^{81}Zn yield shown in Fig. 3 is actually a lower limit as it was deduced assuming 100% absolute branching ratio for the 351 keV gamma ray. Probably this branching ratio is indeed rather high (>50%), else a more pronounced difference in ^{81}Ga intensity in the "laser on" and "laser off" spectra should be observed. One also observes a significant ^{80}Ga activity in the ^{81}Zn decay spectrum which suggests a P_n value >10% for ^{81}Zn, which would be higher than the previously measured $P_n = 7.5(30)\%$ [9].

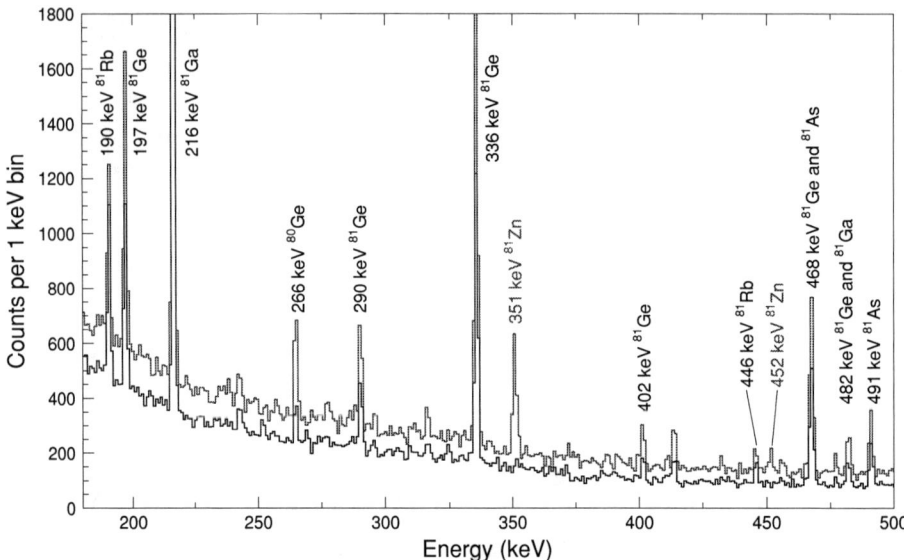

FIGURE 5. Part of the gamma ray spectrum measured on mass 81 in the "laser on" (upper red histogram) and "laser off" (lower blue histogram) mode respectively.

CONCLUSION AND OUTLOOK

It has been shown that clean and intense beams of neutron-rich Zn isotopes can be produced at ISOLDE by combining resonant laser ionization with a neutron converter. The beam purity was sufficient to easily perform decay spectroscopy up to ^{81}Zn and Coulomb excitation measurements up to ^{78}Zn. To proceed to still heavier Zn isotopes, the beam purity has to be further increased without compromising the Zn beam intensity. A possible way is the combination of the present measures with a chemically selective transfer line. In preparatory off-line thermochromatography experiments by the TARGISOL collaboration it has been found that a transfer line made from quartz or different types of glass could help to retain Ga and Rb without delaying Zn too much [16]. A test of such a prototype target at ISOLDE combining a chemically selective transfer line with resonant laser ionization is planned soon. It should make it possible to use in the future ^{80}Zn for Coulomb excitation experiments and ^{82}Zn or even ^{83}Zn for $\beta\gamma$ decay spectroscopy.

ACKNOWLEDGMENTS

Supported by the EU-RTD project TARGISOL (HPRI-CT-2001-50033). We thank the IS386 collaboration for "squatting" their γ-spectroscopy set-up for the measurement of the $^{77-81}$Zn yields and the IS412 collaboration for providing figure 4.

REFERENCES

1. H. Ravn, S. Sundell, L. Westgaard, and E. Roeckl, *J. Inorg. Nucl. Chem.*, **37**, 383–393 (1975).
2. F. Wenander, J. Lettry, and M. Lindroos, *Nucl. Instr. Meth. B*, **204**, 261–266 (2003).
3. M. Oinonen, et al., *Hyp. Int.*, **127**, 431–436 (2000).
4. A. Jokinen, et al., *Nucl. Instr. Meth. B*, **126**, 95–99 (1997).
5. U. Köster, *Eur. Phys. J. A*, **15**, 255–263 (2002).
6. U. Bergmann, et al., *Nucl. Instr. Meth. B*, **204**, 220–224 (2003).
7. U. Köster, et al., *Nucl. Instr. Meth. B*, **204**, 347–352 (2003).
8. I. Dillmann, et al., *Eur. Phys. J. A*, **13**, 281–284 (2002).
9. K. Kratz, et al., *Z. Phys. A*, **340**, 419–420 (1991).
10. U. Köster, V. Fedoseyev, and V. Mishin, *Spectrochimica Acta B*, **58**, 1047–1068 (2003).
11. J. Lettry, et al., *Rev. Sci. Instr.*, **69**, 761–763 (1998).
12. M. Bernas, et al., *Nucl. Phys. A*, **725**, 213–253 (2003).
13. R. Catherall, J. Lettry, S. Gilardoni, U. Köster, and the ISOLDE Collaboration, *Nucl. Instr. Meth. B*, **204**, 235–239 (2003).
14. J. Lettry, et al., *Nucl. Instr. Meth. B*, **126**, 130–134 (1997).
15. D. Verney, et al., *Brazilian J. Phys.*, **34**, 979–982 (2004).
16. C. Jost, *Thermochromatographische Experimente zur Ionenquellenentwicklung*, Diplomarbeit, Johannes-Gutenberg-Universität, Mainz (2005).

NEW FACILITIES—II

High Intensity Beams Of Fission Fragments At SPIRAL 2

Marek Lewitowicz

for the SPIRAL 2 project group

Grand Accélérateur d'Ions Lourds (GANIL)
BP 55027
14076 Caen Cedex
France

Abstract. The future plans of the GANIL/SPIRAL facility related to the SPIRAL~2 project are presented. In the frame of this project a new superconducting linear driver (LINAG) will deliver a high intensity, 40 MeV deuteron beam as well as a variety of heavy-ion beams with mass over charge ratio equals to 3 and energy up to 14.5 AMeV. Using a carbon converter and the 5 mA deuteron beam, a neutron-induced fission rate is expected be up to 10^{14} fissions/s for high-density UC_x target. In the contribution different schemes of production of intense beams of fission-fragments as: low and high neutron-induced fission, photo-fission and charge-particle-induced fission proposed by several Radioactive Nuclear Beams projects are shortly compared.

Keywords: Neutron-induced fission, properties of nuclei, reactions induced by unstable nuclei, linear accelerators, cyclotrons.

PACS: 25.85.Ec; 21.10.-k; 25.60.-t; 29.17.+w; 29.20.Hm

MOTIVATION

New exotic shapes and excitation modes, halo-like and molecular structures, and new modes of nuclear decay have been recently observed, while tests of fundamental symmetries, testing and refinement of the Standard Model of fundamental interactions, and exploration of the 'magic' numbers of protons and nuclei in very exotic nuclei are all enticing avenues of discovery. The study of radioactive nuclei, involved as they are in nucleosynthesis in the stars, leading to the creation of nuclei, atoms, molecules and all other complex structures our world is made of, has until now been strongly restricted by their short lifetimes and the limited production yields. The quest for Rare Isotope Beams (RIBs), which are orders of magnitude more intense than those currently available, is the main motivation behind the SPIRAL 2 project.

Thanks to the high-energy and high-intensity neutron flux available at SPIRAL2, the facility will offer also a unique opportunity for material irradiations and cross-section measurements, both for fission-related (notably accelerator driven systems (ADS) and Gen-IV fast reactors) and fusion-related research, tests of various detection systems and of resistance of electronics components to irradiations, etc.

The conceptual design and, more recently, the technical design of SPIRAL2 have been developed by more than 20 laboratories from 10 countries between 2001 and 2005. More than two hundred scientists and engineers have contributed to the scientific case for the project and to the design of the facility.

The French Ministry of Research took a decision on the construction of SPIRAL 2 at the end of May 2005. Its construction cost, estimated to 130 M€ (including personnel and contingency), will be shared by French funding agencies CNRS/IN2P3 and CEA/DSM, the local authorities in the Basse-Normandie region and the European partners. The construction will last about five years.

WHAT IS SPIRAL 2 ?

The SPIRAL 2 facility (figure 1) is based on a high-power, superconducting linear accelerator (LINAG), which will deliver a high-intensity, 40-MeV deuteron beam as well as a variety of heavy-ion beams with a mass-to-charge ratio equal to 3 and energies up to 14.5 MeV/nucleon. Using a carbon converter, a 5-mA deuteron beam and a uranium carbide target, the fast-neutron-induced fission rate expected reach 10^{14} fissions/s. The RIB intensities in the mass range from A=60 to A=140 will be of the order of 10^6 to 10^{11} particles per second (pps), surpassing by one or two order of magnitude the existing facilities anywhere in the world. For example, the intensities should reach 10^9 pps for ^{132}Sn and 10^{10} pps for ^{92}Kr. A direct irradiation of the UC_2 target with beams of deuterons, ^{3,4}He, ^{6,7}Li, or ^{12}C can be used if higher excitation energy leads to a higher production rate for a specific nucleus of interest.

The neutron-rich fission RIB could be complemented by beams of light nuclei and nuclei near the proton drip-line provided by fusion-evaporation or transfer reactions. For example, a production of up to 8×10^4 atoms of ^{80}Zr per second using a 200-µA ^{24}Mg^{8+} beam on a ^{58}Ni target should be possible. Similarly, the heavy- and light-ion beams from LINAG on different production targets can be used to produce high-intensity light RIB with ISOL technique.

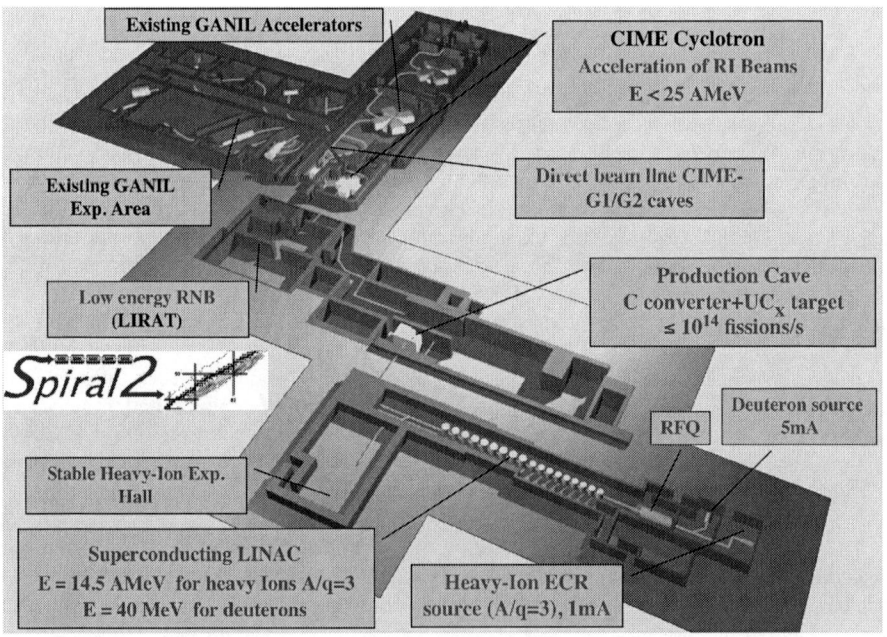

FIGURE 1. Schematic layout of the SPIRAL 2 facility at GANIL.

The extracted RIB will subsequently be accelerated to energies of up to 20 MeV/nucleon (typically 6–7 MeV/nucleon for fission fragments) by the existing CIME cyclotron.

Thus, using different production mechanisms and techniques SPIRAL 2 would allow us to perform experiments in a wide range of neutron- and proton-rich nuclei far from the line of stability (figure 2).

One of the important features of the future GANIL/SPIRAL/SPIRAL 2 facility will be the possibility to deliver up to five stable or radioactive beams simultaneously in the energy range from keV to several tens of MeV/nucleon. For example, let's assume that using the LINAG accelerator and the adequate uranium target one produces high intensity radioactive beams of fission fragments. After ionization the beam of a given mass can be used in the very low energy (several tens of keV/nucleon) experimental area. At the same time, the mass separator will deliver another beam of different mass for the further acceleration in the CIME cyclotron. This beam, thank to the dedicated beam line from CIME to the G1 and G2 caves, can be used for experiments with EXOGAM and VAMOS.

Simultaneously, the standard GANIL beams can be accelerated and used in the IRRSUD facility (stable beams at about 1 AMeV), at the SME beam line (stable beams at 8-10 AMeV) and in the existing experimental area (50-100 AMeV stable or radioactive beams produced in-flight). Thus the parallel operation of the current GANIL facility and SPIRAL 2 will allow for doubling of the beam time available for users. A full version of the SPIRAL 2 Design Study Report is available through the GANIL Web page [1].

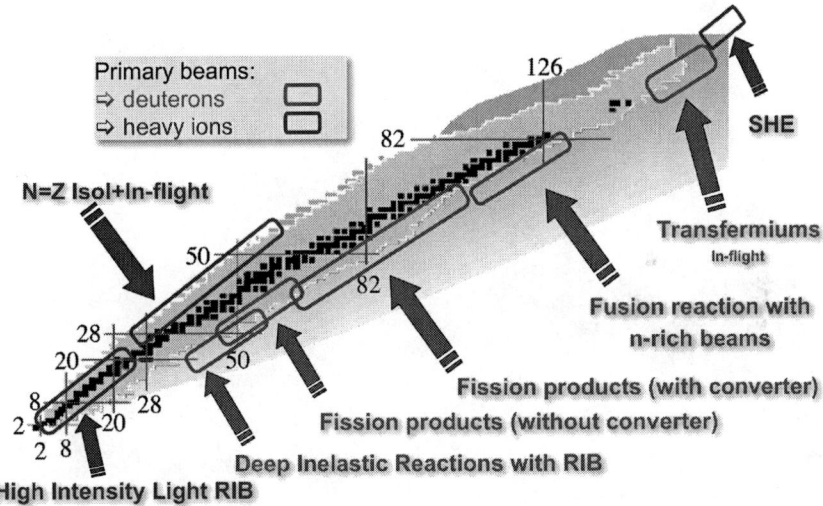

FIGURE 2. Regions of the chart of nuclei accessible for research of nuclei far from stability at SPIRAL2.

PHYSICS AT SPIRAL 2

A full description of the nuclear physics research and interdisciplinary applications, which might be covered with the SPIRAL 2 beams, is beyond the scope of this contribution. In the following only few examples will be given. A draft of the SPIRAL 2 physics case, elaborated by more than hundred physicists from Europe, US and Japan can be found on the GANIL Web page [1].

As it is shown in figure 2, both high-intensity stable and fission-fragment radioactive beams accelerated to several MeV/nucleon can be used to access a broad range of nuclei very far from stability. The facility will provide also a very high flux of fast neutrons (up to ~ 10^{15} n/s/cm^2). These opportunities will allow SPIRAL 2 to contribute to the following research area:

- Structure of exotic nuclei
- Dynamics and thermodynamics of N/Z asymmetric nuclei and nuclear matter
- Origin of elements
- Fundamental interactions: searching for physics beyond the Standard Model
- Physics with fast neutrons
- Interdisplinary research as:
    o Material irradiation studies
    o Fast ion-slow ion collisions
    o Probing nuclear properties with atomic measurements on ion beams
    o The nuclear radioactive probes for solid-state physics
    o Radiation damage and radiation chemistry
    o Radiobiology

and others.

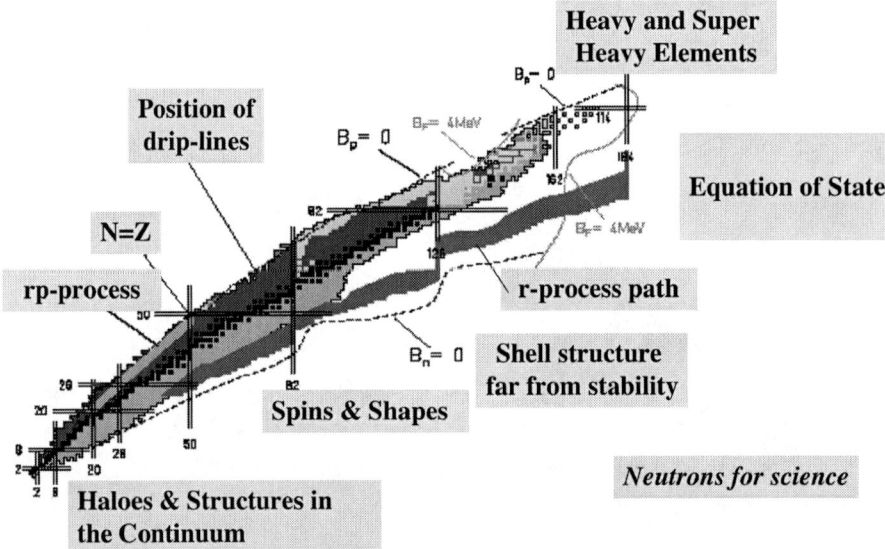

FIGURE 3. Physics with SPIRAL 2.

The following examples of experiments are illustrating a variety of topics that might be studied with the SPIRAL 2 beams.

Example 1: The binding energies of the light nuclei and 3-body forces

With the intense beams of exotic nuclei produced by SPIRAL2, and by using reactions involving transfer of nucleons or charge-exchange transformation of protons into neutrons, it will be possible to conduct studies aiming at the clarification of the binding energies of light nuclei. Of particular interest are beams of ^{6}He and ^{8}He, containing respectively 4 and 6 neutrons, but only 2 protons. For example, SPIRAL2 is expected to deliver an ^{8}He beam with an intensity of 10^9 ions/s on target. This will allow to perform unprecedented reaction studies, to thus gain new insight into the structure of light exotic nuclei and to seek new (unbound) nuclear systems. In particular, a significant constraint on the theory of the nuclear forces could come from systems containing neutrons exclusively, like the quadri-neutrons or the hexa-neutrons, even if they exist only in the form of resonances.

Example 2: Identification, mass measurements and radioactivity studies of exotic nuclei at SPIRAL2

At SPIRAL2, ground-state properties of exotic nuclei can be studied either in the low-energy experimental hall (LIRAT) or after post-acceleration in the high-energy area. The latter method is chosen if the nuclear state of interest is too short-lived to be released from the target/ion source system including, in particular, unbound nuclei, or is produced there in too low quantity due to the low release efficiency for (short-lived) nuclei of certain elements. Accelerated beams of radioactive nuclei delivered by CIME at specific energies of the order of 3 to 20 MeV/nucleon can be used to induce reactions for producing and investigating nuclei that are even further away from stability (and thus more short-lived) than the isotopes delivered by the target/ion source system. Experiments in the high-energy area will use multinucleon-transfer and fusion-evaporation reactions, yielding neutron-rich and neutron-deficient isotopes, respectively. Examples of CIME beams that are relevant for inducing, e.g., fusion-evaporation or multinucleon-transfer are ^{6}He, ^{60}Zn, ^{62}Zn, ^{80}Zn, ^{74}Kr, ^{90}Kr and ^{140}Xe, their intensities ranging from 10^6 to 10^{11} ions/s.

Example 3: Coulomb excitation and inelastic scattering measurements

Knowledge of the excitation energies and transition strengths of low-lying collective states can be obtained by Coulomb excitation. The projectile is excited by a high-Z target, and the decay γ-rays are detected. Such experiments can be performed with very low beam intensities (down to 1000 ions/s) and require a highly efficient granular γ-ray detector such as EXOGAM or, even better, AGATA. Long isotopic chains that could be investigated are, for example, ^{72}Kr to ^{94}Kr, the chain ^{112}Xe to ^{145}Xe, and from ^{125}Sn to ^{134}Sn.

The combination of the VAMOS spectrometer with the state-of-the-art light charged-particle and γ-ray arrays such as MUST2, TIARA and AGATA will permit one to carry out experiments with the aim to determine the contributions of protons and neutrons to nuclear excitation.

Example 4: s-process studies at SPIRAL2

SPIRAL2 will be very well suited to produce the samples of unstable isotopes required for studying s-process branching points. These samples could be subsequently irradiated by intense neutron fluxes in order to determine their (n,γ) cross-section. A dedicated neutron time of flight beam-line can be envisaged at the SPIRAL2 facility (see below).

For some selected cases when the final reaction product is radioactive (2n,γ) cross-section studies are feasible via the neutron activation technique. Nevertheless, the neutron-energy spectrum has to be scaled from 14 MeV down to about 100 keV to mimic the neutron energy in stellar conditions. The determination of neutron-capture cross-sections using the radioactive nuclei ^{85}Kr and ^{60}Fe could be considered as the first key experiments for the SPIRAL2 facility. An alternative means to study (n,γ) cross-sections would be to use indirect methods such as the (d,p) reaction.

Example 5: Neutron Time-of-Flight (ToF) measurements

The SPIRAL2 linac might also serve as a driver accelerator for a neutron time-of-flight facility in which the neutrons are produced by a 40 MeV deuterium beam incident on a thick graphite target. The beam will be pulsed at 500 kHz leaving 2 μs between bursts. The burst width is a few hundred picoseconds. Two flight paths of 5 and 10 m are proposed. Flux estimates were done with MCNPX and amount to $8.8(2.2) \times 10^6$ n/(cm^2 s) for 5(10) m. About 6×10^{12} neutrons are produced per second and at this pulse rate the deuteron average current is ~28 μA.

This facility would offer favourable conditions for measurements of cross-sections for γ-rays emitted in inelastic and (n,xn) reactions to provide a coherent study of several isotope chains (Si, Fe/Ni, Zr/Mo, W, Pb, Th/U) throughout the mass table to assist nuclear model development and can be focussed at key materials of interest to applications (Si, Cr, Fe, Ge, Pd, ...). Also the cross-section measurements of the (n,xp) and (n,xα) reactions are of particular importance to support radiation damage studies. A well-known problem area for which measurements are particularly scarce is the range from 5 to 35 MeV, which is exactly in the range covered by SPIRAL2.

CONCLUSIONS

Because of the importance of Rare Isotope Beams for a "broad programme of research in fundamental nuclear physics and astrophysics, as well as in applications of nuclear science", NuPECC (Nuclear Physics European Collaboration Committee - an expert committee of the European Science Foundation) has – in its 'roadmap' for the future – recommended the construction of two new-generation complementary RIB facilities in Europe, one based on the in-flight fragmentation (IFF) method and the other on isotope-separation on-line (ISOL) method.

The IFF method will be extensively employed at the FAIR facility proposed at GSI (Darmstadt, Germany). In ISOL method, a thick target is bombarded with a primary light or heavy-ion beam, or with a secondary neutron beam in case of fission targets. The produced radioactive nuclei are post-accelerated to energies from keV to tens of MeV. The ISOL method provides high-intensity beams of high optical quality comparable to stable beams. The ultimate-generation RIB ISOL facility will be realised when the EURISOL facility is built. However, a full engineering design study and the necessary R&D programme imply that the expected beginning of operation of the EURISOL facility would be around 2020. Because of the time-line for EURISOL, NuPECC recommends the construction of intermediate-generation facilities that will benefit the EURISOL project in terms of R&D and give the community opportunities to perform research and applications with RIBs. Among the proposed intermediate facilities, SPIRAL2 meets the criteria of European dimension in terms of physics potential, site and the size of the investment.

In March 2005, the ESFRI (European Strategy Forum on Research Infrastructures) published the "List of Opportunities", which includes 23 Research Infrastructure projects corresponding to major needs of the European scientific community in the coming years. FAIR and SPIRAL 2 are among the selected projects and were recommended as Research Infrastructures necessary to maintain Europe's position at the cutting-edge of world research.

SPIRAL 2 will reinforce European leadership in the field of exotic nuclei and will serve a community of about 600 scientists. In this very competitive domain, Japan has invested more than 500M€ in its RIKEN-RIB Facility and the US are proposing to build a "Rare Isotope Accelerator". It is of the outmost importance that EU maintain its present leadership and prepare its future in the field, with SPIRAL2 at GANIL for the low-energy domain (few keV - 20 MeV/nucleon) and FAIR at GSI for the complementary high-energy side (100-1000 MeV/nucleon).

SPIRAL2 has also a remarkable potential for neutron-based research both for fundamental physics and various applications. In particular, in the neutron energy range from a few MeV to about 35 MeV this research facility would have a leading position for the next 10-15 years, if compared to other neutron facilities in operation or under construction worldwide.

ACKNOWLEDGMENTS

We would like to acknowledge all collaborators, physicists, engineers and technicians in France and numerous European and overseas countries actively participating in the elaboration of the physics case and the technical design of the SPIRAL 2 facility.

REFERENCES

1. http://www.ganil.fr

Accelerator Studies for an ADS within the European Project EUROTRANS

Alex C. Mueller

CNRS-IN2P3
Institut de Physique Nucléaire, 91406 Orsay, France

Abstract. Accelerator-Driven Systems (ADS) for transmutation of nuclear waste require multi-megawatt-power proton-accelerators of exceptional reliability. A study for such a machine has been recently performed as part of the FP5 EC contract PDS-XADS, "Preliminary Design Study of an Experimental ADS". A reference solution, based on a linear superconducting accelerator with its associated doubly achromatic beam line was elaborated. For high reliability, the design is intrinsically fault tolerant, relying on highly modular "derated" components associated to a fast digital feedback system. As next step on the proposed roadmap towards construction, an R&D phase was identified. It will be performed within he FP6 contract EUROTRANS of which it is one of the major work packages. The present article gives a short review on all these aspects

Keywords: Nuclear Waste Incineration, Accelerator Driven Systems, high power Accelerators, Reliability of Accelerators
PACS: 28.41.Kw, 28.50.Dr, 28.50.Ft, 29.17.+w, 29.25.Dz

INTRODUCTION

While nuclear power is advocated for its negligible contribution to global warming, it is also heavily debated because of the long-term environmental burden from the present-generation light-water reactors. Therefore, transmutation of their long-lived radioactive waste is of high interest, both in critical and sub-critical reactors. For dedicated transmutation systems, critical reactors loaded with fuel containing large amounts of minor actinides (Americium and Curium) pose safety problems caused by unfavorable reactivity coefficients and small delayed neutron fractions. A sub-critical system using externally provided additional neutrons is very attractive: it allows maximum transmutation rate while operating in a safe manner. Coupling a proton accelerator, a spallation target and a sub-critical core, the name **ADS**, for Accelerator Driven System, is used for such a reactor.

A few years ago, a **T**echnical **W**orking **G**roup (TWG) was created by the advisors to several European Research Ministers in order to report on ADS technology. The TWG concluded by presenting in 2001 a "European Roadmap" towards ADS technology [1], proposing the construction of an experimental facility, "XADS", for the 2015 horizon. Triggered by an initiative of the TWG members, the project PDS-XADS, **P**reliminary **D**esign **S**tudy of an e**X**perimental **ADS**, was funded by the European Commission in 2002 (Contract N° FIKW-CT-2001-00179). Performed by 25 participating organisations PDS-XADS [2,3] contained 5 Working Packages (WP), which studied 3 versions of an eXperimental ADS: a molten-metal (eutectic Pb-Bi) and a gas cooled ADS of 100 MW$_{th}$ class, and a smaller-scale ($\approx$50 MW$_{th}$) system based on the MYRRHA project of SCK Mol in Belgium. WP1, "Global coherence" assured the overall approach to the project, WP2 concerned the safety of these hybrid nuclear systems, WP3 elaborated the accelerator, WP4 the design of core and target, and WP5 the system integration. The PDS-XADS contract arrived at term in October 2004. Below we first give giving a short summary as far as the accelerator is concerned, and then present the required R&D, part of the FP6 project EUROTRANS that started at the time of writing (April 2005).

MAIN SPECIFICATIONS FOR THE XADS ACCELERATOR

The main specifications for the XADS accelerator are summarized in Table 1. Evidently, this machine belongs to the category of "HPPA" (high-power proton accelerators). HPPA are presently very actively studied (or even under construction) for a rather broad use in fundamental or applied science [4, 5]. Compared to other HPPA, many specifications are similar, but it is to be noted that the reliability requirement, i.e. the number of unwanted "beam-trips", is rather specific to ADS. The WP3 studies had to integrate this stringent requirement from the very beginning, since this issue could be a potential "show-stopper" for ADS technology in general.

TABLE 1. PDS-XADS Specifications for the proton beam characteristics

Proton beam parameters	Nominal Values
Max. beam intensity	6 mA CW on target (10 mA rated)
Proton energy	600 MeV (including 800 MeV upgrade study)
Beam entry	Vertically from above
Number of Beam Trips	Less than 5 per year (exceeding 1 second)
Beam Stability	Energy: $\pm$ 1 %, Intensity: $\pm$ 2 %, Size: $\pm$ 10 %
Beam footprint on target	Gas-cooled XADS: circular $\varnothing$ 160 mm LBE-cooled XADS: rectangular 10$\times$80 mm MYRRHA: circular, "donut" $\varnothing$ 72 mm
Intensity modulation	0.2 ms "interruptions" in CW beam for neutronics measurements, repetition frequency 0.01-1 Hz

Beam Energy

In the TWG report, an energy of about 1 GeV was selected in view of the physics of spallation reactions and the energy deposition in the target. Yet, since the mission of the (100 MW$_{th}$) XADS plant is a global demonstration of the operation and safety and not industrial waste transmutation, cost considerations favour a lower energy. In any case, a lower limit of about 600 MeV can be set in order to have a reasonable efficiency in neutron production and an affordable beam load on the target window. The XADS-accelerator was designed accordingly, but with the requirement that the concept should be upgradeable in energy and keeping in mind that the smaller-scale ADS only requires 350 MeV protons.

Beam Intensity and Time Structure

The beam intensity can be deduced from the multiplication factor of the sub-critical assembly and the thermal power of the ADS reactor. The PDS-XADS studies showed, that for a power level of P_{max} = 100 MW, and a core load with MOX type fuel and with a multiplication factor $k_0=k_{max}=0.98$, the proton beam current needed at 600 MeV is about 2 mA average. Considering a core multiplication factor of $k_0=0.90$ as an extreme lower limit, the current would rise up to about 10 mA. Higher currents had at least conceptually also be considered for the XADS accelerator in order to demonstrate also the feasibility of industrial operation.

Concerning the beam time structure, operation in CW (RF) mode was chosen because of its numerous advantages, e.g. lower peak power for the same average power, vanishing mechanical stress from Lorentz forces in the superconducting cavities, significantly lower R&D effort, and strongly reduced thermo-mechanical stress on the beam window, the target and the sub-critical assembly. However, a pulsed operation of the accelerator is possible for time scales that are shorter than the thermal inertia of the different components of the target and the reactor. Further, such pulsing enables on-line measurements of the sub-criticality level through dynamic neutron flux analysis [7], important in order to ensure a safe operation. Therefore, the XADS accelerator uses CW operation with the RF continuously applied on the RF structures, while the beam intensity can be modulated by macro-pulsing.

CHOICE OF THE BASIC ACCELERATOR CONCEPT

The reliability requirements are essentially related to the number of allowable beam trips, because frequently repeated, they can significantly damage the reactor structures, the target or the fuel and, also, decrease the plant availability. Therefore, beam trips in excess of 1 second duration should not occur more frequently than 5 per year.

Up to now, only sector-focused cyclotrons and linear accelerators (LINACs) are able to provide beam currents in the mA domain. The 600 MeV cyclotron of PSI [8] delivers about 2 mA on a routine basis. However, reaching up to 10 mA is more questionable, and might require two cyclotrons with the beams being funneled together. A (given) cyclotron also cannot be expanded in energy, so that boosting the energy from 600 to 800 MeV, as specified by WP1, would require the full replacement of the final and main stage, an absolutely not cost effective operation. The industrial transmuter needing about 1 GeV, the intrinsic limit of the very working principle of a cyclotron is reached, because the proton is becoming too relativistic. Even for a well optimized extraction system (losses well below 10^{-3}), the hot spot created by µA of lost protons is a serious radiation hazard in contrast to the quest for hands-on maintenance. Furthermore, the requirement to provide pulses for sub-criticality monitoring is a major difficulty for a cyclotron of such power. None of all these limitations are present in a LINAC which can reach 100 mA intensities without an intrinsic energy limit.

The chosen strategy to implement reliability relies on *over-design, redundancy and fault-tolerance* [9]. This approach requires a highly modular system where the individual components are operated substantially below their performance limit. In contrast to a cyclotron, a superconducting LINAC, with its many repetitive accelerating sections grouped in "cryomodules", conceptually meets this reliability strategy. The activation of the structures is rather low (large diameter opening for the beam), important for radioprotection and maintenance issues. For all these reasons, WP3 concluded that the cyclotron solution for an XADS presents a number of difficulties if not impossibilities: funneling, pulsing, beam trips, double-machine scheme, intrinsic current limitation, activation, energy upgrading that precludes this solution. Therefore, the reference solution discussed below is a superconducting LINAC [10]. Further, this assessment is corroborated by the one of OECD/NEA [11].

THE XADS REFERENCE ACCELERATOR

The proposed reference design for the XADS accelerator, optimized for reliability, is shown in Figure 1. It and the associated beam-line are discussed below.

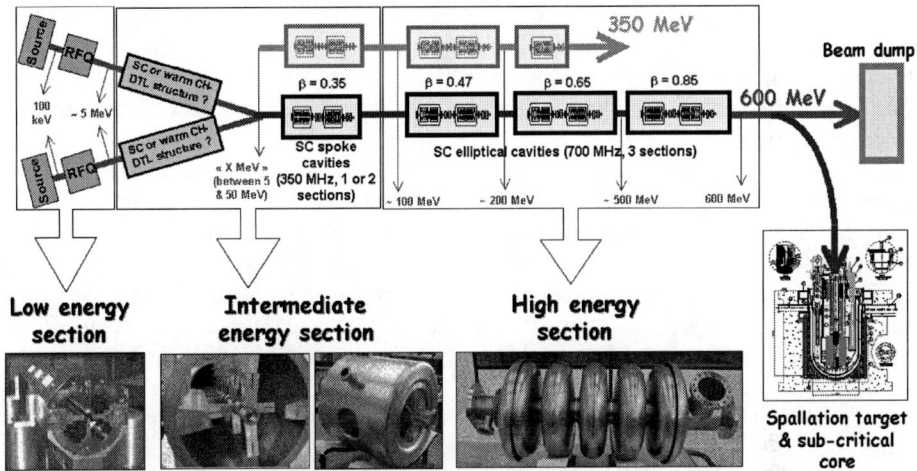

FIGURE 1. XADS reference accelerator layout: a doubled injector accelerator is followed by a fully modular spoke and elliptical cavity superconducting linac. Photos of typical cavity prototypes are shown in the lower part. From left to right: RFQ, CH structure, Spoke, Elliptical 5-cell.

The LINAC

For the injector, an ECR source with a normal conducting RFQ is used, followed by warm IH-DTL or/and superconducting CH-DTL structures up to a transition energy. Then a fully modular superconducting LINAC accelerates the beam up to the final energy. Below the transition energy, fault-tolerance is guaranteed by means of a "hot stand-by" spare. Above this energy, spoke and, from 100 MeV on, elliptical cavities are used. Beam dynamic calculations showed that an individual cavity failure can be handled without loss of the beam. Another remarkable feature of the concept is its validity for a very different output energy range: 9 cryomodules of $\beta=0.65$ cavities for 350 MeV; for 600 MeV, simply 10 more cryomodules have to be added (7 with β=0.65 and 3 with β=0.85) and 12 additional (β=0.85) boost the energy to 1 GeV. Therefore, already the small-scale XADS accelerator is fully demonstrative not only of the 600 MeV XADS (and could be converted to it), but even for an industrial machine. The performance of first prototypical cavities has been measured to exceed the specifications for the XADS by a very comfortable *over-design* safety margin [12].

The Beam Transport Line

The objective of the transport line is to safely inject the proton beam onto the spallation target with the specified footprint. A doubly achromatic module composed of two 45 degrees bending dipoles and three focusing quadrupoles has been designed. With scaled magnetic rigidity, the same layout can be used at 350 MeV or 1 GeV. Overall, such a double-achromat is non-dispersive. Thus, the beam position at the target is independent on energy variations, and the beam size independent on the energy spread. Thus, spread and central energy fluctuations from the accelerator will have no effect at the target. The system is however dispersive in the region situated between the two dipoles. Position and size monitors located in this region will be able to provide information on proton energy variations, and to trigger a feedback system. The footprint is obtained by raster scanning. The pencil-like beam is deflected by fast steering magnets operated at frequencies of 50 to a few hundreds Hertz, and acting in the two transverse directions. Various shapes (rectangular, circular) and various particle distributions (uniform, parabolic...) are achievable. Four raster magnets will be operated synchronously and independently so that the beam will be always moved at the target if one magnet fails. Redundant fault detection circuits will monitor the magnet current and the magnetic field to ensure proper operation and to shut down the beam in case of necessity. Similar systems are used in cancer therapy where they meet the stringent requirements for medical use.

Maintenance strategy and radiation protection

Proper maintenance has to guarantee that the *over-design* margin does not deteriorate and that equipment redundancy is regularly restored if partial failures have occurred. The supervising control system regularly must undergo a complete performance check to ensure that it can replace components-at-fault by readjusting the operational parameters of the overall machine. This requests the development of an expert system, able, while the accelerator is running and delivering nominal beam, to precisely identify and locate equipment that has started to loose rated performance, and/or that is out-of-order and to be replaced or repaired. This system provides the database for the scheduled maintenance periods of repair and/or replacement of deteriorated or faulty equipment. This is also in-line with the ALARA (As Low As Reasonably Achievable) principle for the concerned personnel. Indeed, many conditions for the ALARA principle, like enough working place and quick disconnection of sub equipments are actually the very much the same that are asked by an optimization of the reliability. The shielding calculations for the XADS accelerator had to be in line with the radiation protection philosophy [13] adopted by the

European decree Euratom/96/29. Thus, the goal of the shielding design was to guarantee that, under normal operational conditions, the added integrated dose to personnel will be extremely low, i.e. smaller than the natural background of 1 mS/y. To obtain this goal, the calculations were made using conservative (= pessimistic) assumptions, and assuming an "occupancy factor" of 1, 2000 hours-per-year presence just behind the shield wall where maximum dose rates exist. The design dose rate was 0.5 µSv/h. Some more details are given in [14] and refs quoted therein.

R&D PROGRAMME AND ROADMAP

The broad field of applications covered by HPPA accelerators is at the origin of remarkable R&D effort presently underway world-wide. The study, design and testing of the main components of these new generation linear accelerators have contributed to a good synergy by developing complementary activities between many laboratories. The XADS accelerator can profit from this general background and even built on it quite directly. However, a dedicated R&D programme is needed for the requirement of an extremely low number of beam trips. In this spirit, the participants to WP3 have elaborated such a program, focused on reliability and fault tolerance design. It is included the 4 year 6th FP project EUROTRANS, that started in April 2005.

Injector and Intermediate Energy Section

Concerning the injector section, a thorough campaign to test the reliability of every component of the injector, operated over a long period of time (e.g. a continuous run of many weeks) will be performed "full-scale" with IPHI. Presently under construction in France [15], IPHI, consisting of an already operational ECR source and a 6 m long RFQ, will deliver, in 2006, its 3 MeV high intensity beams (10 – 100 mA). Some basic R&D is required for the subsequent sections, up to 100 MeV, in order to assess a solution simultaneously reliable and economical. While superconducting components should in principle be deployed from the lowest possible energies on, room-temperature-structures have nevertheless to be studied and prototyped: the transition energy to the superconducting structures might be higher than the RFQ output. Secondly, while room-temperature structures have large RF losses, their development risk is low (well established technology). The superconducting resonators considered here are short and modular in view of the reliability strategy. First SC cavity prototypes are presently successfully tested [16]. It is therefore important to push these developments for spoke- and CH- structures, by adding helium tanks and power couplers. The final aim of all these developments is to asses the best technical option for the intermediate section of the XADS accelerator based on established demonstrated performance. It might well be a combination of several technologies.

High-Energy Section and RF System

While the R&D on 704 MHz superconducting elliptical cavities is well advanced in Europe [12], the demonstration of the full technology is not yet accomplished. Besides the development of the bare superconducting cavity, it is important to prototype each auxiliary system needed for the cavity operation in a real environment (power coupler, RF source, power supply, RF control system, cryogenic system, cryostat…). The construction of a full-scale module with a given beta value (e.g. $\beta = 0.5$) can be considered as a rather general proof-of-principle of the technology, since the higher beta modules are very similar. Moreover, tests with RF at nominal power (although without beam), could be done and used for specific studies, like the RF control system procedure in case of a cavity failure. Indeed, the reaction speed for retuning the whole accelerator, to nominal beam conditions must be less than 1 second for fault-tolerance. Digital techniques are necessary to meet the speed and software configuration requirements.

Possible Roadmap

The roadmap for the accelerator is given in figure 2. Consecutive to the advanced technology R&D phase, the construction of the accelerator of the XADS facility typically may take 7 years. This estimate was based on industrial studies and experience gained from the construction of similar facilities like ESRF, SOLEIL and SNS. Concerning the cost, prior to routine operation, about 300 M€ have been estimated for the 600 MeV machine, including in-house and external man-power as well as infrastructure investments of 30 M€. Considerable savings are possible if the energy is lower, and the injector not doubled. As concerning operation costs, the concept of using superconducting technology from a very low energy range on is most cost effective. It is hoped that the R&D program within EUROTRANS will not only provide answers related to the accelerator performance in the critical area of reliability/availability, but may even allow to relax on some of the presently rather conservative specifications, with positive impact on the price.

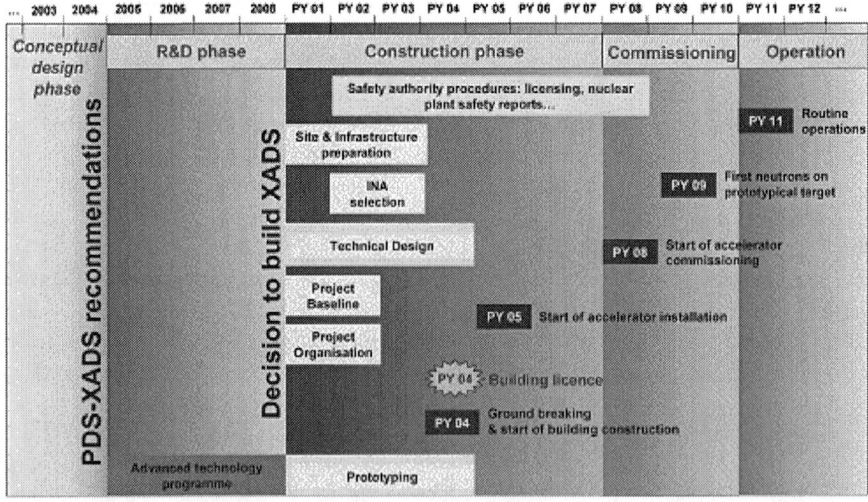

Figure 2. The Roadmap for the Accelerator for an experimental ADS demonstrator

CONCLUDING REMARKS

Within the 3 year contract PDS-XADS, a generic and robust technical solution for the XADS accelerator has been developed. A superconducting LINAC, with its associated doubly-achromatic beam line, has been found to constitute an optimal technical solution. It is also representative for an industrial machine. It is reliable through the rigorous implementation of a highly modular system with de-rated components operated in a fault-tolerant way. The well-controlled beam dynamics behaviour also allows a straightforward concept for the radioprotection. The required R&D is identified, and a technological validation programme, focused on reliability will be performed within the FP6 project EUROTRANS.

ACKNOWLEDGMENTS

The author, coordinator of WP3 of PDS-XADS, acknowledges the work of all his colleagues of the WP3 collaboration on which this short summary paper is based.

REFERENCES

1. Rubbia, C. et al., "A European Roadmap for Developing Accelerator Driven Systems (ADS) for Nuclear Waste Incineration", ENEA, Rome (2001), ISBN 88-8286-008-6.
2. Abderrahim, H.A., Carluec, B. and Mueller, A.C. "The European Programme", in ref. [4], 3rd meeting
3. Giraud, B. "Principle Concepts of the XADS Designs in the 5th FP", in ref. [6]
4. *OECD Nuclear Energy Agency, International Workshops on Utilization and Reliability of High Power Proton Accelerators (HPPA)*:
   1st Meeting, October 13-15, 1998, Mito, Japan, ISBN 92-64-17068-5
   2nd Meeting, November, 22-24, 1999, Aix-en-Provence, France, ISBN 92-64-18749
   3rd Meeting, May, 12-16, 2002, Santa Fe, NM, USA, ISBN 92-64-10211-6
   4th Meeting, May,16-19, 2004, Daejon, Republic of Korea, in press.
5. *International Workshop on Physics with a multi-MW proton source*, CERN, Geneva, 25-27 May, 2004, http://physicsatmwatt.web.cern.ch/physicsatmwatt
6. Proceedings of ICRS 10/RPS2004, May 2004, Funchal, Madeira Island, Portugal, http://www.itn.mces.pt/ICRS-RPS and Radiation Protection Dosimetry (2005) in press.
7. Rimpault, C. et al. "Core instrumentation and reactivity Control of the experimental ADS", in *International Workshop on P&T and ADS Development*, SCK Mol, Belgium (2003), http://www.sckcen.be/sckcen_en/activities/conf/conferences/20031006/cd/
8. Stammbach, Th. et al., *Nucl. Inst. Meth.* **B113** 1 (1996).
9. Pierini, P. "ADS Reliability Activities in Europe, in ref. [5], 4th meeting.
10. Safa, H. "Design of a XADS linear accelerator", in proceedings quoted ref. [7].
11. "Summary of Working Group Discussion on Accelerators", in ref. [5], 3rd Meeting.
12. Junquera, T., "Status and Perspectives of the R&D programs for the XADS accelerator", in proceedings quoted in ref. [9]
13. Recommendations of the International Commission on Radiological Protection, ICRP Publication 60. Annals of the ICRP 21(1-3) Pergamon Press, Oxford, 1991
14. Mueller, A.C., "The PDS-XADS Reference Accelerator and its radioprotection issues", in ref. [6]
15. Beauvais, P.Y., "Recent Evolutions in the Design of the French High Intensity Proton Injector (IPHI)", *European Conf. on Particle Accelerators*, EPAC 2004, Luzern, Switzerland, http://accelconf.web.cern.ch/AccelConf/e04/PAPERS/TUPLT053.PDF
16. Olry, G. et al., see proceedings quoted in ref. [15] ,/TUPKF019.PDF

** POSTERS **

Major Actinide diffusion (U, Pu) in oxidised zirconium Application to nuclear fuel cladding tubes

N. Bérerd [1], Y. Pipon [1], N. Moncoffre [1], A. Chevarier [1], H. Faust [2], H. Catalette [3].

[1] Institut de Physique Nucléaire de Lyon, IN2P3/CNRS, Université Claude Bernard, F-69622, Villeurbanne, France.

[2] Institut Laue Langevin, B.P. 156, F-38042 Grenoble Cedex 9, France.

[3] EDF Les Renardières, Route de Sens, Ecuelle, F-77818 Moret sur Loingt, France.

Abstract: Nuclear fuel devices of Pressurised Water Reactors are composed of uranium oxide pellets which are enclosed in zircaloy cylinders. During reactor operation, major actinides are created among which plutonium 239. Further on, energy deposition of fission recoils leads to sputtering of uranium and plutonium onto the inner surface of the cladding material. Thus, sputtered Pu and U ions start to migrate outwards. Experiments to investigate the migration process were performed at the Institut Laue Langevin (Grenoble) on the Lohengrin mass spectrometer. A model is proposed to deduce an apparent uranium and plutonium diffusion coefficient in zirconia (ZrO_2) from the energy distribution broadening of a selected fission product.

Keywords: actinide, uranium, plutonium, diffusion, irradiation.
PACS: 61.80 Lj, 61.82 Ms, 66.30.-h.

1. INTRODUCTION

During reactor operation, the fuel cladding inner surface is oxidised and contaminated with fission products and actinides [1]. Actinide contamination is due to successive recoil effects induced by fission products and alpha emission. The main actinide contaminants are uranium and plutonium. Recently, we have studied the migration behaviour of uranium in ZrO_2 under fission product irradiation [2]. In the present paper, we have enlarged this study to the plutonium diffusion in zirconia and the plutonium apparent diffusion coefficient thus obtained is compared to the uranium one.

2. EXPERIMENTAL SET-UP

The experiment was performed using an in-pile position at the Institut Laue Langevin (ILL) in Grenoble. The ILL is equipped with a high neutron flux nuclear reactor ($\Phi = 5 \times 10^{14}$ n cm^{-2} s^{-1}). The experiments have been performed on the H9 beam line, on which the Lohengrin mass spectrometer is located. The residual pressure in the beam tube is close to 5×10^{-3} Pa. The line is composed of three main parts: the target, the Lohengrin spectrometer and an ionisation chamber as detector. The target is made of a titanium support covered with a thin platinum deposit on which a 162 µg cm^{-2}

plutonium oxide layer enriched up to 98% with ^{239}Pu has been deposited. A 3.1 µm thick x 3.5 cm^2 zirconia foil (obtained from the irradiation enhanced oxidation of a 2 µm thick zirconium foil as described in [3]) is in contact with the ^{239}PuO$_2$. This device is positioned in the ILL thermal neutron flux. The corresponding fission fragment (FF) flux generated by the ^{239}PuO$_2$ film is estimated to be 1.5x10^{11} particles cm^{-2} s^{-1}. Plutonium diffusion in the zirconia target will occur under neutron and FF irradiation.

The principle of our experiment is the measurement of the energy distribution evolution of a selected FF using the Lohengrin spectrometer. It is schematized in figure 1.

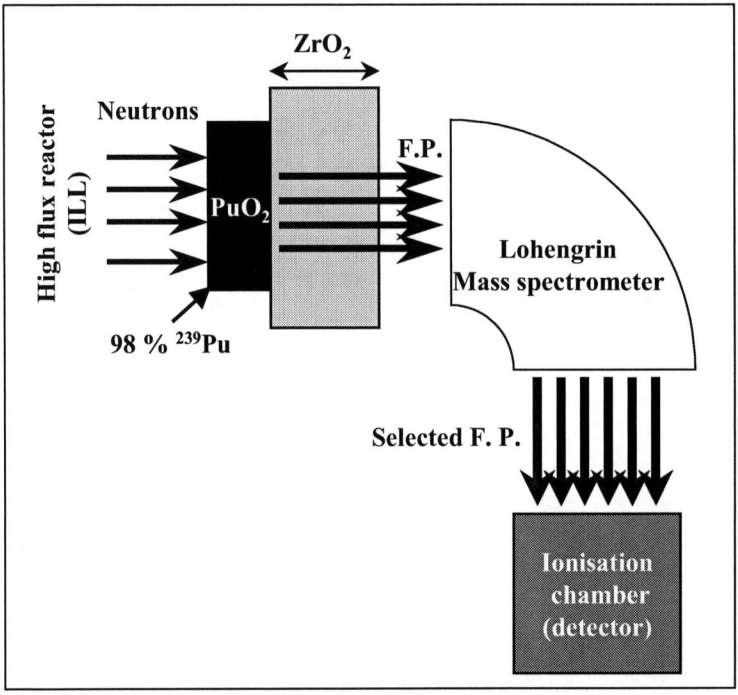

FIGURE 1: Schematic representation of the ILL experiment.

High energy FF are emitted from the ^{239}Pu thermal fission and pass through zirconia before being detected. Their mass and kinetic energy are analysed in three successive magnetic and electric fields. At the exit point, all FF having the same M/q ratio (M is the FF mass and q is its ionic charge state) and the same velocity are selected. The separated fragments are identified using a high resolution ionisation chamber [4]. We have chosen to detect the mass 89 with the most probable charge state (q=20). At regular time intervals, the energy spectra of the chosen fission fragments are measured. All the spectra are normalised to a relative intensity equal to 1 at the spectra maximum.

The target temperature was measured during irradiation using a specific infra-red pyrometer to be equal to 540°C.

The major drawback of this experiment is that it is impossible to collect the sample after irradiation, because of the very high radioactivity level. However, it has been shown in a GANIL experiment [5] that around 500°C at the air pressure of 5×10^{-3} Pa and under 50 MeV Xe irradiation, the zirconium oxidation is accelerated and the obtained zirconia structure is a mixture of tetragonal phase (50%) and monoclinic phase (50%). This composition is stable for xenon fluences close to the FP fluence used in this experiment. Therefore, in this study, we consider that the zirconia sample structure corresponds to a stable ZrO_2 structure.

3. RESULTS

Figure 2 displays the evolution of the FF kinetic energy distribution for two diffusion times (t=0, t=1000 h) and at 540°C.

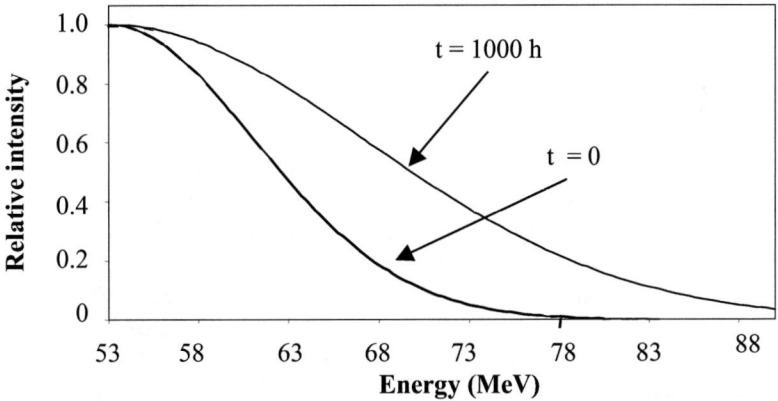

FIGURE 2: Evolution of the kinetic energy distribution (M = 89) as a function of time.

Each distribution is fitted with gaussian-like curves. As previously observed with uranium [2], the kinetic energy distributions broaden while the mean kinetic energy value remains constant (53 MeV for M=89). Only the right part of the distribution, which is correlated to plutonium diffusion in zirconia, is considered since the low energy part corresponds to diffusion in the platinum deposit of the target support.

In order to deduce an apparent diffusion coefficient of plutonium in zirconia under FF irradiation (called D*), we have simulated the evolution of the energy spectra as a function of time using the model developed for the uranium diffusion study [2]. This model is based on the second Fick's law, and in our conditions, the analytical solution is:

$$C(x, t) = C_S \, \mathrm{erfc}\left(\frac{x}{2\sqrt{D^* t}}\right) \quad (3)$$

where t is the irradiation time, x the diffusion depth, C(x,t) the plutonium concentration and C_S the plutonium concentration at x = 0. The model is described in detail elsewhere [2]. The result of the simulation is presented in figure 3.

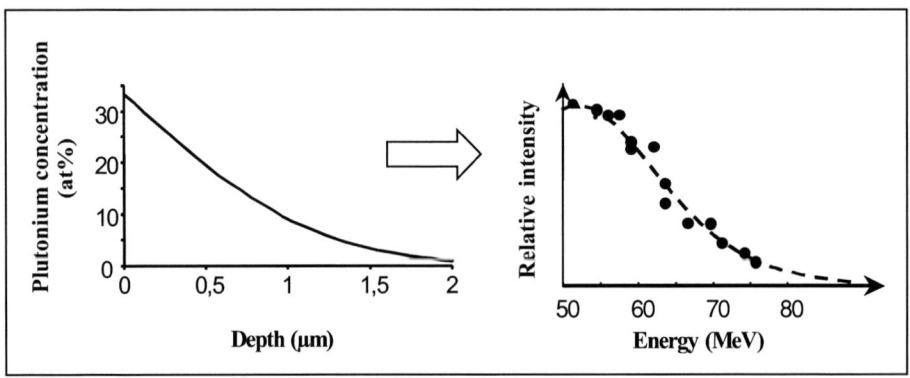

FIGURE 3: (a) Concentration profile of plutonium in zirconia at t = 1000 h, and (b) comparison of similated profile (dot line) and experimental data (dots)

The apparent plutonium diffusion coefficient D* in ZrO_2 deduced from this model at 540°C and for a flux of 1.5×10^{11} FF cm^{-2} s^{-1} has been determined to be equal to $(2 \pm 1) \times 10^{-15} cm^2 s^{-1}$. The error bar is determined from the diffusion model fit.
This value has to be compared with that obtained for U diffusion under FF irradiation in ZrO_2 at 480°C. It was found to be to equal to 10^{-15} cm^2 s^{-1}. Consequently, U and Pu apparent diffusion coefficients are very close and we can assume that D* is temperature independent. The similar behaviour of these actinides in ZrO_2 can be explained by the dominant influence of defect evolution induced by bombardment with FF [6].

REFERENCES

[1] T. Hirabayashi, T. Sato, C. Sagawa, N. M. Masaki, M. Saeki, T. Adachi, J. of Nucl. Mat., 174 (1990) 45.
[2] N. Bérerd, A. Chevarier, N. Moncoffre, Ph. Sainsot, H. Faust, H. Catalette, accepted to publication in Nucl. Instr. and Meth. in Phys. Res. B, lycen 2005-15 (2005).
[3] N. Bérerd, H. Catalette, A. Chevarier, N. Chevarier, H. Faust, N. Moncoffre, Surface and Coating Technology, 158-159 (2002) 473.
[4] J. P. Bocquet, R. Brissot, H. Faust, Nucl. Instr. and Meth. in Phys. Res. B A267 (1988) 466.
[5] N. Bérerd, A. Chevarier, N. Moncoffre, H. Jaffrézic, E. Balanzat, H. Catalette, Accepted for publication in J. of Appl. Physics, Lycen 2004-24 (2004).
[6] Hj. Matzke, Rad. Effects, 64 (1982) 3.

Binary fission-fragment yields from the reaction ^{251}Cf(n$_{th}$, f)

E. Birgersson[*†], S. Oberstedt[*], A. Oberstedt[†], F.-J. Hambsch[*], D. Rochman[¶], I. Tsekhanovitsch[‡]

[*] EC-JRC IRMM, B-2440 Geel, Belgium,
[†] Dept. of Natural Sciences, Örebro University, SE-70182 Örebro, Sweden
[¶] Brookhaven National Laboratory, National Nuclear Data Center (NNDC), USA
[‡] ILL, F-38042 Grenoble Cedex, France

Abstract. The recoil mass spectrometer LOHENGRIN of the Laue-Langevin Institute, Grenoble has been used to measure the light fission-fragment mass yield and kinetic energy distributions from neutron-induced ^{252}Cf*, using ^{251}Cf as target material.

Keywords: Californium-251, neutron-induced fission, fission fragment spectroscopy.
PACS: 25.85.Ec, 28.20.-v, 29.30.Ep, 24.75.+i

INTRODUCTION

The interpretation of fission-fragment properties in terms of so-called fission modes has been successfully applied in the actinide region to describe mass yield and total kinetic energy distributions as a function of incident neutron energy. From investigating fission-fragment characteristics as a function of excitation energy the mass-asymmetric standard I (S1) and standard II (S2) modes as well as the mass-symmetric super-long (SL) mode have been used to consistently describe the mass- and total kinetic energy (TKE) distributions [1-4]. In case of spontaneous fission of ^{252}Cf, which is considered as standard reaction in nuclear fission, the number of traditional fission modes is not sufficient to properly describe the experimentally obtained fission-fragment distributions. Additional theoretically obtained fission modes have to be included into the analysis to improve the description of the data. In order to achieve experimental confirmation of the number of fission modes present in ^{252}Cf the fission-fragment properties and mean kinetic energies were measured for the reaction ^{251}Cf(n$_{th}$, f) at thermal excitation.

EXPERIMENT

The experiment was performed at the recoil mass spectrometer LOHENGRIN of the Institute Laue-Langevin. The target consisted of Cf$_2$O$_3$ deposited on a titanium backing. The active spot had a diameter of 4 mm and a thickness of 87.7 μg/cm^2. The initial isotopic composition of the californium batch was the following: ^{249}Cf (18 %),

^{250}Cf (35 %), ^{251}Cf (46 %) and ^{252}Cf (1 %). In order to avoid loss of material due to sputtering, when the target is heated in the high thermal neutron flux of $5 \cdot 10^{14}$ neutrons/cm^2/s, the target was covered by a 0.25 µm thick nickel foil. The nickel foil is supported by an acrylic layer, which is supposed to evaporate before the measurements starts.

In LOHENGRIN the separation of fission-fragments is done according to their A/q and E/q ratios, where A, E and q are the mass number, kinetic energy and ionic charge state of the fission fragments, respectively. The fission fragments are detected after separation with an ionisation chamber, which has a segmented anode to serve as a ∆E-E telescope.

MEASUREMENTS AND DATA ANALYSIS

The emission yield Y(A) of a fission fragment of mass A is determined according to:

$$Y(q,E) = \frac{Y(A)}{\sqrt{2\pi}\sigma_E} e^{-\frac{(E-<E>)^2}{2\sigma_E^2}} \frac{1}{\sqrt{2\pi}\sigma_q} e^{-\frac{(q-<q(E)>)^2}{2\sigma_q^2}}, \quad (1)$$

where <q(E)> is the mean ionic charge as a function of energy and calculated according to Ref. [5] but with parameters refitted using the experimental results. For each mass an energy distribution at the mean ionic charge was measured and at the mean kinetic energy an ionic charge distribution was measured. However several corrections have to be made to the raw data before yield and kinetic energy may be determined.

Due to the high neutron-induced reaction cross-sections for ^{251}Cf and the extremely high neutron flux the decrease of target material, the so-called burn-up, has to be taken into account and monitored. This was done by measuring mass A = 100 at ionic charge q = 22 and kinetic energies E_k = 80 to 115 MeV in steps of 5 MeV several times per day. The resulting intensity and mean kinetic energies for the different burn-up measurements as a function of time is shown in Fig. 1. The burn-up is described in the usual way by a two-exponential function, where the fast component accounts for material losses during initial heating of the target. As it may be depicted from Fig. 1 the burn-up data show quite some structure, which has to be attributed to drifts in the electric field between the condenser plates of LOHENGRIN. However, based on only those burn-up runs performed directly after each day formation, performed on the electronic condenser plates, proper corrections to these instabilities could be established. Eventually, all burn-up measurements were used to correct the raw data for the electronic instabilities.

The increase in mean kinetic energy, see right part of Fig. 1, is described by a two exponential, where one of the time constants is given the same value as in another experiment performed at LOHENGRIN on ^{245}Cm which had a similar fission rate [6]. The initial strong increase of the mean kinetic energy is due to the acrylic layer on the nickel foil, which is supposed to evaporate before starting the experiment. Since this foil was mounted upside down at ILL, the acrylic layer evaporated much slower. The function describing the increase in kinetic energy as shown in Fig. 1 was used to estimate the kinetic energy at t = 0, when the target properties were known. Energy

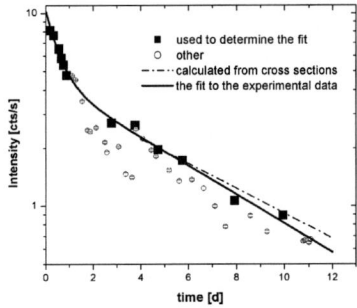 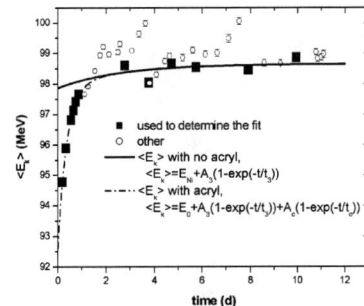

FIGURE 1. **Left:** Burn-up measurements performed to monitor the decrease of fissile material. The fit is based on the squares. The open circles and the squares are used to monitor electronic drifts. **Right:** The increase in mean kinetic energy described by a two-exponential is based on the squares. The open circles and squares are used to monitor the electronic drifts. Also shown is the one-exponential with time dependence according to Ref. [6].

losses in the target and the Ni-foil were then calculated using SRIM2003 [7].

The composition of the target changes during the experiment. At the end of the experiment 25% of the fission-fragments come from ^{249}Cf. The other isotopes contribute to less then 10% and are not taken into account. Therefore a correction to the burn-up function was made since it was measured at A = 100 and the relative contributions from ^{249}Cf and ^{251}Cf are not the same. Once these corrections to the obtained data have been performed the extraction of the contribution from the ^{249}Cf can be calculated. The relative mass yield of ^{251}Cf is given by

$$y_1(A) = \frac{Y(A)}{f_1(t(A))} \cdot \frac{1+\sum_{A'=80}^{124}\frac{f_9(t(A'))}{f_1(t(A'))}y_9(A')}{\sum_{A'=80}^{124}\frac{Y(A')}{f_1(t(A'))}} - \frac{f_9(t(A))}{f_1(t(A))}y_9(A), \qquad (2)$$

where $f_1(t(A))$ and $f_9(t(A))$ are describing the fission fragments coming from ^{251}Cf and ^{249}Cf, respectively, and $f_1(t(A))+f_9(t(A)) = 1$. The $y_1(A)$ and $y_9(A)$ are the relative mass yield for ^{251}Cf and ^{249}Cf, respectively. Y(A) is the burn-up corrected yield.

RESULTS AND DISCUSSION

The obtained post-neutron fission-fragment distribution of ^{252}Cf* at thermal excitation is shown (square symbol) in Fig. 2 together with an evaluation from Ref. [8] and data from the spontaneous fission of ^{252}Cf [9]. The data from the fission at thermal excitation are in good agreement with the evaluation. The distribution appears to be broader than in the case of spontaneous fission. For a more realistic comparison the LOHENGRIN data have been 5-point smoothed to mimic the resolution of the SF data measured with the 2E-technique. The emission yields around A = 115 is enhanced and diminished around A = 105. In terms of fission modes the more compact standard I mode seems to be enhanced compared to the more deformed mode standard II. This should lead to an increase of the mean kinetic energy for the mass region above A =

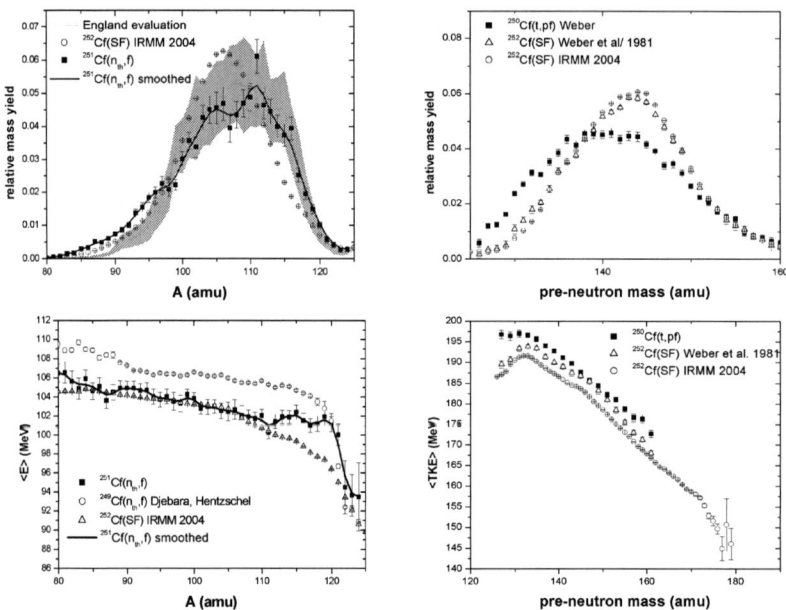

FIGURE 2. Top-left: Light fission-fragments mass distribution from thermal neutron induced fission of ^{252}Cf* (squares) compared to an evaluation from Ref. [8], where the grey-shaded area indicates lower and upper limits. Circles are data from the spontaneous fission of ^{252}Cf [9]. **Top-Right:** Heavy pre-neutron fission-fragment mass distribution from the fission of ^{252}Cf* (squares, [10]) following the reaction ^{250}Cf(t, p) together with data from the spontaneous fission of ^{252}Cf (open symbols, [9,10]). **Bottom-left:** Mean kinetic energy as a function of the fission-fragment mass A: ^{251}Cf(n_{th}, f) (squares, this work), ^{249}Cf(n_{th}, f) (circle [11,12,13]) and ^{252}Cf(SF) (triangles, [9]). **Bottom-Right:** Total kinetic energy (TKE) as a function of the fission-fragment mass prior to prompt neutron emission [9, 10].

110, which is indeed the case (Fig. 2, lower part). This observation confirms findings in Ref. [10] using the reaction ^{250}Cf(t, p) at a slightly higher mean excitation energy.

REFERENCES

1. P. Siegler, F.-J. Hambsch, S. Oberstedt, J. P. Theobald, Nucl. Phys. **A594** 45 (1995)
2. S. Oberstedt, F.-J. Hambsch, F. Vives, Nucl. Phys. **A644** 289 (1998)
3. F. Vivès, F.-J. Hambsch, H. Bax, S. Oberstedt, Nucl. Phys. **A662** 63 (2000)
4. F.-J. Hambsch, F. Vivès, P. Siegler, S. Oberstedt, Nucl. Phys. **A679** 3 (2000)
5. V. S. Nikolaev, and Dmitriev, *Physics Letters* **28A**, 277-278 (1968).
6. T. Friedrichs, *PhD thesis,*. Braunschweig: unpublished (1998).
7. Computer code SRIM 2003.26, available from J.F. Ziegler, IBM Research, Yorktown, NY-10598, USA and J. P. Biersack, Hahn-Meitner Institut, 1 Berlin 39 (2003)
8. T. R. England, and B. F. Rider, *LA-UR-94-3106,ENDF-349* (1994).
9. F. -J. Hambsch, *Private communication* (2004).
10. J. Weber, H. C. Britt, and J. B. Wilhelmy, *Phys. Rev.* **C23**, 2100 (1981).
11. M. Djebara et al., *Nucl. Phys* **A496,** 346 (1989).
12. R. Hentzschel, *Nucl. Phys* **A571,** 427 (1994).
13. R. Hentzschel, *PhD thesis*. Mainz University: unpublished (1992).

The Neutron Induced Fission Cross-Section of ^{240}Pu, ^{243}Am and ^{nat}W in the Energy Range 1 - 200 MeV

A.B. Laptev[1,*], A.Yu. Donets[2], A.V. Fomichev[2], A.A. Fomichev[3], R.C. Haight[4], O.A. Shcherbakov[1], S.M. Soloviev[2], Yu.V. Tuboltsev[5], A.S. Vorobyev[1]

[1] *Petersburg Nuclear Physics Institute, Gatchina, 188300, Russia*
[2] *V.G. Khlopin Radium Institute, St. Petersburg, 194021, Russia*
[3] *Saint Petersburg State University, St. Petersburg, 198504, Russia*
[4] *Los Alamos National Laboratory, Los Alamos, NM, 87545, USA*
[5] *A.F. Ioffe Physico-Technical Institute, St. Petersburg, 194021, Russia*

Abstract. The neutron-induced fission cross-section ratios relative to ^{235}U has been measured for ^{240}Pu, ^{243}Am and ^{nat}W in a wide energy range of incident neutrons from 1 MeV to 200 MeV at the GNEIS facility. The measurements were performed using the multiplate ionization chambers and time-of-flight techniques. The results obtained in this measurement are presented in comparison with the other data.

Keywords: Fission; Cross-section; Measurement; Plutonium-240; Americium-243; Tungsten; Fast neutrons; Time-of-flight method.
PACS: 25.85.Ec; 29.30.Hs; 27.90.+b; 27.70.+q

INTRODUCTION

There is a long-standing need for information about fission reactions of heavy nuclei induced by particles at intermediate energies required for such applications as accelerator-driven transmutation of nuclear waste, energy generation, fundamental physics, *etc* [1]. Recently the fission cross-sections of ^{233}U, ^{238}U, ^{232}Th, ^{237}Np, ^{239}Pu, ^{nat}Pb and ^{209}Bi have been measured at the GNEIS facility [2] in the energy range 1-200 MeV [3]. As the next step of this research the nuclei of ^{240}Pu, ^{243}Am and ^{nat}W have been chosen for investigation. Existing data sets are very incomplete for these nuclides. For ^{240}Pu above 20 MeV, there is only the single data set obtained at the "white" neutron source of the WNR/LANSCE facility [4]. There are no available data for ^{243}Am above 40 MeV. In case of ^{nat}W the data consist of one point obtained by Goldanskiy et al. [5] and several energy points obtained at the cyclotron of Uppsala University [6]. The new fission cross-sections of ^{240}Pu, ^{243}Am and ^{nat}W in the energy range 1 – 200 MeV are presented in this paper. The preliminary results of this measurement have been published earlier [7].

---

[*] Present address: *Japan Nuclear Cycle Development Institute, Tokai-mura, Ibaraki 319-1194, Japan,* E-mail: *laptev@jnc.go.jp*

EXPERIMENTAL PROCEDURE

The fission cross-section ratios for ^{240}Pu, ^{243}Am and ^{nat}W relative to ^{235}U have been measured using the neutron time-of-flight spectrometer GNEIS. This facility is based on the 1 GeV proton synchrocyclotron of PNPI and has average intensity $3·10^{14}$ n/s and a burst duration 10 ns. The flight path was 48.5 m. A schematic layout of the experimental set-up is shown in Fig. 1.

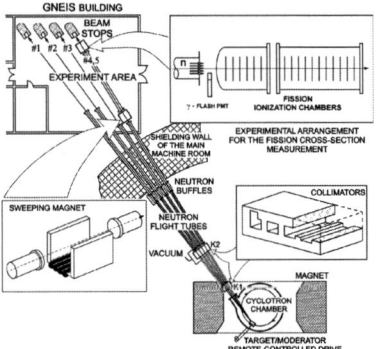

The fission reaction rate was measured using two fast ionization chambers. Both chambers (actinides and ^{nat}W) have several sections containing one pair of cathode and anode plates spaced by 5 mm. The painting technique has been used for targets production. The data acquisition system was based on a 100-MHz flash-ADCs. The start signal was provided by gamma flash-detector, a bare PMT placed in the neutron beam.

FIGURE 1. Schematic layout of the GNEIS facility and experimental set-up for fission cross-section measurements.

DATA PROCESSING

Identification of start (gamma flash) and stop (fission) signals on the background has been made by a method of digital filtering [8]. The raw data reduction included composing pulse height and TOF information into a 2-dimensional matrix. The example of this matrix in case of ^{240}Pu is shown in Fig. 2. The total pulse height spectrum in case of ^{nat}W and its fission and background components are shown in Fig. 3. The calibration of energy scale was made using the positions of lead resonances and the weak gamma-flash peak observed in the TOF-spectra (v. Fig. 4).

In the data reduction, the fission event counting rates were corrected for (1) background events, (2) the neutron flux attenuation for investigated and reference

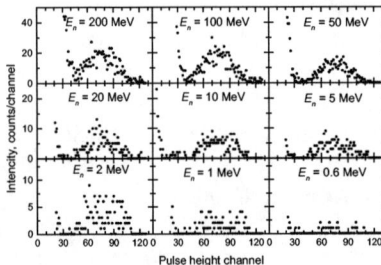

FIGURE 2. Pulse height spectra of ^{240}Pu measured at different neutron energies.

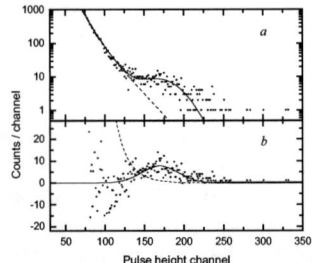

FIGURE 3. (*a*) Measured pulse height spectrum for one foil of ^{nat}W; the solid curve denotes fission fragments and the dash curve shows background events; (*b*) pulse height spectrum for fission fragments after background subtraction.

nuclei, (3) fragment losses in the targets, and (4) neutron momentum transfer and angular anisotropy of fission fragments. The attenuation correction was calculated using the neutron total cross-sections from ENDF/B-VI. A correction for the energy dependent linear-momentum and angular-momentum effects were calculated following a method of G. Carlson [9]. The fission cross-section ratios for investigated nuclei and the reference nucleus ^{235}U obtained in the "shape" measure-

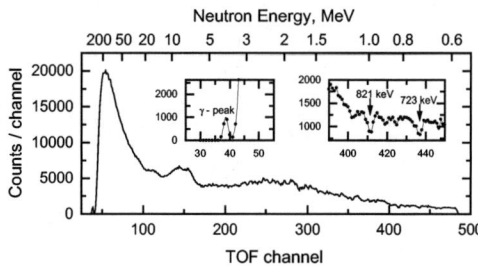

FIGURE 4. Time-of-flight spectra (10 ns channel width) obtained for ^{235}U target after background subtraction. Inserts show TOF-spectra in regions of gamma-flash and lead resonances in more detail.

ments have been normalized using the target thicknesses. An accuracy of this normalization is 5%. Finally, the normalized fission cross-section ratios have been converted to the cross-sections using the recommended data [10,11].

RESULTS AND CONCLUSIONS

Results of the present measurements for ^{240}Pu and ^{243}Am are shown in Figs. 5 and 6 and compared with other experimental and theoretical data. The error bars of our data represent the statistical errors only and these uncertainties are about 2%.

For ^{240}Pu (Fig. 5), there is good enough agreement between the present data and those of Staples and Morley [4]. Some small differences are within stated errors. On our opinion, most of the differences are in normalization rather than shape. Such differences vanish after normalization of both data sets at neutron energy 14 MeV. Smooth curves present cross section values from ENDF B-VI and JENDL-3.3 libraries, calculations of Maslov et al. [12] and systematics of Fukahori et al. [13]. All these evaluations show a rough agreement with our data for ^{240}Pu.

For ^{243}Am (Fig. 6), a comparison of the present data with other data sets shows good agreement of the present data with that of Behrens et al. and Goverdovskiy et al. There are no available previous data for the fission cross section of ^{243}Am above ~40 MeV. In case of ^{243}Am there are significant disagreements between previous data. The cross section values from the libraries ENDF/B-VI and JENDL-3.3 and the calculations of Maslov et al. [14] and the evaluation of Ignatyuk et al. [15] were based, where possible, on data these previous data. Our data would be in rather good agreement with these works if our results were normalized to them at 14 MeV.

The results of present measurements of neutron-induced fission cross-sections in case of ^{nat}W and ^{209}Bi are shown in Figs. 7 and 8. For ^{nat}W uncertainties vary from 19% at 100 MeV to 7% at 200 MeV. There is generally good agreement between our data and those of Refs. [5,6] for ^{nat}W, except for a possible discrepancy in the 90-100 MeV region. In our work the neutron-induced fission cross-section of ^{nat}W has been measured for the first time with a "white" neutron source. Comparison of fission

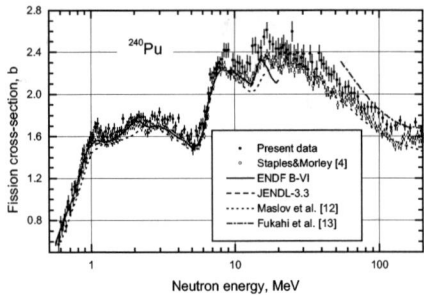
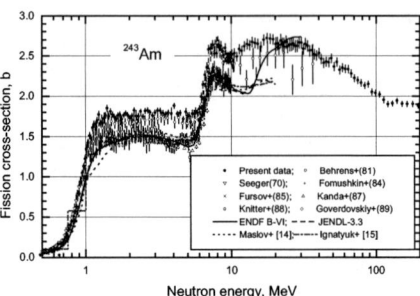

FIGURE 5. Fission cross-section of ^{240}Pu in the energy range 0.5 – 200 MeV.

FIGURE 6. Fission cross-section of ^{243}Am in the energy range 0.5 – 200 MeV. Experimental data references can be found in the EXFOR database.

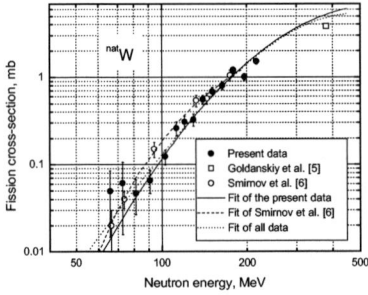
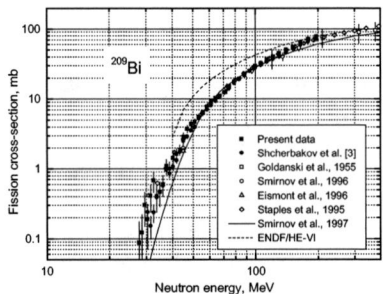

FIGURE 7. Measured fission cross-section of ^{nat}W in comparison with other data.

FIGURE 8. Measured fission cross-section of ^{209}Bi in comparison with the result of our previous experiment [3] and other data. References can be found in [3].

cross section of ^{209}Bi measured in the present experiment with that obtained in our previous measurement [3] shows excellent agreement.

REFERENCES

1. Salvatores, M., *J. Nucl. Sci. Tech.*, Suppl. 2, 4-12 (2002).
2. Abrosimov, N.K., Borukhovich, G.Z., Laptev, A.B., et al., *Nucl. Instr. Meth.* **A 242**, 121-133 (1985).
3. Shcherbakov, O., Donets, A., Evdokimov, A., et al., *J. Nucl. Sci. Tech.*, Suppl. 2, 230-233 (2002).
4. Staples, P., and Morley, K., *Nucl. Sci. Eng.* **129**, 149-163 (1998).
5. Goldanskiy, V.I., Tarumov, E.Z., and Penkina, V.S., *Dokl. Akad. Nauk* **101**, 1027 (1955).
6. Smirnov, A.N. , Eismont, V.P., Filatov, N.P., et al., *Phys. Rev.* **C 70**, 054603 (2004).
7. Laptev, A.B., Donets, A.Yu., Fomichev, A.V., et al., *Nucl. Phys.* **A 734S**, E45-E48 (2004).
8. Burgess, D.D., and Tervo, R.J., *Nucl. Instr. Meth.* **214**, 431-434 (1983).
9. Carlson, G.W., *Nucl. Instr. Meth.* **119,** 97-100 (1974).
10. Nakagawa, T., Shibata, S., Chiba, S., et al., *J. Nucl. Sci. Tech.,* **32**, 1259-1271 (1995).
11. Carlson, A.D., Chiba, S., Hambsch, F.-J., et al., Report INDC(NDC)-368, Vienna: IAEA,1997, p.23.
12. Maslov, V. Porodzinskij, Yu., Baba, M., et al., *J. Nucl. Sci. Tech.*, Suppl. 2, 80-83 (2002).
13. Fukahori, T., and Pearlstein, S., Report BNL-45200 (1991).
14. Maslov, V., Sukhovitskij, E.Sh., et al., Report INDC(BLR)-006, Vienna: IAEA, 1996.
15. Ignatyuk, A.V., Blokhin, A.I., Lunev, V.P., et al., *VANT, Ser. Jadernye Konstanty*, issue 1,25(1999).

A Novel High-Resolution Time-of-Flight Spectrometer with Tracking Capabilities for Photo-Fission Fragments and Beams of Exotic Nuclei

N. Nankov[*,†], E. Grosse[*,**], A. Hartmann[*], A. R. Junghans[*], K. Kosev[*], K. D. Schilling[*], M. Sobiella[*] and A. Wagner[*]

Institute of Nuclear and Hadron Physics, Forschungszentrum Rossendorf, P.O.Box 510119, 01314 Dresden, Germany
†*Institute for Nuclear Research and Nuclear Energy, 72 Tzarigradsko Chaussee Blvd., 1784 Sofia, Bulgaria*
**Institute of Nuclear and Particle Physics, Technical University of Dresden, Mommsenstr. 7, 01062 Dresden, Germany*

Abstract. Bremsstrahlung photons produced at the superconducting electron accelerator ELBE at FZ Rossendorf will be used for the production of neutron-rich nuclei by photon-induced fission. The properties of such exotic nuclei will be studied by decay spectroscopy. The mass and charge identification plays a key role in such experiments and is based on a double time-of-flight (TOF) method for both fission fragments. The reaction products are registered using secondary electrons emitted by thin foils and detected by position-sensitive MCP detectors. Currently a position resolution of 1.8(0.3) mm (FWHM) in both x and y directions and a time resolution of about 330 ps (FWHM) have been deduced using a source of α-particles.

Keywords: nuclear structure, nuclear astrophysics, bremsstahlung, photon-induced fission, time-of-flight spectrometry
PACS: 25.85.Jg,26.30.+k,29.30.Aj,29.40.Gx

PRODUCTION OF FISSION FRAGMENTS

The superconducting electron accelerator ELBE at the Forschungszentrum Rossendorf can deliver electron beams of up to 40 MeV energy and average beam currents of up to 1 mA with high duty cycles. The micro-pulse frequency can be varied between 100 kHz and 260 MHz with bunch widths less then 10 ps [1, 2]. The electron beam, delivered by ELBE is used to produce a high-intensity photon beam via bremsstrahlung. Photon fluxes of up to 10^8 cm^{-2}s^{-1}MeV^{-1} can be achieved. Neutron-rich exotic fragments can be produced via photon-induced fission in heavy nuclei by the bremsstrahlung photon beam. After proper identification their structure can be studied by the means of decay-spectroscopy. The β-decay will be measured by an array of CsI-scintillators and the γ-ray detection will be performed by an array of HPGe (cluster) detectors [3].

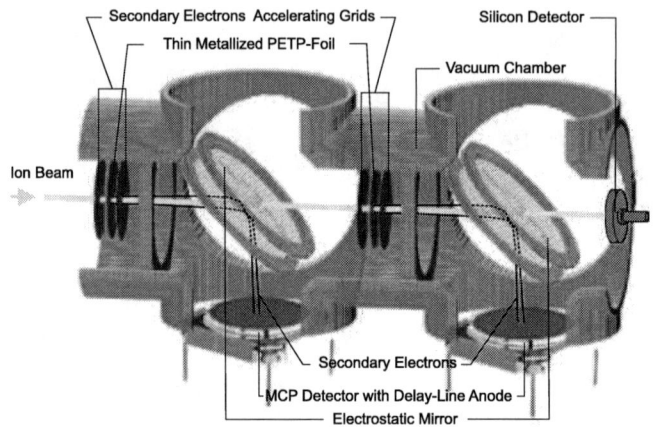

FIGURE 1. Mechanical design of the TOF spectrometer for heavy fragments and its working principle.

EXPERIMENTAL SETUP

In the double arm TOF spectrometer both the charge and the mass of each fragment are determined by two independent TOF measurements [4]. In order to study and optimize its characteristics an experimental device was build and tested (see Fig. 1). It consists of an α-source and two identical TOF detectors. The secondary electrons (SEs) are knocked out from a 300 $\mu g/cm^2$ thin Al-metallized polyethylene terephthalate (PETP) foil by the ions passing through and accelerated up to 6 keV by an electric potential between the foil and the grids. After that the SEs are deflected at 90° by an electrostatic mirror towards a position-sensitive micro-channel plate (MCP) detector from RoentDek [5]. The second detector is positioned 270 mm behind the first one. In our present setup a mixed α-source (^{226}Ra, ^{222}Rn, ^{210}Po, ^{218}Po, ^{214}Po) was used. It was positioned 160 mm in front of the first foil. At the end of the second detector a small Si detector was installed to detect only those α-particles passing through. Both TOF detectors are used in coincidence to minimize the number of the random events on the MCPs.

EXPERIMENTAL RESULTS

Detection Efficiency

For the efficiency measurement only one MCP detector and the silicon detector were used with discrimination levels set as low as possible. The efficiency is taken to be the ratio of the electron counts on the MCP detector over the silicon detector energy peak counts. Efficiency measurements for the forward as well as for the backward emitted secondary electrons were made. The results are shown in Fig. 2. The stable detection efficiency has been reached at an accelerating potential of about 1 kV. The ratio between the forward and backward efficiency taken in the saturation area is 1.92.

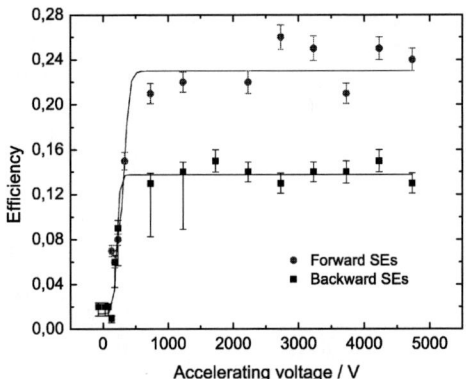

FIGURE 2. Detection efficiency for forward and backward emitted SEs as a function of the accelerating potential.

Position and Time Resolution

In order to determine the position resolution a mask with several 2 mm apertures, separated by 4 mm (center to center) was installed in front of the foil. Only those particles that have passed through the mask holes were recorded. In such a way the electrostatic mirror formed an image of the mask holes on the position sensitive MCP detector (see Fig. 3(left)). From our position resolution and efficiency measurements an initial transverse momentum of approx. 1 eV/c can be deduced. This spread is the main limitation for the position resolution, which is currently 1.8(0.3) mm. The measured TOF spectrum of the α-particles is shown on Fig. 3(right) for a distance of 80 mm between the foils. From the spectrum one can derive an apparent time resolution of about 330 ps (FWHM) for the TOF detector. The actual resolving power doubles as the distance between the foils increases from 27 to 80 mm and additionally the position information from the delay-lines is used in order to correct the TOF spectrum for different flight paths due to the initial angular distribution of the particles. Calculations have shown [6] that using an isochronous transport of the secondary electrons can improve the time resolution. This simple condition can be expressed as following:

$$d/(L_1 + L_2) = 0.236 \left(\Delta V / \Delta V_f \right), \qquad (1)$$

where L_1 and L_2 are the distances between the mirror and the foil and between the mirror and the MCP respectively, d is the mirror height, ΔV_f and ΔV are the foil acceleration potential and the mirror deflection potential. Fulfilling this condition one can expect considerable improvement of the time resolution.

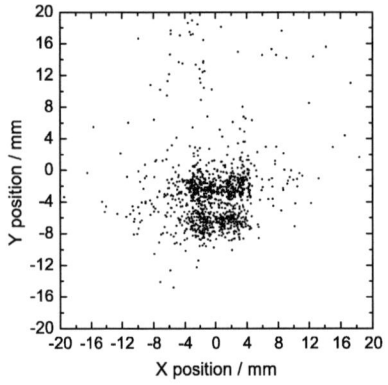 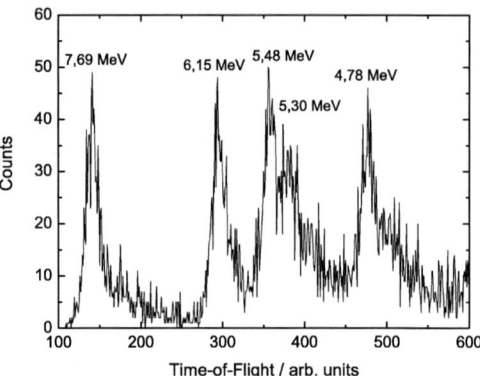

FIGURE 3. Isometric mapping of 4 apertures onto the delay-line anode (left) and the time-of-flight spectrum of alpha-particles with different energies coming from a mixed alpha source (right).

CONCLUSIONS

The first working prototype of the novel TOF spectrometer for fission fragments and exotic nuclei has been built and successfully tested. It has an acceptable efficiency and time and position resolution. The position resolution is limited by the initial energy spread of the SEs. The time resolution is limited until now by the detector readout electronics. The position information form the delay-lines and the isochronous transport of the secondary electrons could significantly improve the time resolution. Online tests with heavy fragments are planed at FZ Rossendorf (ELBE) and at GSI.

ACKNOWLEDGMENTS

This project is supported by the GSI under contracts DR-DON, and DR-WAG.

REFERENCES

1. F. Gabriel, et al., *Nucl. Instr. Meth. Phys. Res. B*, **161-163**, 1143 (2000).
2. J. Teichert, et al., *Nucl. Instr. Meth. Phys. Res. A*, **507**, 354 (2003).
3. A. Wagner, et al., The new bremsstrahlung facility at the superconducting electron accelerator ELBE, submitted to: Journal of Physics G.
4. H. Sharma, et al., "Planned Photofission Experiments at the New ELBE Accelerator in Rossendorf," in *Seminar on Fission, Pont d'Oye V-2003*, edited by C. Wagemans *et al.*, World Scientific, Singapore, 2004, pp. 201–208.
5. *RoentDek GmbH;* http://www.roentdek.com.
6. N. Nankov, et al., *Wiss.-Techn. Berichte FZR*, **423**, 25 (2004).

On The Double And Triple-Humped Fission Barriers And Half-Lives Of Actinide Elements

G. Royer, C. Bonilla

Laboratoire Subatech, UMR: IN2P3/CNRS-Université-Ecole des Mines, 4 rue A. Kastler, 44307 Nantes Cedex 03, France

Abstract. The potential energy of a deformed nucleus has been determined within a Generalized Liquid Drop Model taking into account the proximity energy, the microscopic corrections and compact and necked shapes. Multiple-humped potential barriers appear. A third minimum and third maximum exist in specific exit channels where one fragment is close to a magic spherical nucleus while the other one varies from oblate to prolate shapes. The heights of the fission barriers and half-lives of actinides are in agreement with the experimental results.

Keywords: Fission, Actinides, Liquid Drop Model, Half-lives.
PACS: 24.75.+i, 21.60.Ev, 27.90.+b.

INTRODUCTION

The fission probability, the angular distribution of the fission fragments and the low energy α decay in some actinides support the hypothesis of hyperdeformed states lodging in a third well in several Th and U isotopes [1] confirming the pioneering work of Blons et al [2]. It is even also advocated that this third minimum could be the true ground state of the heaviest elements [3]. The potential barriers governing the actinide fission have been determined [4] within a Generalized Liquid Drop Model taking into account both the proximity energy when a neck exists, an accurate nuclear radius, the mass asymmetry and the microscopic corrections. The path leading rapidly to the formation of a deep neck in compact shapes has been selected and the ellipsoidal deformations of the separated fragments have been taken into account.

FISSION BARRIERS

The proximity forces included in this GLDM allow to strongly lower the deformation energy in the quasi-molecular shape path leading rapidly to separated spherical fragments and allow to obtain the experimental fission barrier heights in the whole mass range, even for the Se, Br, Mo, In and Tb nuclei [5,6]. The α and cluster emission, the highly deformed rotating state [7] and fusion data can also be described within this unified approach.

In this work the coaxial ellipsoidal deformations have been taken into account since the limitation to spherical fragments leads to actinide fission barriers higher of some MeV than the experimental ones. The dependence of the potential barriers on the two-body shapes and microscopic corrections is displayed in Fig. 1. The shell effects generate the deformation of the ground state and increase the height of the first peak which appears already macroscopically. The proximity energy flattens the potential energy and will explain with the microscopic effects the formation of a second minimum lodging the superdeformed isomeric states for the heavier nuclei. In the two-sphere exit channel the rupture of the bridge of matter between the nascent fragments occurs before reaching the barrier top. The transition between one-body and two-body shapes is more sudden when the ellipsoidal deformations are allowed. It corresponds to the passage from a quasi-molecular one-body shape with spherical ends to two touching ellipsoidal fragments. The introduction of the microscopic energy still lowers the second peak ans shifts it to an inner position. It even leads to a third minimum and third peak. The heaviest fragment is a magic nucleus and remains almost spherical while the non magic fragment was born in an oblate shape. When the distance between mass centers increases the proximity energy tends to keep close the two tips of the fragments and the lightest one reaches a spherical shape which corresponds to a maximum of the shell energy, which is at the origin of the third peak. Later on, the proximity forces maintain in contact the fragments and the shape of the smallest one becomes prolate. Finally, a plateau exists at larger distances and much below the ground state when the proximity forces can no more compensate for the Coulomb repulsion and the fragments go away.

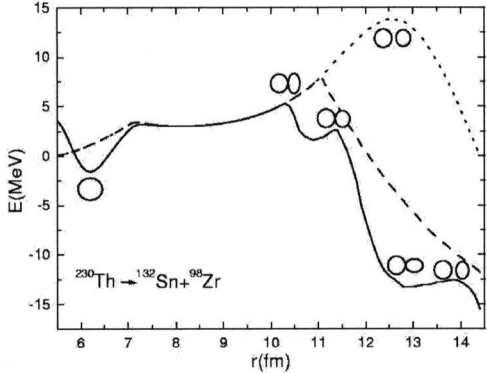

FIGURE 1. Fission barrier of a ^{230}Th nucleus emitting a doubly magic ^{132}Sn nucleus. The dotted and dashed lines correspond to the macroscopic energy within the two-sphere approximation and the ellipsoidal deformations for the two-body shapes. The solid line includes the microscopic corrections.

This third barrier appears only in the asymmetric decay channels and for some specific nuclei. In the symmetric mass exit path the proximity and Coulomb energies counterbalance the small shell effects and the two fragments remain in contact, one fragment being prolate while the other one is oblate before becoming both prolate at larger distances.

The whole reproduction of the heights of the inner and outer fission barriers which are almost constant from Th to Am isotopes is a very difficult task for all the theoretical approaches (liquid drop or droplet models, asymmetric two-center shell model, Hartree-Fock-Bogoliubov or Relativistic mean field theories,..). The theoretical and experimental energies of the maxima and minima of the potential barriers are compared in table 1. The choice of the most probable fission path is difficult for some elements since it exists a true degenerescence in energy between several mass asymmetry, particularly for the heaviest elements where the symmetric path seems more probable. The agreement with the experimental data is quite correct. For the heaviest nuclei the external barrier disappears.

TABLE 1. Experimental (e) and theoretical (t) first E_a, second E_b and third E_c peak heights and energies E_2 and E_3 of the second and third minima relatively to the ground state energy (in MeV).

Reaction	$E_a(e)$	$E_a(t)$	$E_2(e)$	$E_2(t)$	$E_b(e)$	$E_b(t)$	$E_3(t)$	$E_c(t)$
$^{231}_{90}Th \to ^{132}_{50}Sn + ^{99}_{40}Zr$	-	5.5	-	5.2	6.5	7.1	3.9 5.6(e)	6.9 6.3(e)
$^{233}_{90}Th \to ^{132}_{50}Sn + ^{101}_{40}Zr$	-	5.6	-	5.1	6.8	7.0	5.0 5.2(e)	7.8 6.8(e)
$^{232}_{92}U \to ^{134}_{52}Te + ^{98}_{40}Zr$	4.9	4.5	-	3.2	5.4	5.0	4.2	5.1
$^{234}_{92}U \to ^{131}_{50}Sn + ^{103}_{42}Mo$	5.6	5.0	-	4.4	5.5	5.9	3.7 3.1(e)	5.6
$^{235}_{92}U \to ^{131}_{50}Sn + ^{104}_{42}Mo$	5.7	5.7	2.5	4.9	5.8	6.6	5.4	6.9
$^{236}_{92}U \to ^{132}_{50}Sn + ^{104}_{42}Mo$	5.6	5.5	2.3	4.8	5.5	6.2	3.1 3.1(e)	4.4
$^{237}_{92}U \to ^{132}_{50}Sn + ^{105}_{42}Mo$	6.1	6.1	2.5	5.3	5.9	6.5	3.6	6.2
$^{238}_{92}U \to ^{132}_{50}Sn + ^{106}_{42}Mo$	5.7	5.5	2.6	4.5	5.7	5.6	4.1	5.6
$^{238}_{94}Pu \to ^{130}_{50}Sn + ^{108}_{44}Ru$	5.6	5.2	2.7	3.6	5.0	4.5	3.2	3.6
$^{239}_{94}Pu \to ^{130}_{50}Sn + ^{109}_{44}Ru$	6.2	5.8	2.6	4.1	5.5	5.0	4.1	5.6
$^{240}_{94}Pu \to ^{130}_{50}Sn + ^{110}_{44}Ru$	5.7	5.3	2.4	3.3	5.1	4.6	-	-
$^{241}_{94}Pu \to ^{131}_{50}Sn + ^{110}_{44}Ru$	6.0	6.1	1.9	4.4	5.5	5.6	5.1	5.5
$^{243}_{94}Pu \to ^{132}_{50}Sn + ^{111}_{44}Ru$	5.9	6.3	1.7	4.6	5.4	5.2	3.2	4.6
$^{242}_{95}Am \to ^{131}_{50}Sn + ^{111}_{45}Rh$	6.5	6.8	2.9	5.1	5.4	5.7	4.1	5.1
$^{244}_{95}Am \to ^{132}_{50}Sn + ^{112}_{45}Rh$	6.3	7.0	2.8	5.3	5.4	5.7	2.4	4.2
$^{243}_{96}Cm \to ^{130}_{50}Sn + ^{113}_{46}Pd$	6.4	6.0	1.9	3.6	4.2	4.2	2.4	2.7
$^{245}_{96}Cm \to ^{130}_{50}Sn + ^{115}_{46}Pd$	6.2	6.0	2.1	3.1	4.8	3.7	-	-
$^{248}_{96}Cm \to ^{130}_{50}Sn + ^{118}_{46}Pd$	5.7	5.3	-	2.0	4.6	3.0	-	-
$^{250}_{97}Bk \to ^{130}_{50}Sn + ^{120}_{47}Ag$	6.1	6.4	-	2.6	4.1	3.7	-	-
$^{250}_{98}Cf \to ^{125}_{49}In + ^{125}_{49}In$	5.6	4.9	-	0.1	-	1.7	-	-
$^{256}_{99}Es \to ^{128}_{50}Sn + ^{128}_{49}In$	4.8	5.9	-	0.8	-	2.4	-	-
$^{255}_{100}Fm \to ^{127}_{51}Sb + ^{128}_{49}In$	5.7	5.5	-	0.3	-	1.9	-	-

HALF-LIVES

Within this asymmetric fission model the decay constant is simply the product of the assault frequency by the barrier penetrability. Our theoretical predictions are compared with the experimental data [8,9] in table 2. There is a correct agreement on 24 orders of magnitude, except for the lighest U isotopes.

TABLE 2. Experimental and theoretical spontaneous fission half-lives of actinide nuclei.

Reaction	$T_{1/2,\exp}(s)$	$T_{1/2,th}(s)$
$^{232}_{92}U \rightarrow ^{134}_{52}Te + ^{98}_{40}Zr$	2.5×10^{21}	3.6×10^{16}
$^{234}_{92}U \rightarrow ^{131}_{50}Sn + ^{103}_{42}Mo$	4.7×10^{23}	8×10^{19}
$^{235}_{92}U \rightarrow ^{131}_{50}Sn + ^{104}_{42}Mo$	3.1×10^{26}	7.7×10^{23}
$^{236}_{92}U \rightarrow ^{132}_{50}Sn + ^{104}_{42}Mo$	7.8×10^{23}	1.0×10^{22}
$^{238}_{92}U \rightarrow ^{132}_{50}Sn + ^{106}_{42}Mo$	2.6×10^{23}	5.3×10^{22}
$^{238}_{94}Pu \rightarrow ^{130}_{50}Sn + ^{108}_{44}Ru$	1.5×10^{18}	2.6×10^{19}
$^{239}_{94}Pu \rightarrow ^{130}_{50}Sn + ^{109}_{44}Ru$	2.5×10^{23}	4.8×10^{22}
$^{240}_{94}Pu \rightarrow ^{130}_{50}Sn + ^{110}_{44}Ru$	3.7×10^{18}	4.8×10^{19}
$^{243}_{95}Am \rightarrow ^{133}_{51}Sb + ^{110}_{44}Ru$	6.3×10^{21}	1.1×10^{23}
$^{243}_{96}Cm \rightarrow ^{122}_{48}Cd + ^{121}_{48}Cd$	1.7×10^{19}	3×10^{21}
$^{245}_{96}Cm \rightarrow ^{130}_{50}Sn + ^{115}_{46}Pd$	4.4×10^{19}	3×10^{20}
$^{248}_{96}Cm \rightarrow ^{130}_{50}Sn + ^{118}_{46}Pd$	1.3×10^{14}	7.7×10^{15}
$^{250}_{98}Cf \rightarrow ^{140}_{55}Cs + ^{110}_{43}Tc$	5.2×10^{11}	4.9×10^{11}
$^{250}_{98}Cf \rightarrow ^{132}_{52}Te + ^{118}_{46}Pd$	5.2×10^{11}	1.2×10^{10}
$^{255}_{99}Es \rightarrow ^{128}_{50}Sn + ^{127}_{49}In$	8.4×10^{10}	8×10^{9}
$^{256}_{100}Fm \rightarrow ^{121}_{47}Ag + ^{135}_{53}I$	1.0×10^{4}	82
$^{256}_{102}No \rightarrow ^{128}_{51}Sb + ^{128}_{51}Sb$	110	0.9×10^{-2}
$^{256}_{102}No \rightarrow ^{116}_{46}Pd + ^{140}_{56}Ba$	110	0.3×10^{-1}

REFERENCES

1. Krasznahorkay, A., et al, *Phys. Rev. Lett.* **80**, 2073 (1998).
2. Blons, J., Mazur, C., Paya, D., Ribrag, M., and Weigmann, H., *Nucl. Phys. A* **414**, 1 (1984).
3. Marinov, A., Gelberg, S., Kolb, D., Brandt, R., and Pape, A., *Int. J. Mod. Phys. E* **12** No. 5 1 (2003).
4. Bonilla, C., and Royer, G., *Heavy-Ion Physics*, 2005, in print.
5. Royer, G., and Remaud, B., *J. Phys. G* **10**, 1541 (1984).
6. Royer, G., and Zbiri, K., *Nucl. Phys. A* **697**, 630 (2002).
7. Royer, G., Bonilla, C., and Gherghescu, R. A., *Phys. Rev. C* **67**, 34315 (2003).
8. Björnholm, S., and Lynn, J. E., *Rev. Mod. Phys.* **52**, 725 (1980).
9. Wagemans, C., *The nuclear fission process*, Boca Raton, CRC press, 1991.

Experimental Study of Energy Dependence of Proton Induced Fission Cross Sections for Heavy Nuclei in the Energy Range 200-1000 MeV.

A. A. Kotov[1], Yu. A. Gavrikov[1], L. A. Vaishnene[1], V. G. Vovchenko[1], V. V. Poliakov[1], O. Ya. Fedorov[1], T. Fukahori[2], Yu. A. Chestnov[1], A. I. Shchetkovskiy[1]

[1] *Petersburg Nuclear Physics Institute, Gatchina, Leningrad district, Orlova roscha 1, 188300, Russia*
[2] *Japan Atomic Energy Research Institute, Tokai-mura, Ibaraki 319-1195, Japan*

Abstract. The results of the total fission cross sections measurements for ^{nat}Pb, ^{209}Bi, ^{232}Th, ^{233}U, ^{235}U, ^{238}U, ^{237}Np and ^{239}Pu nuclei at the energy proton range 200-1000 MeV are presented. Experiments were carried out at 1 GeV synchrocyclotron of Petersburg Nuclear Physics Institute (Gatchina). The measurement method is based on the registration in coincidence of both complementary fission fragments by two gas parallel plate avalanche counters, located at a short distance and opposite sides of investigated target. The insensitivity of parallel plate avalanche counters to neutron and light charged particles allowed us to place the counters together with target immediately in the proton beam providing a large solid angle acceptance for fission fragment registration and reliable identification of fission events. The proton flux on the target to be studied was determined by direct counting of protons by scintillation telescope. The measured energy dependence of the total fission cross sections is presented. Obtained results are compared with other experimental data as well as with calculation in the frame of the cascade evaporation model.

Keywords: fission cross sections, intermediate energy protons, lead, bismuth 209, thorium 232, uranium 233, uranium 235, uranium 238, neptunium 237, plutonium 239.
PACS: 25.85.Ge; 29.40.Cs.

1. Introduction

The need for the information concerning fission reactions induced in heavy nuclei by intermediate energy projectiles has been obvious. The interest in this process emerges from both fundamental and applied problems of nuclear physics. In spite of extensive experimental efforts, the fission process of nuclei induced by intermediate energy projectiles remains insufficiently understood in many aspects. The proposed measurements of the energy dependence of total fission cross sections of heavy nuclei induced by intermediate energy protons will add to our understanding of the fission process in terms of nuclear properties of highly excited nuclei, such as temperature dependence of level density and fission barriers of excited nuclei. For physics applications, the nuclear data are required for new energy production concepts with the help of accelerator driven system (ADS), for nuclear waste transmutation technologies, for accelerator and cosmic device radiation shields. All the above-mentioned problems require fission cross section data of high accuracy and reliability.

Unfortunately, most of experimental data have been obtained in various experiments by using different methods of registration. That is why available data on fission cross section are dispersed in the range which exceeds declared accuracy of measurements, not allowing to establish reliable energy dependence of fission cross section on proton energy. High accuracy of the fission cross section measurements may be only achieved by use of the modern electronic methods of the registration of the both fission fragments in coincidence. together with high precision monitoring of a proton flux on the studied target. In the present experiment the method based on use of the gas parallel plate avalanche counters (PPAC) for registration of complementary fission fragments in coincidence and the telescope of scintillation counters for direct counting of the incident protons on the target has been used. This method allowed us to measure the absolute proton induced fission cross sections of ^{239}Pu, ^{237}Np, ^{238}U, ^{235}U, ^{233}U, ^{232}Th, ^{209}Bi and ^{nat}Pb in the energy range from 200 to 1000 MeV with step 100 MeV and results on the energy dependence of total fission cross sections are presented.

2. Experiment

The experiment was carried out at the 1 GeV proton synchrocyclotron of Petersburg Nuclear Physics Institute (Gatchina). Proton beams in the energy range 200 - 900 MeV were obtained by decreasing the energy of the primary 1 GeV proton beam by copper absorbers. The system for transportation and formation of the external proton beam, consisting of a bending magnet and quadrupole magnetic lenses, represents a magnetic spectrometer with an energy resolution of <20 MeV in the energy range 200 - 1000 MeV.

The detector for recording coincident fission fragments represents an assemblage consisting of two identical gas parallel plate avalanche counters and the target to be studied. The good timing properties of PPAC and the possibility to discriminate particles with low energy losses allow us to use this device directly in the incident proton beam, providing a solid angle acceptance of nearly 4 π and 100 % detection efficiency. The proton beam with a diameter of 1 - 2 cm and an intensity of 10^5 -10^6 protons per second after traversing the target was monitored by a telescope of scintillation counters. The detection of both fission fragments and beam protons will be performed by direct absolute experimental methods thus providing a high degree of accuracy of the total fission cross section measurements.

The cross section measurement method are described in detail in Ref. [1].

3. Results

So, using the methods of detection of the binary fission events and beam monitoring, described in the previous sections, the fission cross sections for each target have been measured at nine proton energies.

Energy dependence of the fission cross sections obtained in the present experiment for the nuclei mentioned is shown in Fig. 1. The results of the previous experiments, compiled in papers [2,3], are also given. In paper [2] a parametrization of the all the world fission cross section data was proposed after the critical analysis and data selection for a series of the actinide and pre-actinide nuclei for proton energies up to 10-30 GeV. The results of this fission cross section estimation based on all the world experimental data are also given in Fig .1 by dashed lines.

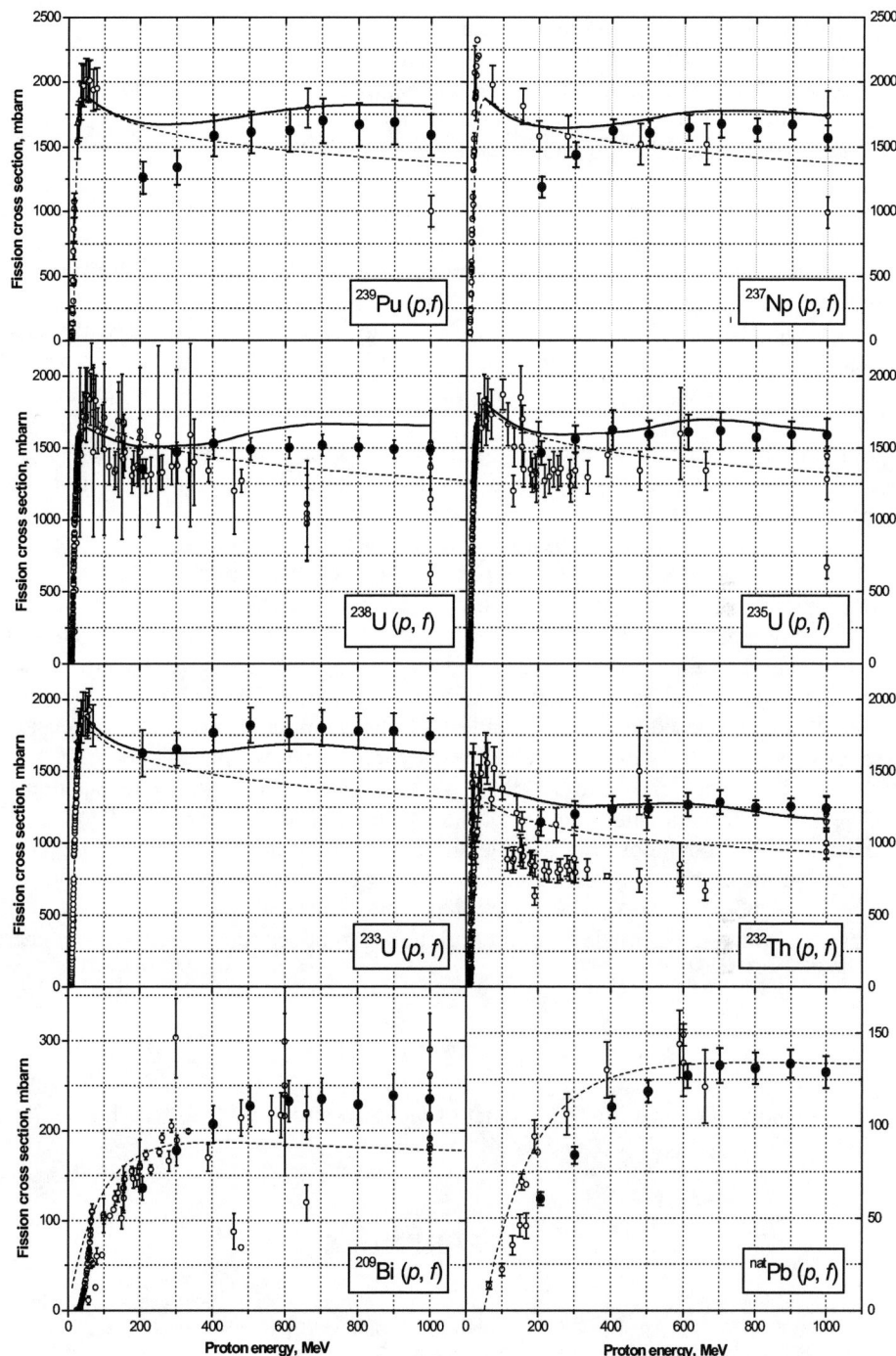

FIGURE 1. Energy dependence of the fission cross sections. Black circles-our data, open circles-data from refs. [2,3]. Dashed line – a parametrization from paper [2], solid line- theoretical calculation.

The cross section calculation procedure comprised the following steps:
1. Background subtraction after the analysis of the bidimensional amplitude distribution of the detected events. The number of the background events amounted to 2-3% of the total number of the detected events.
2. Determination of the solid angle for fission event detection for each assembly. The calculation was done by a Monte-Carlo simulation which took into account: the proton beam profile at the target; detection geometry for the fission fragments, their mass and energy distributions, as well as separation asymmetry of the fragments in the laboratory due to the longitudinal component of the momentum of the fissioning nucleus. The statistical accuracy of the solid angle calculations amounted to 0.1%.
3. The calculations of the undetectable part of the fission events, caused by the energy loss of the fragment in the target, in the backing and in the PPAC electrodes. The undetectable fraction of the events depended on the target nucleus and thickness and amounted to 3-8%.
4. Determination of the integral proton flux through the target, with the account of the scintillation telescope efficiency and the probability of the appearance of more than one proton in a microbunch.

Theoretical calculations of fission cross sections were carried out in the framework of the two-step cascade evaporation model [4]. When calculating the fission cross sections, one needs to vary the nuclear level density parameters at the equilibrium deformation a_n and at the saddle point of the fissioning nucleus a_f, as well as the fission barrier value B_f. It was supposed that at high excitation energy nuclear shell effects in the fission barriers may be ignored, the ratio of the level density parameters a_f/a_n being supposed to be independent of the excitation energy of the decaying nucleus. In our calculations the level density parameter a_n was taken to be $A/10$ MeV^{-1} for all actinide nuclei, the ratio of the level density parameters a_f/a_n being equal to 1.1. The fission barriers calculated in the liquid drop model [5] were taken for B_f. The results of the calculations for actinide nuclei are presented by solid line in Fig. 1 It is seen that the calculations reproduce qualitatively the general cross section behavior in the range 50-1000 MeV. For pre-actinide nuclei the calculations in frame of the cascade evaporation model we used fail to reproduce the cross section energy behavior.

ACKNOWLEDGMENTS

We are grateful to Nuclear Data Center of Japan Atomic Energy Research Institute for support through the ISTC Project -1405.

REFERENCES

1. Chtchetkovski A.I., Gavrikov Y.A., Kotov A.A., Vaishnene L.A., Vovchenko V.G., Fedorov O.Y., Fukahori T., Poliakov V.V., Chestnov Y.A., *Physica Scripta*, **104**, 101-104 (2003).
2. Prokofiev A.V., *Nucl. Instrum. Meth. Phys. Res.* **A463**, 557 (2001).
3. Obukhov A.I., *Phys. Part. and Nucl.* v.1, **32**,162 (2001).
4. Serber R. *Phys.Rev.* **72**,1114(1947).
5. Myers W.D. and Swiatecki W.J., *Ark.Phys.*,36, 343 (1967); *Ann.Phys.*,68, 186 (1974).

Development Of A Digital Technique For The Determination Of Fission Fragments And Emitted Prompt Neutron Characteristics

N. Varapai[1,2], F.-J. Hambsch[2], S. Oberstedt[2], O. Serot[1], G. Barreau[3], N. Kornilov[2,4], Sh. Zeinalov[5]

1 CEA Cadarache, DEN/DER/SPRC/LEPh, F-13108 St. Paul-Lez-Durance, France
2 EC-JRC-Institute for Reference Materials and Measurements, Retieseweg, B-2440 Geel, Belgium
3 CEN Bordeaux Gradignan, 33175 Gradignan Cedex, France
4 Institute for Physics and Power Engineering, 249033 Obninsk, Kaluga oblast, Russia
5 Joint Institute for Nuclear Research (JINR), 141980 Dubna, Moscow Region, Russia

Abstract. The present work demonstrates the application of the digital technique for nuclear measurements. This method has been implemented for measurements of promptly emitted fission neutrons in coincidence with fission fragments from ^{252}Cf(sf). A double Frisch-grid ionization chamber is used as fission fragment detector. The promptly emitted neutrons are detected by a NE213 liquid scintillation detector. The experimental set-up is installed at the Institute for Reference Materials and Measurements. Preliminary results are presented.

Keywords: Waveform Digitizer, Fission Fragment Spectroscopy, Neutron Spectroscopy, Scintillator, Spontaneous Fission of ^{252}Cf.

1. INTRODUCTION

In recent years digital processing technology is slowly replacing traditional analogue techniques for nuclear physics application [1, 2]. This technology gives the possibility to simplify the analogue technique when separate units for the separation and storage of the information are used. The main advantage is the digitalization of the signal shape which contains all the information on the particle, i.e. kinetic energy, emission angle and mass. This information is actually determined off-line by digital signal processing and opens up the possibility to modify the analysis procedure without repeating the experiment.

To test the digital technique and to verify the method of off-line analysis a ^{252}Cf(sf) source is used since here both the fission fragment and the prompt emitted neutron properties are well known from literature.

2. EXPERIMENTAL SET-UP

A double Frisch-grid ionization chamber (IC) was used as a fission fragment detector. As it is shown in Ref. [3], it is possible to determine the fragment kinetic energy, the emission angle and to obtain the fragment masses by measuring simultaneously the pulse-height of the signals from the chamber electrodes for each of the fragments. The ^{252}Cf(sf) target is mounted at the centre of the common cathode. The chamber was operated with a gas flow of 0.1 l/min. The counting gas was 90% Ar + 10% CH$_4$. The bias voltage applied to the cathode was -1700V, and to the anodes +1000V. The grids were grounded.

As neutron detector a 15 cm diameter and 6 cm thick NE213 liquid scintillator coupled to a 10 cm diameter XP2041 photomultiplier (PM) was used. It was on loan from CEN Bordeaux Gradignan and placed at a distance of 1.5 m from the IC. The anode signal of the PM was used for the discrimination between γ-rays and neutrons by means of pulse shape analysis.

The data acquisition is based on 2 waveform digitizers (WFD) of FAST-ComTec (12 bit resolution). One WFD (100 MHz, 2 input channels) was used to record the fission fragment waveforms. The anode signal of the neutron detector was summed with the cathode signal of the IC and then registered with the second WFD (200 MHz). The digitized signals are stored on the hard disk. A schematical view of the experimental set-up is shown in Fig.1.

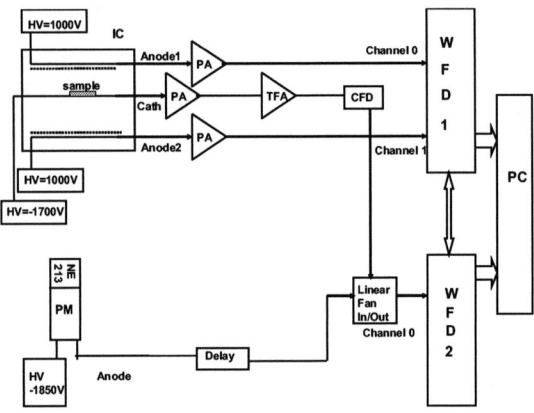

FIGURE 1. Schematical view of the experimental set-up using charge sensitive preamplifier (PA).

3. DATA ANALYSIS AND RESULTS

Since all information is contained in the signal shape, it is possible to apply any off-line method for signal data analysis.

Typical waveforms of the ionization chamber and neutron detector are presented in Fig. 2. The waveforms (left figure) represent the fragment pulses after passing charge sensitive pre-amplifiers (PA). The height of the pulse is proportional to the particle kinetic energy. From the slope of the signal the emission angle of the fission fragment may be extracted. To obtain the angular distribution, the centre of gravity of the current signal (dQ/dt, where Q(t) is the signal in the left part of Fig.2) is determined. Then the center of gravity is transformed into the cosine of the emission angle as explained in Ref. [3] and plotted as a function of the pulse height. Figure 3 shows the corresponding two-dimensional distributions.

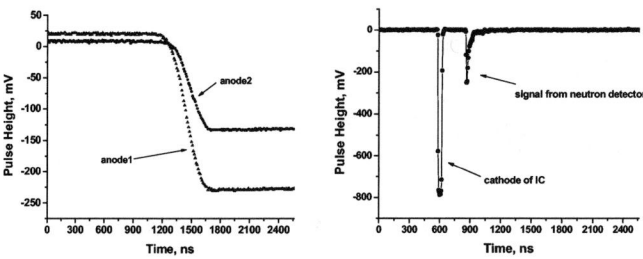

FIGURE 2. Typical waveforms of the anode signals from the ionization chamber (left) and anode from the neutron detector (right).

The pre-neutron mass distribution of the fission fragments for the ^{252}Cf(sf) is obtained after correcting for energy loss in the sample and backing material of the target and the pulse height defect in the counting gas as well as for the emission of prompt neutrons. The resulting distribution presented in Fig. 4 is in good agreement with literature values: $<A_L>$=109.05 amu, $<\sigma_L>$=6.6 amu, $<A_H>$=142.95 amu, $<\sigma_H>$=6.6 amu ($<A_L>$=108.6 amu, $<A_H>$=143.5 amu: taken from Ref. [4]).

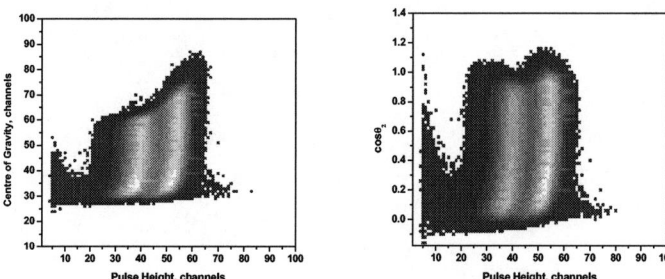

FIGURE 3. Centre of gravity versus pulse height (left) and cosθ versus pulse height for the sample side of the target (right).

The signal in the right part of Fig. 2 is the sum of two signals: the cathode signal from the IC after passing a Constant Fraction Discriminator (CFD) giving the start-time (fission) and the anode signal from the neutron detector giving the stop-time (neutron, γ-rays). Sending both signals ("start" and "stop") in one channel of the WFD we avoid the possible problem of the synchronization of different channels. Both time stamps were determined by the positions of the centre of gravity: for the signal from the neutron detector an interval of 50 ns around the minimum was taken into consideration; for the signal from the cathode the position of the centre of gravity was found by the derivation of the left part of the rectangular signal. This method of determination of the start and the stop positions gives a time resolution 4.2 ns for the Time-of-Flight spectrum (see left part of Fig. 5).

In the right part of Figure 5 the quality of the neutron-γ separation is shown. A two integral method was applied for the pulse shape analysis.

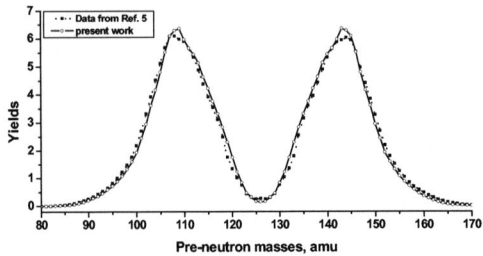

FIGURE 4. Fission fragment mass distribution for the ^{252}Cf(sf).

The time windows for the slow and fast components are 250 ns and 25 ns, respectively.

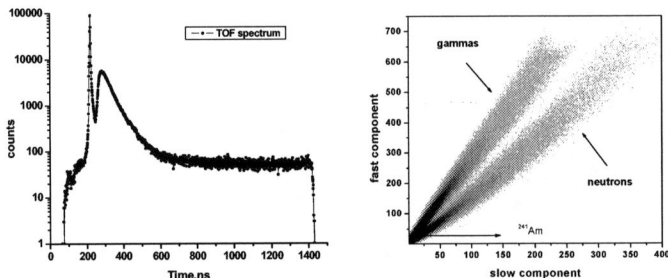

FIGURE 5. The Time-Of-Flight spectrum (1ns channel width) (left) and 2-dimensional distribution used for the n-γ discrimination for ^{252}Cf(sf) (right).

CONCLUSION

In this work we demonstrated the application of the WFD technique to nuclear experiments. This technique was successfully applied for the determination of fission fragments and prompt emitted neutron characteristics in case of spontaneous fission of ^{252}Cf. The results for the fission fragment mass distribution are comparable with the traditional analogue technique (see Fig. 4 and Ref. 5). A reasonably good n-γ separation using the pulse shape analysis method has been obtained, too.

REFERENCES

1. Kornilov, N. V., et al., *Nucl. Inst. Meth, Phys. Res. A,* vol. 497, p.467 (2003).
2. Bardelli, L., et al., *Nucl. Inst. Meth, Phys. Res. A,* vol.521, p.480 (2004).
3. Budtz-Jorgensen, C., et al., *Nucl. Inst. Meth, Phys. Res. A,* vol.258, p.209 (1987).
4. Gönnenwein, F., "The Nuclear Fission Process", edited by C. Wagemans, CRC Press, Boca Raton, 1991, 323ff.
5. Hambsch, F.-J., Oberstedt, S., *Nuclear Physics A,* vol.617, Issue 3, p. 347-355 (1997).

Calculation Of Fission Fragment Yields At Low And Intermediate Energy Fission

S.Yavshits*, O.Grudzevich[¶]

*V.G.Khlopin Radium Institute, St.-Petersburg, Russia
[¶]Technical State University, Obninsk, Russia

Abstract. The model for fission fragment mass distributions and results of calculations for low energy fission and fission induced by intermediate energy nucleons is presented. Formation of mass distributions is considered as a result of oscillations on mass asymmetry degree of freedom in the potential well calculated in the macroscopic-microscopic approach. For intermediate energy fission the distribution of fissioning nuclei is taken into account with detailed reaction calculations including direct, preequilibrium and statistical reaction stages.

Keywords: Fission fragment mass spectra calculations, collective model, temperature dependence, nucleon-induced fission in wide energy region.
PACS: 25.85.Ec, 25.85.Ge

INTRODUCTION

Formation of fragments in nuclear fission is closely tied with important and still incompletely studied nuclear matter fragmentation for both low excited and excited states in the cases of intermediate and high energy nucleon-induced fission.

It is clear at present that formation of fission fragment mass distributions is first of all connected with the properties of potential energy surface on the stage of saddle-to-scission descent. Dynamical effects have a less influence on the mass spectra and manifest themselves in widths of mass distributions mainly. The model used in the given work originates from a notion of nuclear shape oscillations on mass-asymmetry degree of freedom in the potential well calculated in macroscopical-microscopical approach (Strutinsky' prescription) in one-center shape parameterization. Solution of one-dimensional Schrödinger equation in such a potential let us possibility to define fragment mass spectra and introduce temperature dependence of fragment yields naturally through temperature dependence of collective potential and population of states in the well.

THEORETICAL MODEL

One of the important points in the study of nuclear configurations near the scission point is the choice of shape parameterization. We use parameterization proposed by V.Pashkevich [1] for axially symmetrical configurations. Here nuclear shape is defined in the orthogonal coordinate system where basic family of surfaces are deformed Cassinian ovaloids allowing to describe both oblate and prolate shapes including strongly prolate ones up to division of nucleus on two fragments.

Let us restrict the consideration by three main parameters of deformations: $\{\alpha\} = (\alpha, \alpha_1, \alpha_4)$ where α is the lemniscate parameter, α_1 defines mirror symmetry of nuclear shape and α_4 is the parameter of hexadecapole deformations. At small values of the lemniscate parameter the shape of nucleus looks like ovaloid while values $\alpha > 0.9$ correspond to configurations with developed neck.

We use one-dimensional Schrödinger equation for description of the collective motion in the mass-asymmetry coordinate:

$$\left\{-\frac{\hbar^2}{2\sqrt{B_1}}\frac{\partial}{\partial \alpha_1}\frac{1}{\sqrt{B_1}}\frac{\partial}{\partial \alpha_1} + V_T(\alpha_1;\alpha_{sc};\alpha_4^{min})\right\}\psi_v(\alpha_1) = E_v\psi(\alpha_1). \quad (1)$$

Here B_1 is the mass parameter for mass-asymmetry mode α_1, $V_T(\alpha_1;\alpha_{sc};\alpha_4^{min})$ is the temperature dependent collective potential energy of deformation as function of α_1 at the scission point $\alpha = \alpha_{sc}$, the potential energy being minimized on the parameter of hexadecapole deformation $\alpha_4 = \alpha_4^{min}$, where α_4^{min} is taken from the condition of potential energy minima, $\psi_v(\alpha_1)$ are collective wave functions and E_v is the energy spectrum of collective states. We used liquid drop value for the mass parameter and Strutinsky' prescription [2] for potential energy calculation:

$$V_T(\{\alpha\}) = E_{ld}(\{\alpha\}) + f(T)\delta E(\{\alpha\}), \quad (2)$$

where E_{ld} is a liquid drop energy, δE takes into account shell correction and pair energy, $\delta E = E_{shell} + E_{pair}$, and function $f(T)$ reflects melting of nuclear structure effects with nuclear temperature increase. We used Fermi function $f(T)$,

$f(T) = 1/(1 + \exp(T - T_{cr})/a)$, where T_{cr} and a are model parameters.

Probability to find configuration with given value of α_1 can be expressed as follows:

$$Y(\alpha_1) \propto \sum_v |\Psi_v(\alpha_1)|^2 e^{-E_v/T} d\alpha_1, \quad (3)$$

and fragment mass spectrum will have the following form

$$Y(A_1(\alpha_1)) \propto Y(\alpha_1)\frac{dA_1}{d\alpha_1}, \quad (4)$$

where A_1 is the fragment mass number linked with parameter α_1 by volume integration to the left (or right) from the point of minimal neck radius.

Similar model has been proposed earlier in the work [3]. However, two-center configuration in the scission point (configuration of nascent fragments) used in [3] requires correct account of fragment interaction including nucleon exchange between fragments [4] which was not done in this work. The one-center model used in the present work is free from these weaknesses and is naturally connected with the saddle-to-scission descent stage.

RESULTS

The scission point potential energy as function of fragment mass for the case of ^{239}Pu fission is shown in the Fig.1.

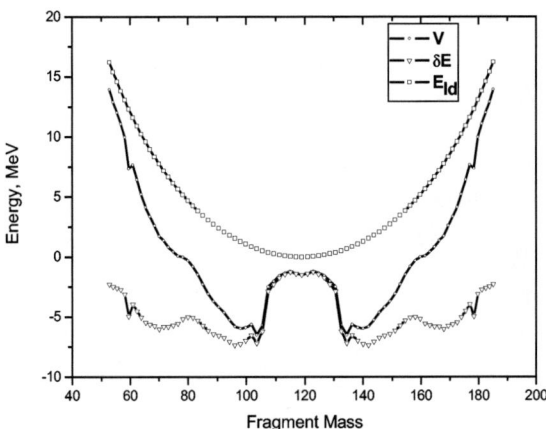

FIGURE 1. Potential energy of deformation and its components in the scission point for the case of ^{239}Pu fission as function of fragment mass number for zero nuclear temperature

To describe observed fragment mass spectra for actinides we reduced liquid drop stiffness at scission point approximately twice as compared with the stiffness in the ground state considering it as a model parameter because behavior of coulomb and surface energies in the strongly deformed nuclear state with thin neck in the scission point may be different from equilibrium ground state.

Our results for low-energy fission fragment mass distributions are presented in the left part of Fig.2. For intermediate-energy fission one should take into account the formation of wide distribution of fissioning nuclei on mass and charge numbers and excitation energies A_f, Z_f, E_f^* due to prefission particle emission at each stage of nucleon-induced reaction, i.e. at direct, preequilibrium and equilibrium (multichance fission) stages. The final fragment mass distribution could be found as a superposition of fragment yields for each nucleus with corresponding weights $W(A_f, Z_f, E_f^*)$ defined by the reaction mechanism:

$$Y \propto \sum_{A_f, Z_f, E_f} W(A_f, Z_f, E_f^*) Y_f, \qquad (5)$$

where weights $W(A_f, Z_f, E_f^*)$ were calculated with MCFx code developed earlier [5].

In the right part of Fig.2 the calculated fission fragment mass distributions are shown for 50 MeV neutron-induced fission of ^{238}U and ^{209}Bi(p,f) reaction at proton energy 475 MeV. One can see that in all cases our results reproduce experimental data rather well. So, we can conclude that the model developed in link with MCFx code can be used for generation of data files on the fission fragment mass distributions at low and intermediate energy fission.

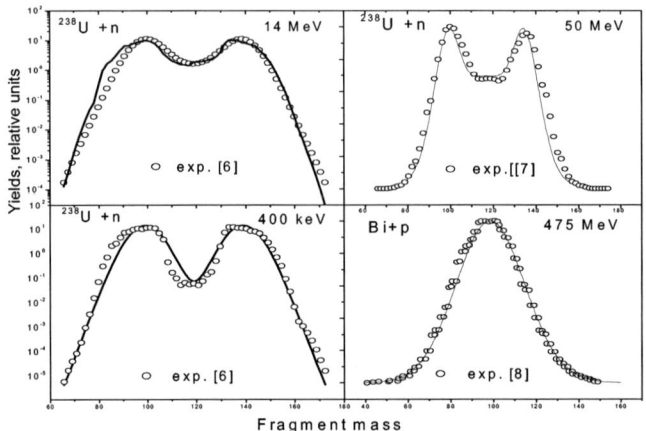

FIGURE 2. Comparison of calculated and experimental fission fragment yields

REFERENCES

1. V.V Pashkevich., *Nucl.Phys. A*, **169**, 275 (1969).
2. V.M.Strutinsky, *Nucl.Phys. A*, **122**, 1 (1968).
3. J.Maruhn, W.Greiner, *Phys. Rev. C*, 13, 2404 (1976).
4. V.A.Rubchenya, S.G.Yavshits, *Sov. J. Nucl. Phys.*, **40**, 649 (1984).
5. S.Yavshits, V.Ippolitov, A.Goverdovskii, O.Grudzevich, in *ND-2001*, *Proc.* of the International Conference on Nuclear Data for Science and Technologies, Tsukubo, Oct.7-12, 2001, Japan , p. 104; S.Yavshits, *Report IAEA-NDS-153* (2002) - htpp://www-nds.iaea.org/reports/nds-153.pdf.
6. JEFF 2.2 *Report AEA-1015*, 1993 and htpp://www.nea.fr/Janis
7. C.M.Zoller *Doctoral thesises* TH Darmstadt, 1995.
8. Z.Fraenkel et al., *Phys., Rev. C*, **41**, 1050 (1990).

List of Participants

LASTNAME	FIRSTNAME	ORGANIZATION	CITY	COUNTRY	EMAIL
AICHE	Mourad	CEN	Bordeaux-Gradignan	France	aiche@cenbg.in2p3.fr
AL MAHAMID	Ilham	LBL	Berkeley	USA	AlMahamid@lbl.gov
ARMBRUSTER	Peter	GSI	Darmstadt	Germany	-
ASTIER	Alain	CSNSM	Orsay	France	astier@csnsm.in2p3.fr
BAO	Zongyu	China Institute of Atomic Energy	Beijing	China	baozongyu0440@sina.com.cn
BARREAU	Gérard	CEN	Bordeaux-Gradignan	France	barreau@cenbg.in2p3.fr
BAUGE	Eric	CEA/DAM/DIF	Bruyères-le-Châtel	France	eric.bauge@cea.fr
BERERD	Nicolas	IPN	Lyon	France	bererd@ipnl.in2p3.fr
BERGER	Jean-François	CEA/DAM/DIF	Bruyères-le-Châtel	France	jean-françois.berger@cea.fr
BERNAS	Monique	IPN	Orsay	France	bernas@ipno.in2p3.fr
BHATTACHARYYA	Sarmishtha	GANIL	Caen	France	sarmi@ganil.fr
BIRGERSSON	Evert	EC-JRC IRMM	Geel	Belgium	evert.birgersson@cec.eu.int
BONNEAU	Ludovic	LANL	Los Alamos	USA	bonneau@lanl.gov
BOUDARD	Alain	CEA Saclay	Gif sur Yvette	France	aboudard@cea.fr
CARJAN	Nicolae	CEN	Bordeaux-Gradignan	France	carjan@in2p3.fr
CHABOD	Sébastien	CEA Saclay	Gif sur Yvette	France	schabod@cea.fr
CHADWICK	Mark	LANL	Los Alamos	USA	mbchadwick@lanl.gov
CHAU	Pierre	CEA/DAM/DIF	Bruyères-le-Châtel	France	huu-tai.chau@cea.fr

CHBIHI	Abdu	GANIL	Caen	France	chbihi@ganil.fr
DASSIE	Danielle	CEN	Bordeaux-Gradignan	France	dassie@cenbg.in2p3.fr
DAYRAS	Roland	CEA Saclay	Gif sur Yvette	France	rdayras@cea.fr
DE FRENNE	Denis	University of Gent	Gent	Belgium	denis.defrenne@Ugent.be
DELONCLE	Isabelle	CSNSM	Orsay	France	deloncle@csnsm.in2p3.fr
DORÉ	Diane	CEA Saclay	Gif sur Yvette	France	ddore@cea.fr
DUARTE	Helder	CEA/DAM/DIF	Bruyères-le-Châtel	France	helder.duarte@cea.fr
DUCAUZE-PHILIPPE	Marie	CEA/DAM/DIF	Bruyères le Chatel	France	marie.ducauze@cea.fr
DUPONT	Emmeric	CEA/DEN	Cadarache	France	edupont@cea.fr
FAUST	Herbert	ILL	Grenoble	France	faust@ill.fr
FIONI	Gabriele	CEA Saclay	Gif sur Yvette	France	gabriele.fioni@cea.fr
GALY	Jean	EU-JRC-Inst. for Transuranium Elements	Karlsruhe	Germany	jean.galy@itu.fzk.de
GENEVEY	Janine	LPSC	Grenoble	France	genevey@lpsc.in2p3.fr
GIACRI-MAUBORGNE	Marie-Laure	CEA Saclay	Gif sur Yvette	France	mlgiacri@cea.fr
GÖNNENWEIN	Friedrich	Physikalisches Institut	Tübingen	Germany	goennenwein@uni-tuebingen.de
GOUTTE	Dominique	GANIL	Caen	France	goutte@ganil.fr
GOUTTE	Héloïse	CEA/DAM/DIF	Bruyères-le-Châtel	France	heloise.goutte@cea.fr
GRIGORIEV	Yuri	JINR	Dubna	Russia	griguv@mail.ru
HAAS	Bernard	CEN	Bordeaux-Gradignan	France	haas@cenbg.in2p3.fr
HAMBSCH	Franz-Josef	EC-JRC IRMM	Geel	Belgium	franz-joseph.hambsch@cec.eu.int

Surname	First name	Institute	City	Country	Email
HANAPPE	François	PNTPM Université Libre de Bruxelles	Bruxelles	Belgique	fhanappe@ulb.ac.be
HEINRICH	Sophie	CEA/DAM/DIF	Bruyères-le-Châtel	France	heinrics@bruyeres.cea.fr
HEYSE	Jan	University of Gent	Gent	Belgium	jan.heyse@ugent.be
ITKIS	Mikhail	FLNR, JINR	Dubna	Russia	morozova@nrmail.jinr.ru
JURADO	Beatriz	CEN	Bordeaux-Gradignan	France	jurado@cenbg.in2p3.fr
KALININ	Valery	V.G. Khlopin Radium Institute	St Petersburg	Russia	kalinin@atom.nw.ru
KELIC	Aleksandra	GSI	Darmstadt	Germany	a.kelic@gsi.de
KOSTER	Ulli	CERN	Genève	Switzerland	Ulli.Koster@cern.ch
LACHKAR	Jean	CEA	Paris	France	j.lachkar@cea.fr
LEDOUX	Xavier	CEA/DAM/DIF	Bruyères-le-Châtel	France	xavier.ledoux@cea.fr
LEE	YongDeok	KAERI	Daejon	Korea	ydlee@kaeri.re.kr
LEMAIRE	Sebastien	LANL	Los Alamos	USA	lemaire@lanl.gov
LERAY	Sylvie	CEA Saclay	Gif sur Yvette	France	sleray@cea.fr
LETOURNEAU	Alain	CEA Saclay	Gif sur Yvette	France	aletourneau@cea.fr
LEWITOWICZ	Marek	GANIL	Caen	France	lewitowicz@ganil.fr
MARCHIX	Anthony	GANIL	Caen	France	marchix@ganil.fr
MOREL	Pascal	CEA/DAM/DIF	Bruyères-le-Châtel	France	pascal.morel@cea.fr
MORILLON	Benjamin	CEA/DAM/DIF	Bruyères-le-Châtel	France	benjamin.morillon@cea.fr
MORJEAN	Maurice	GANIL	Caen	France	morjean@ganil.fr
MUELLER	Alex	IPN	Orsay	France	mueller@ipno.in2p3.fr
NANKOV	Nikolai	Inst of Nucl. and Hadronphysics	Dresden	Germany	n.nankov@fz-rossendorf.de

NICHOLS	Alan	IAEA		Vienna	Austria	a.nichols@iaea.org
OBERSTEDT	Andreas	Dept of Natural Sciences		Orebro	Sweden	Andreas.Oberstedt@nat.oru.se
OBERSTEDT	Stephan	EC-JRC IRMM		Geel	Belgium	stephan.oberstedt@cec.ev.int
PEREZ	Sara	Universidad Autonoma		Madrid	Spain	Sara.Perez@uam.es
PETROV	Gennady	PNPI		Gatchina	Russia	gpetrov@pnpi.spb.ru
PINSTON	Jean-Alain	LPSC		Grenoble	France	pinston@lpsc.in2p3.fr
PORQUET	Marie-Geneviève	CSNSM		Orsay	France	porquet@csnsm.in2p3.fr
PREVOST	Aurélien	CSNSM		Orsay	France	prevost@csnsm.in2p3.fr
PRIEELS	René	U.C.L.		Louvain-la-neuve	Belgium	R.Prieels@fynu.ulc.ac.be
QUENTIN	Philippe	CEN		Bordeaux-Gradignan	France	quentin@cenbg.in2p3.fr
REJMUND	Fanny	GANIL		Caen	France	frejmund@ganil.fr
RICCIARDI	Maria Valentina	GSI		Darmstadt	Germany	m.v.ricciardi@gsi.de
RIDIKAS	Danas	CEA Saclay		Gif sur Yvette	France	ridikas@cea.fr
ROBLEDO	Luis	Universidad Autonoma		Madrid	Spain	luis.robledo@uam.es
ROMAIN	Pascal	CEA/DAM/DIF		Bruyères-le-Châtel	France	pascal.romain@cea.fr
ROYER	Guy	Subatech		Nantes	France	royer@subatech.in2p3.fr
RUGAMA	Yolanda	OECD-NEA		Issy les Moulineaux	France	rugama@nea.fr
SCHMITT	Christelle	IPN		Lyon	France	c.schmitt@ipnl.in2p3.fr
SCHMIDT	Karl-Heinz	GSI		Darmstadt	Germany	k.h.schmidt@gsi.de
SEROT	Olivier	CEA/DEN		Cadarache	France	olivier.serot@cea.fr

SIDA	Jean-Luc	CEA/DAM/DIF	Bruyères-le-Châtel	France	jean-luc.sida@cea.fr
SIMPSON	Gary	ILL	Grenoble	France	simpson@ill.fr
STASZCZAK	Andrzej	Institut of Physics	Lublin	Poland	stas@tytan.umcs.lublin.pl
STUTTGé	Louise	Ires	Strasbourg	France	stuttge@in2p3.fr
THOMAS	Jean-Charles	GANIL	Caen	France	thomasjc@ganil.fr
TOCCOLI	Cécile	CEA/DAM/DIF	Bruyères-le-Châtel	France	cecile.toccoli@cea.fr
TOVESSON	Fredrik	LANL	Los Alamos	USA	tovesson@lanl.gov
TUDORA	Anabella	University of Bucharest	Bucharest	Romania	anabella@olimp.fiz.infim.ro
/AYSHNENE	Larisa	PNPI	Gatchina	Russia	vaishnen@mail.pnpi.spb.ru
VARAPAI	Natallia	CEA/DEN	CADARACHE	France	natallia.varapai@cea.fr
VERMOTE	Sofie	University of Gent	Gent	Belgium	sofie.vermote@ugent.be
VOLANT	Claude	CEA Saclay	Gif sur Yvette	France	cvolant@cea.fr
VOROBYEV	Alexander	PNPI	Gatchina	Russia	alexander.vorobyev@pnpi.spb.ru
WAGEMANS	Cyriel	University of Gent	Gent	Belgium	cyrillus.wagemans@ugent.be
WEISS	Béatrice	LRSAE	Nice	France	bweiss@unice.fr
YAVSHITS	Sergey	V.G. Khlopin Radiium Institute	ST Petersburg	Russia	yav@mail.rcom.ru, yavshits@atom.nw.ru

AUTHOR INDEX

A

Ahmad, I., 19
Aiche, M., 19
Alahari, N., 263
Al Mahamid, I., 11, 157
Armbruster, P., 37, 45, 49
Audouin, L., 45

B

Bacri, C.-O., 263
Bao, Z., 221
Barreau, G., 19, 263, 369
Bauge, E., 19
Beghini, S., 305
Behera, B. R., 305
Behrens, T., 315
Benlliure, J., 37, 45, 49, 263
Bérerd, N., 345
Berger, J.-F., 69
Bernas, M., 37, 45, 49
Berthoumieux, E., 19
Bessolaz, N., 232
Bigot, B., xvii
Billebaud, A., 19
Birgersson, E., 27, 349
Blandin, C., 11, 57
Bogatchev, A. A., 305
Bokov, P., 277
Bonilla, C., 361
Bonneau, L., 77, 297
Bouchat, V., 305
Boudard, A., 37, 45, 49
Boyer, S., 19
Bringer, O., 11

C

Carjan, N., 123
Casajeros, E., 37
Casarejos, E., 45, 263
Catalette, H., 345
Cederkall, J., 131
Chabod, S., 11, 57

Chadwick, M. B., 31, 167, 213
Chartier, F., 11, 57
Chaves de Jesus, C., 273
Chestnov, Y. A., 365
Chevarier, A., 345
Clausen, C., 315
Corradi, L., 305
Covello, A., 145
Czajkowski, S., 19, 45, 49

D

Dassié, D., 19
David, J.-C., 31, 277
Delahaye, P., 315
De Maesschalck, A., 131
De Witte, H., 131
Dobaczewski, J., 93
Donets, A. Y., 353
Doré, D., 31, 277
Dorvaux, O., 305
Ducret, J. E., 45
Dushin, V. N., 255

E

Egido, J. L., 103
Enqvist, T., 37, 45, 49, 263

F

Faust, H., 11, 145, 221, 232, 345
Fedorov, D., 131
Fedorov, O. Y., 365
Fedoseyev, V. N., 131, 315
Fernandez-Dominguez, B., 45
Filipescu, D., 3
Fioni, G., 11, 57
Fioretto, E., 305
Fomichev, A. A., 353
Fomichev, A. V., 353
Foucher, Y., 11, 57
Fraile, L. M., 315
Franchoo, S., 131

Frankland, J., 263
Fritsch, V., 27
Fukahori, T., 365
Fynbo, H. O. U., 131

G

Gadea, A., 305
Gagarski, A. M., 115
Galy, J., 269
Gamboni, T., 273
Gargano, A., 145
Gavrikov, Y. A., 365
Geerts, W., 273
Geltenbort, P., 111, 182
Genevey, J., 137, 145, 149
Georg, U., 131
Gernhäuser, R., 315
Giacri, M.-L., 277
Giacri-Mauborgne, M.-L., 31
Giles, T. J., 315
Gönnenwein, F., 115
Gogny, D., 69
Gorska, M., 131
Goutte, H., 69, 123
Greene, J. P., 19
Grosjean, C., 19
Grosse, E., 357
Grudzevich, O., 373
Guiral, A., 19
Gunsing, F., 19

H

Haas, B., 19
Haight, R. C., 353
Hambsch, F.-J., 3, 27, 247, 255, 349, 369
Hanappe, F., 305
Hartmann, A., 357
Henriksson, H., 61
Heyse, J., 111, 182
Huyse, M., 131

I

Ignatyuk, A. V., 239
Ionan, A., 315

Itkis, I. M., 305
Itkis, M. G., 305

J

Jakovlev, V. A., 255
Jandel, M., 305
Janssens, R. V. F., 19
Jesinger, P., 115
Jonsson, O., 131
Junghans, A. R., 357
Jurado, B., 19, 45, 263

K

Kalinin, V. A., 255
Kawano, T., 167, 213
Kelić, A., 178, 239, 263
Kliman, J., 305
Knyazheva, G. N., 305
Kojouharov, I., 115
Kondratiev, N. A., 305
Kopatch, Y. N., 115
Kornilov, N., 27, 369
Kosev, K., 357
Kotov, A. A., 365
Kozulin, E. M., 305
Kröll, T., 315
Kruglov, K., 131
Krupa, L., 305
Köster, U., 131, 315
Kudryavtsev, Y., 131

L

Lammer, M., 285
Laptev, A. B., 255, 353
Latina, A., 305
Ledoux, X., 277
Lee, Y. D., 190
Lee, Y. O., 190
Legrain, R., 37, 45
Lemaire, S., 213
Leray, S., 37, 45, 49
Letourneau, A., 11, 57
Lewitowicz, M., 327
Lubkiewics, E., 115

Lövestam, G., 27, 273
Lyapin, V. G., 305

M

MacFarlane, R. E., 167
Mach, H., 315
Madland, D. G., 213
Magill, J., 269
Marie, F., 11, 57
Marsh, B., 315
Materna, T., 305
Mezentseva, Z., 115
Michel-Sendis, F., 19
Mikhailov, I. N., 297
Mishin, V. I., 131
Mittig, W., 263
Moncoffre, N., 345
Montagnoli, G., 305
Morillon, B., 247
Morjean, M., 194, 263
Mueller, A. C., 334
Mueller, W. F., 131
Mustapha, B., 45, 49
Mutterer, M., 115
Mutti, P., 11

N

Nankov, N., 357
Napolitani, P., 37
Nazarewicz, W., 93
Nezvishevsky, V., 115
Nichols, A. L., 285

O

Oberstedt, A., 27, 349
Oberstedt, S., 3, 27, 247, 273, 349, 369
Oganessian, Y. T., 305
Orlandi, R., 145
Osmanov, B., 19

P

Pauwels, D., 131
Pereira, J., 37, 45
Pérez, S., 85

Perrot, L., 19
Petit, M., 19
Petrov, B. F., 255
Petrov, G. A., 115, 205
Pinston, J. A., 137, 145, 149
Pipon, Y., 345
Plukiene, R., 277
Plukis, A., 277
Pokrovsky, I. V., 305
Poliakov, V. V., 365
Prasad, N. V. S. V., 131
Pravikoff, M., 45, 49
Prokhorova, E. V., 305

Q

Quentin, P., 77, 297

R

Rejmund, F., 37, 45, 49, 239, 263
Rejmund, M., 263
Ricciardi, M. V., 37, 45, 49, 239, 263
Ridikas, D., 31, 277
Robledo, L. M., 85, 103
Rochman, D., 349
Rowley, N., 305
Royer, G., 361
Rubchenya, V. A., 305
Rugama, Y., 61
Rusanov, A. Y., 305

S

Sagaidak, R. N., 305
Savajols, H., 263
Scarlassara, F., 305
Schaffner, H., 115
Scharma, H., 115
Scherillo, A., 145
Schilling, K. D., 357
Schmidt, K.-H., 37, 45, 49, 178, 239, 263
Schmitt, C., 263, 305
Seliverstov, M., 315
Serot, O., 111, 123, 182, 369
Shaughnessy, D. A., 157